Invisible Labour in Modern Science

Global Epistemics

In partnership with the Centre for Global Knowledge Studies (*gloknos*)

Titles in the Series:

Imaginaries of Connectivity: The Creation of Novel Spaces of Governance
Edited by Luis Lobo-Guerrero, Suvi Alt and Maarten Meijer

Mapping, Connectivity, and the Making of European Empires
Edited by Luis Lobo-Guerrero, Laura Lo Presti and Filipe dos Reis

Invisible Labour in Modern Science
Edited by Jenny Bangham, Xan Chacko, and Judith Kaplan

Invisible Labour in Modern Science

Edited by
Jenny Bangham, Xan Chacko, and
Judith Kaplan

ROWMAN & LITTLEFIELD
Lanham • Boulder • New York • London

Published by Rowman & Littlefield
An imprint of The Rowman & Littlefield Publishing Group, Inc.
4501 Forbes Boulevard, Suite 200, Lanham, Maryland 20706
www.rowman.com

86-90 Paul Street, London EC2A 4NE

British Library Cataloguing in Publication Information Available

Library of Congress Cataloging-in-Publication Data Available

ISBN 9781538159958 (cloth) | ISBN 9781538159965 (ebook)

∞™ The paper used in this publication meets the minimum requirements of American National Standard for Information Sciences—Permanence of Paper for Printed Library Materials, ANSI/NISO Z39.48-1992.

Contents

Series Editor's Note

It has been a great privilege to accompany Jenny Bangham, Xan Chacko, and Judith Kaplan in the final stages of this important collaborative project they began many years ago. *Invisible Labour in Modern Science* returns science to the multitude of epistemic labourers whose voices, perspectives, experiences, practices, concerns, and values have been erased from our consciousness and memory. It does so by reconstituting, without ever simplifying them, the complex contexts and dynamics through which modern science becomes constituted and operative as an autonomous but socially situated field of action, and those through which its authoritative knowledge is called upon to provide meaning and arbitration in various areas of human life.

Invisible Labour in Modern Science is also conceived as a purposeful intervention with pedagogical intent. It will likely become a referential volume for scholars and students of the history and sociology of science, but its contribution to critical pedagogy and to the ethos of reflexive scholarship extends well beyond these fields. Because of the place and authority of modern science in contemporary society and academia, the elucidation of the erasures that accompany its formation and deployment is important to all of us. At the very least, the volume should prompt us to interrogate what similar or other erasures and invisibilities we ourselves inherit, ignore, and consolidate as we deploy our epistemic practice as researchers and teachers.

Inanna Hamati-Ataya
Cambridge, January 31, 2022

Acknowledgements

This book is the result of many years of collaborative dialogue. It first took shape at the Max-Planck-Institut für Wissenschaftsgeschichte (MPIWG), Germany, where we were given the opportunity to convene a local workshop in June 2015 for colleagues whose interests converged on invisible labour in the human sciences. For their contributions to the workshop, we wish to thank Josh Berson, Mirjam Brusius, Eric Llavaria Caselles, Iris Clever, Samuël Coghe, Josephine Fenger, Donna Germanese, Sally Gregory Kohlstedt, Petter Hellström, Myriam Klapi, Nina Lerman, Nina Ludwig, Birgitta von Mallinckrodt, Johanna Gonçalves Martin, Minakshi Menon, Christine von Oertzen, Jenny Reardon, Helga Satzinger, and Kathleen Vongsathorn. The workshop was funded by Department II, then directed by Lorraine Daston, and the Independent Research Group "Histories of Knowledge about Human Variation," directed by Veronika Lipphardt.

Later, we expanded the geographical, political and disciplinary scope of the project and in February 2021 convened a second online workshop hosted by Inanna Hamati-Ataya of the Centre for Global Knowledge Studies (gloknos) in Cambridge, UK (an international network of cross-disciplinary academics funded by the European Research Council). We are grateful to Inanna for hosting the 2021 workshop, for her enthusiastic inclusion of this book in the Global Epistemics series, for the valuable insights she brought to the project, and the careful attention she paid to the manuscript. We thank the anonymous reviewers solicited by Rowman & Littlefield and the patient and efficient editorial team, which has included Isobel Cowper-Coles, Dhara Snowden, Frankie Mace, Deni Remsberg, and Mary Wheelehan. We thank Nate Freiburger for the thoughtful and comprehensive index. Jenny Bangham wishes to acknowledge support from the Wellcome Trust during this stage of the project (grant number: 200299/Z/15/Z). For comments on the drafts of the introduction, we thank Rohan Deb Roy, Nick Hopwood, Boris Jardine, Daniel Midena,

Gabriela Soto Laveaga, Simon Schaffer, Laura Stark, and many of the participants of the 2021 workshop. Our thanks also go to Susannah Chapman, Iris Clever, Petter Hellström, Lara Keuck, Ahmed Ragab, and May Ee Wong, who helped us to refine our summarizing blurb of the book.

Because this is a collaborative venture, and to save precious word quotas, we suggested that authors leave out their individual thanks to the authors and editors of the volume; but on their behalf we want to acknowledge the collegiate and cooperative spirit that all participants brought to one another's contributions. We are mindful that editing is, itself, a representational practice, and during the project we have had the opportunity to reflect on how invisible (to some) racialized, gendered, national, and institutional power structures affect who takes part, or not, in books like this one—we hope we can bring those lessons to bear on future projects.

Introduction

Jenny Bangham, Xan Chacko, and Judith Kaplan

> It came suddenly, splendid and complete, into my mind. I was alone, the laboratory was still, with the tall lights burning brightly and silently. . . . "One could make an animal—a tissue—transparent! One could make it invisible! All except the pigments. I could be invisible!" I said. . . . "To do such a thing would be to transcend magic. And I beheld, unclouded by doubt, a magnificent vision of all that invisibility might mean to a man—the mystery, the power, the freedom . . ."[1]
>
> —H. G. Wells, *The Invisible Man*, 1897

> I am an invisible man . . . I am a man of substance, of flesh and bone, fibre and liquids—and I might even be said to possess a mind. I am invisible, understand, simply because people refuse to see me . . . they see only my surroundings, themselves, or figments of their imagination—indeed, everything and anything except me.
>
> —Ralph Ellison, *Invisible Man*, 1952

Science fiction writer H. G. Wells put invisibility at the heart of a psychological thriller. In his 1898 novel, *The Invisible Man*, a young scientist discovers how to alter the human body's refractive index so that it neither absorbs nor reflects light, making it transparent. Elated by the possibilities of escaping scrutiny, he eagerly tests the technique on himself, but fails to reverse the effects, and his optimism soon fades. Despite attempts to hide his predicament by wearing clothes over his invisible form, he appears as a phantom to those he encounters. The terror that ensues turns the scientist into an agent of desperate destruction as he attempts to find a secluded place to hide and search for an antidote. Wells explores invisibility as freedom from surveillance, which (while it lasts) gives his protagonist liberty and power, and an ability to observe events that no one else can.[2]

Half a century after Wells' novel, another book with a strikingly similar title took on invisibility from a different perspective. Rather than the literal invisibility of Wells' protagonist, the narrator and central figure in Ralph Ellison's debut novel *Invisible Man* is made invisible by personal, social, and political forces that deny his existence. A poignant commentary on racism and the promise of Black Nationalism in the United States at mid-century, the novel follows the life of an unnamed Black man who leaves the American South for New York City. At every turn the narrator faces and negotiates the difficulty of being unseen, "being bumped against by those of poor vision," with the anguish of constantly needing to convince himself that he exists.[3] While the protagonists in both novels are invisible to others, their juxtaposition highlights important differences in the ways that invisibility is created. Wells wrote of a particular man, who through intellect and hubris has ended up in a situation of his own creation. By contrast, Ellison (with a title that purposefully omits the definite article) wrote of the abject powerlessness that comes from being continually ignored: the protagonist is rendered invisible by the unseeing gaze of others.[4] Together, the novels capture the ambiguities of invisibility in everyday life. They also explore the ways that surveillance and subordination can shape and define a person's actions and examine how (in)visibility is a property of position and perspective.

Invisible Labour in Modern Science explores the layers of invisibility that conceal, eclipse, or anonymize people and practices in scientific research. The themes of surveillance, subordination, and perspective raised by Ellison and Wells connect invisibility to labour.[5] Modern science is constituted by a workforce of diverse and specialized roles, only some of which are part of the public face of science.[6] In varied ways, translators, curators, experimental subjects, citizen scientists, and ethics review boards, for example, are often absent in formal publications and omitted from stories of discovery. Reports of scientific methods are held to ideals of transparency, yet "objective" representations are the result of careful judgements about who and what to reveal.[7] Meanwhile, the credibility of professional scientists has often traded on their apparent self-elimination, even when they themselves are celebrated personalities.[8] This book focuses on such omissions in an array of contexts, periods, and geographic spaces to examine how the concealment of identities and practices shapes knowledge.

Invisibility can empower as well as subordinate.[9] It can make someone a privileged observer, but it can also undermine credibility.[10] Power structures built on hierarchies of race, gender, class, and nation frame what, and how much, can be seen.[11] What is visible, to whom, depends on an observer's position.[12] For some, invisibility can be a deliberate political strategy, or a necessary part of a paid role; for others, it is a consequence of marginalization. Either way, there can be profound epistemological consequences to the omission of people and practices from scientific representations. *Invisible Labour in Modern Science* turns invisibility into a guide for exploring some of the hidden processes, moral sensibilities, and politics of twentieth and twenty-first century science.

INVISIBILITY AND LABOUR IN THE HISTORY OF SCIENCE

Invisibility hides in plain sight as a leitmotif in the historiography of modern science. Studies over the past century have shown how politics, economy, forms of labour, and power have shaped the creation and dissemination of scientific knowledge. As the profession of history of science expanded, dialogue across the humanities and social sciences brought into view a rich and varied cast of scientific characters and emphasized the politics of their representation.[13] Steven Shapin and Sharon Traweek, for example, drew attention to the "invisible technicians" and graduate students of the laboratory.[14] Scholars highlighted the significance of "craft" and "tacit knowledge" in science—those actions and meanings that are invisible (to some) because they cannot be communicated in writing.[15] The "turn to practice" encouraged descriptions of what scientists actually do as they work at the laboratory bench, manage projects, and write papers, and of how scientific labour is organized.[16] In concert with this, scholars examined the strategies by which researchers constructed scientific knowledge by deliberately alienating it from the idiosyncrasies and geographies of its producers.[17]

Meanwhile, historians of fieldwork practices highlighted the crucial mediation and translation work of field "assistants."[18] Studies of science in colonial contexts showed how forms of administrative control profit from the erasure, or extractive alienation, of Indigenous knowledge.[19] Critical race scholars showed that the construction of "universal" knowledge is predicated on a specific reference point: the Western Man—the basis for what philosopher Sylvia Wynter calls the Western world's "referent-we."[20] Historians of science revealed how the creation of such universalizing norms has often nevertheless depended on the bodies of human experimental subjects marked through colonization and enslavement.[21] Revealing the many systematic exclusions that result from the racial invisibilities inherent in science, scholars such as Katherine McKittrick, Alondra Nelson, and Ruha Benjamin have propelled the fields that study science and scientists towards justice.[22] Feminist interventions in science studies moved beyond the binary of marginalization and power to propose that invisibility is intersectional—that the functioning of invisibility often operates together with and cuts across social categories of race, gender, nation, class, and ability.[23] A different strand of scholarship has explored the history of ignorance, secrecy, and forgetting in science (including their virtuous ramifications).[24] In these manifold respects, the recovery of the invisible has been at the heart of the history of science and STS (science and technology studies) projects for many decades; this book centres invisibility and asks how it is made and how it functions in science.

In using the term "invisible," we emphasize the ocular. Donna Haraway sensitized audiences to the partial and situated nature of vision; what a person can observe is always conditioned by their position, their experience, and their habits of seeing.[25] One visual metaphor suggests others, such as the term "perspective" to emphasize that what is visible always depends on where one is standing and how and by what means our eyes are focused. Ellison in his sophisticated use of the visual metaphor of invisibility was careful to acknowledge the situatedness of vision. To clarify the origin of the social invisibility experienced by the narrator of *Invisible Man*, Ellison wrote, "That invisibility to which I refer occurs because of a peculiar disposition of the eyes of those with whom I come in contact. A matter of the construction of their inner eyes, those eyes with which they look through their physical eyes upon reality."[26] Ludwik Fleck, writing just a few years earlier about epistemology in science, made a related point about social positioning: "We look with our own eyes, but we see with the eyes of the collective body."[27] Fleck was observing that the things that we see and recognize in the world are created by our own environments, customs, traditions, and prejudices. The chapters in this book examine the workings of such "inner" and "collective" eyes, and they consider how these betray preconceptions, preferences, and intentions but also promise the possibility of change.

Yet optical metaphors have limits; at a minimum they risk marginalizing the scientific work done with aural and haptic senses. Critics of "ocularcentrism" have worried that the privileging of sight turned our attention away from the interpretation of the observer and point out that the visual always risks seeming self-evident.[28] And other metaphors are available: "voice," for instance, is often used to evoke an individual's agency and the force of personal testimony. Many oral history projects are explicitly political.[29] Bearing that in mind, we have settled on the terms "visibility" and "invisibility" because of their multiple and flexible meanings, but we use them in ways that go beyond the visual.

The term "labour" is built into scientific terminology: the English term "laboratory" comes from the Latin *laboratorium*, or workplace.[30] In everyday speech "labour" evokes the corporeal and manual dimensions of human activity—labour is done with the body, and it shapes the body.[31] But we use "labour" in a capacious sense, that is, to refer to mental, intellectual, emotional, and manual activities, and to encompass efforts that are financially remunerated, as well as those that are not.[32] We also include the donation of body tissues and participation in clinical trials—corporeal contributions that might be termed "bio-labour."[33]

In bringing "invisibility" and "labour" together, we are drawing on feminist scholars who coined the phrase "invisible work" to characterize women's unpaid domestic, reproductive, and volunteer labour, which had been culturally and economically devalued.[34] We are influenced by Margot Lee Shetterly's account of the marginalization of African American women mathematicians in histories of NASA, and the work of Mar Hicks on gendered labour that was gradually displaced in the history of computing.[35] Our choice of the phrase "invisible labour" gestures to the analyses developed since the 1980s of the emotional labour performed by service and health care workers.[36] It acknowledges recent scholarship on the necessary invisibility of infrastructures and the labours that sustain them—especially those of internet and data management.[37] And we also have in mind anthropologist Elizabeth Povinelli's analysis of the failures of Western cultural notions of labour to incorporate forms of Indigenous productivity.[38] As all of these scholarship shows, the (in)visibility of labour operates along the axes of power.

By foregrounding invisibility with respect to scientific labours specifically, the case studies here draw attention to the varied forms of recognition and remuneration in science (or lack thereof). Although many participants exchange their labour for money, rewards also often come in the form of authorship, citations, institutional recognition, awards, political agency, or feelings of virtue or collegiality. Some participate because they are institutionalized in prisons, schools, or hospitals; others volunteer as, for example, blood donors or citizen scientists.[39] In the spirit of the scientific values of disinterestedness and self-denial, many workers accept anonymity as a by-product of their participation, or purposefully cultivate invisibility as part of their jobs. In science, then, "value" has a complex relationship to "visibility," so that the gains promised through the recognition of labour might not be the goal for all.

PEOPLE, PROCESSES, POWER, AND PRACTICES

The case studies in this volume consider "invisibility" and "labour" in twentieth and twenty-first century science in varying geographic locations, political circumstances, and disciplinary contexts. We have organized the chapters into four overlapping domains. The first of these, "People" (with an accompanying commentary by Sabine Clarke), explores how identities are made visible or obscured in ways that make particular individuals credible observers, unbiased conduits, or authentic research subjects. Some of the chapters investigate enterprises that engaged "ordinary" citizens in scientific projects; others examine the practices and expertise of brokers and translators, exploring how hierarchical, gendered, Indigenous, Western, or "local" identities affect visibility. The chapters highlight some of the political conditions under which participants become concealed or obscured, and they draw attention to why many remain in the shadows.

The second section, "Power" (with a commentary by Gabriela Sota Laveaga), ties participation in science to agency. Differences in gender, sexuality, class, race, national, and community identity are leveraged to privilege or underplay contributions to scientific knowledge and control its effects on society. We see how extensive political and epistemological work is required to transform knowledge into scientific data worthy of recognition by state actors and policy-makers. There are stories in this section of communities that recruit and train researchers to secure political visibility in struggles for land, health care, and education.

Invisibility is not fixed, but is rather the outcome of multiple and often conflicting processes. This conjecture is a major theme of the third section, "Process" (accompanied by a commentary by Susan Lindee). The chapters here examine how abstracting and effacing relations are part of the work of collecting specimens, assembling data, and translating interviews. They also explore the perspectival and temporal qualities of visibility and how people and processes come in and out of focus in the practice of research. In several of these chapters, archives and scientific collections take centre stage. Authors reflect on the methodological difficulties of recovering the labour of those who prepare, order, and assemble such collections. They also consider the legal frameworks and changing ethical regimes that condition the continued existence of scientific archives.

As we, as a collaborative group, considered these questions, our attention was perhaps inevitably turned to how invisibilities figure into our own work as historians and STS scholars. What are the moral and political motivations underlying historical recovery projects, and where might such projects lead? How do social and professional identities—racialized, gendered, and disciplinary—affect whose scholarship is included in a volume like this one?[40] The final section of this book, "Practices" (with a commentary by Judith Kaplan), reflects on these questions. The chapters ask whether representation is always an unqualified good—should we always strive to recover invisible lives and experiences, and how should we respect historical desires for privacy?[41] They question how historians' own (varied) social relations affect their narratives, and ask how archivists, curators, museums, and libraries define and shape historians' work. The chapters also discuss how access, visibility, privacy, and protection are constantly being remade in the present, and how these negotiations affect what kinds of histories will be told.

There are many potential pathways through the topic of "invisible labour." The overall structure of the book, and the commentaries within, suggest one such possible path. The rest of this introduction offers another. We start with scientific fieldwork. How might we question the assumed right of fieldworkers to speak on behalf of others, human and non-human? Fieldwork also offers a rewarding opportunity to investigate some of the epistemic consequences of keeping the expertise of interlocutors hidden in formal scientific accounts. Next, we engage the processes of translation—between languages, between forms of inscription and between knowledge systems—and reflect on how and why such processes are often hidden. The following section explores why participants in science might consider it ethical *not* to be included in formal accounts, choosing more anonymous rewards instead. We juxtapose this with struggles to be made visible—specifically in the context of activist "citizen science"—and consider how the experiences of harm can discredit a community's testimony.

Archives are fascinating terrain for understanding how people and events are obscured from the historical record. Several chapters examine the processes through which archives create such silences and reflect on the consequences of this for historical narratives. Others focus on such processes specifically as they relate to biographical narratives. Historical storytelling is all about choices—about what to include and what to leave out—and two chapters consider this with respect to the curation of museum objects. Finally, we study the affective relations that emerge in scientific spaces. Taking "care" as a point of entry, we open the space for scientific practices to engage and create value for relational practices between scientists (broadly conceived) and their multi-species networks.

INVISIBLE FIELDWORK

Some of the best-studied instances of invisibility in science trace the labour of fieldwork assistants, informants, and translators. From explorers to ethnographers, metropolitan researchers who gather data in the "field" have been adept at concealing the efforts and expertise underpinning their work.[42] Attempting to recover lost stories about fieldwork, some historians have shown how to change the frame of reference, from formal scientific outputs to the biographies and practices of "brokers" who facilitate access to the sites under observation.[43] Historian of anthropology Lyn Schumaker achieved such a shift in her history of mid-century anthropologists in central Africa, when she focused her research on the local assistants who guided the work of white British anthropologists in the 1950s and 1960s.[44] In *Africanizing Anthropology*, Schumaker moved away from the history of ideas about functionalism and changes to theory and method, and instead immersed us in administrative practices and the day-to-day labour of fieldwork.

In recovering the motivations and choices of fieldwork "assistants," Schumaker cautioned against reaching for any simple model of exploitation and also encouraged historians to consider assistants' agency and their management of anthropologists for their own ends.[45] Rosanna Dent (chapter 24) does precisely this in her work to recover the labour and expertise of the A'uwe-Xavante community in Brazil. The people in this community have been highly visible as "subjects" since the 1960s, but not as knowledge-makers who guide and drive research. Dent demonstrates the perspectival dimension of that invisibility, and creates a shift in that perspective to show how A'uwe-Xavante people welcomed, enculturated, and guided the research of several generations of U.S. anthropologists, including Dent herself.[46]

Other chapters attend to researchers whose methods go relatively unreported in scientific publications; they also investigate how that invisibility helped to substantiate claims about racial difference. Elise K. Burton (chapter 7) describes the work of three Iranian scientists of the Tehran Pasteur Institute, who were employed by French and U.S. visitors as reliable local informants and interpreters. She draws attention to their relationships with both foreign scientists and the marginalized research subjects that they studied—and examines how power relations became manifest in the linguistic, geographic, and racial categories created by this fieldwork. In a related case study, Ana Carolina Vimieiro Gomes (chapter 9) explores the interests, motivations, and expertise of Brazilian geneticist Eliane Azevedo, who worked with U.S. collaborators in the 1960s–1980s to study the population genetics of Northeastern Brazil. Azevedo's social position and experience of Brazilian society shaped her scientific career and allowed her to make credible judgements about racialized field data.[47]

Also focusing on the making of racial categories, Mihai Surdu (chapter 8) recounts his experiences as a student researcher during the 1990s, when he gathered sociodemographic information about Roma people in Romania. Surdu's chapter examines how sociological fieldwork methods outsourced judgements about race and ethnicity to fieldworkers with minimal formal training. In all three case studies, the relative invisibility of in-the-field judgements helped officials to project an authoritative science of human difference.

Histories of scientific fieldwork, then, offer some first thoughts about the perspectives that are opened up by shifting position. Beyond the lens of authorial credit in scientific publications, they show how we might pay attention to the political or social visibility that "assistants," "collaborators," and "brokers" achieve by recruiting and working with scientists. These histories encourage us to think about how decisions concerning representation shape knowledge and its politics.

TRANSPARENT TRANSLATION

One category of collaborator that has played a particularly elusive role in modern science (including fieldwork) is that of the translator. Science is often figured as a system of direct report. On this view, translations are good and trustworthy so long as they don't get in the way of the "facts"—that is, so long as they are transparent.[48] But this view has been called into question by historians who point out that translation is as integral to scientific practice as measurement or experimentation.[49] Part of the impetus for this historiographic shift has come from an expanded consideration of the human sciences (biomedicine included) where the *in*commensurability of meanings is often explicitly at issue.[50] Literary scholar Lydia Liu framed the problem through reference to processes of modernization, with science playing a key role: translators sometimes confirm the equivalence of meaning across linguistic communities, participating in the "universalizing tendencies of modernity"; other times, they deny such equivalence on the grounds of cultural specificity—terms are left in the original, reifying cultural or phenomenological difference.[51] Consistent with our emphasis on labour, Liu's approach focuses attention on asymmetrical power relations in the global circulation of knowledge.[52] Under these conditions, as a number of the chapters in this volume point out, translators and the world they describe tend to push back. By looking more closely at the structures and institutions that surround the labour of translation, we might begin to recover something of the values attributed to symbols by diverse actors at specific moments in time.

The agency of translators is emphasized in Lan A. Li's biographical sketch of Chinese biochemist and historian Lu Gwei-djen, who worked in China, the United States, and Britain (chapter 2). Lu's life story sits at the intersection of several "alternative models of modernity," and translation featured throughout her career: she inhabited the space between feminist discourses, Confucianism, nationalist impulses, and medicine, even before translating the history of Chinese medicine for English-language audiences more formally. In doing so, Lu grappled with translation between ontological worlds. We see related themes in the essay by Michaela Spencer (chapter 25), who draws attention to her own role as an "epistemic translator"—that is, someone who mediates between different ways of knowing. Spencer reflects on her experiences with "Ground Up," a collaboration between university and community researchers working with Aboriginal communities in northern Australia. During the COVID-19 pandemic, Ground Up took an analytical and collaborative approach to translating Aboriginal narratives about the pandemic into a form that was legible to biomedical health care workers. But the findings produced by this team (findings that might have saved lives) failed to gain traction among public health and government officials. Ultimately, Spencer emphasizes the importance of cultivating receptive audiences rather than commensurable meanings (see also Blacker, chapter 10). The normally invisible work of translation might then be revealed through the creation of communities with shared understanding.

The labour of translation is also manifest in practices of inscription and in attempts to carry meaning across media and different formal registers. María Fernanda Olarte-Sierra (chapter 3) describes how forensic scientists in Colombia carefully moderate their visibility to cultivate anonymity in the media (by hiding their identities) and objectivity in the courtroom (by making visible their professional qualifications). They also *personally* come forward as "translators" of the violence traceable in the bones of those lost to violence for relatives and loved ones in "delivery ceremonies." Here, their task is to explain to bereaved families how the marks on those bones indicate cause of death, translation work that is understood to be crucial to healing. While Olarte-Sierra's focus remains with the scientists who shepherd meaning across communities, Judith Kaplan (chapter 17) queries the suppression of meaning that comes with the shift from oral to textual language—taking as her example the early twentieth-century efforts to record the Wisconsin dialect of the Oneida language.[53] Other chapters (Surdu, chapter 8, and Laemmli, chapter 18) emphasize the loss of meaning that comes when inscriptions travel between qualitative and quantitative formats and from ephemeral cultural expression to decontextualized systems of notation.

Focusing on the politics of translation, these and other chapters engage with meanings that do not carry over easily between languages, cultures, and inscription formats. Understanding takes time and work, which is often at odds with the tempo of administrative initiatives, funding cycles, and professional development. The authors of these chapters situate examples of linguistic, epistemic, and intermedial translation to show how the (in)visibility of translational processes affects credibility.

ANONYMOUS ACTORS

Sometimes people willingly participate in scientific projects without credit, choosing instead more anonymous rewards such as the satisfaction of serving a community or nation, the thrill of discovery or just the reward of contributing "to science."[54] Two chapters describe mid-twentieth century scientific projects in which the identities of participants were effaced through their willing incorporation into scientific bureaucracies. In the 1950s and 1960s, the U.S. National Institutes of Health, a vast research institution that functioned (and still functions) as part of the U.S. bureaucratic state, recruited volunteer research subjects who were willing to participate in obscurity, understanding their own invisibility to be part of their ethic of volunteerism (Stark, chapter 6). During the same decades, British serology and genetics depended on millions of altruistic (often modest, self-effacing) donors who volunteered their blood as commitments to community and nation (Bangham, chapter 15). In both these cases, we see research subjects happily embracing the relative obscurity that came with entering into bureaucratic enterprises.

In a similar vein, but in a very different political context, Elena Aronova (chapter 14) considers "citizen" seismologists in the 1950s USSR—individuals who believed that their partially invisible contributions to earthquake prediction would not only benefit fellow citizens but also help reconcile science with Soviet ideology. In a different intersection between citizenship and state, Alexandra Noi (chapter 5) introduces us to workers who saw in science a way of finding freedom from government surveillance and control. Ex-prisoners of the USSR, formerly incarcerated in Gulag forced-labour camps, willingly participated in physically challenging, extremely remote scientific expeditions to northern Siberia. They sought to carve out a space of invisibility and freedom from Soviet state bureaucracy and, within that space, exert control over their lives. An important lesson from all of these, which applies to a much broader range of scientific projects, is that one's motivations to participate may lie far beyond the notions of credit and authority, and these may be consequential for survival.[55]

In another case study concerning survival, Sarah Blacker (chapter 10) analyses attempts by Indigenous communities in northern Alberta to make their knowledge about environmental damage known to Canadian authorities. These First Nations communities live with serious and ongoing pollution of their land, water, and food sources by the oil industry. Blacker follows a collaborative project led by two First Nations communities to translate their carefully collected evidence of environmental harm into data that could be seen and understood by the federal government. She notes that to protect their privacy, the identities of Indigenous participants have been left anonymous in official reports—an example of how invisibility may be strategically planned for in the interests of justice and agency. But despite this work, First Nations' knowledge claims are yet to adequately influence governance regimes in Canada, and Blacker observes that the Indigenous knowledge-holders experience a double invisibility. Not only are they politically marginalized, but the harm that they experience through environmental pollution can paint them as biased in the eyes of settler-colonial officials, thus further undermining their ability to testify.

The observation that people and communities can be discredited by their embodied experience has been underlined elsewhere by Susan Lindee, who described the painstaking monitoring and organizational work carried out by citizens local to the Fukushima Prefecture in Japan to make visible the radiation experienced following the devastating accident at the Daiichi Nuclear Power Plant.[56] Responding to a serious erosion of public trust caused by the disaster, residents used small, cheap, highly mobile civilian radiation monitoring devices to contest their own invisibility. Like Blacker, Lindee observed the power of radiation to injure not just bodies, but the credibility and visibility of those who have experienced harm. A person suffering the consequences, in this case, of toxicity, is seen by those in power as an unreliable witness. In both cases, "science" was seen as the route to overcome that invisibility, and in both cases, those projects and toxicity are enduring.

Citizens bring motivations, desires, and experiences to the scientific projects in which they participate. Sometimes those correspond to a desire for anonymity and privacy; in other cases, for visibility and political action. Stark suggests how historians' interpretive frames might be sensitively tailored to account for the thoughts, interests, and inclinations of their actors. It is not always straightforward to balance the recovery of marginalized actors with historical desires for privacy and anonymity.

SILENCES IN THE ARCHIVES

Whether riddled with gaps, dust or indecipherable handwriting, archival work continues to be a rite of passage for the aspiring historian, and it entails processes of searching within known documentary repositories for the possibility of serendipitous discovery.[57] Silences and effacements are part of the allure. Anthropologist Michel-Rolph Trouillot observes that the creation and use of archives are predicated on layers of silences. His analysis of how power operates in the creation of historical narratives emphasizes the silences made at every step of their construction: "the moment of fact creation (the making of sources); the moment of fact assembly (the making of archives); the moment of fact retrieval (the making of narratives); and the moment of retrospective significance (the making of history in the final instance)."[58] Archivist and historian Michelle Caswell extends Trouillot's proposal by pointing to the multidimensional ways in which archivists themselves powerfully shape historical storytelling through their processes of record-making, evidence capture, and the organization of personal and institutional archives.[59] Trouillot and Caswell are helpful for thinking through the chapters that reflect on the silences, effacements, and invisibilities created in the construction and use of sources, archives, and narratives.

Several chapters explore what becomes invisible (and how) during the processes of making and assembling sources. Whitney Laemmli (chapter 18) shows how such effacements occur even when a project's aim is to make a highly visible totalizing archive. She recounts the remarkable attempt by American folklorist Alan Lomax between the 1960s and the 1980s to create an archive of film that would capture the modes and patterns of human dance from across the world. But Lomax's attempt at total visibility created its own silences; while the dancers' bodies were highly visible, Lomax's archive omitted their names, motivations, and some aesthetic traits of their work.

Archival materials have the potential to become untethered and alienated from the cultural and political worlds they are seen to represent (see also Kaplan, chapter 17). But they are all situated within national contexts and within institutions, which shape their meanings and uses in ways that are often not immediately evident. The labour and expertise of archivists are crucial to the creation of narrative. Lara Keuck (chapter 21) recounts her variable access to psychiatric archives in Germany. Keuck found that in one institution the records were, above all, understood as historical materials, while at another they remained patient files. Each setting conditioned the way that Keuck encountered and used the material and so helped to define what she could know and write. Bangham (chapter 15), writing about the United Kingdom, notes how present-day scientific knowledge about genetics ultimately affected her access to medical archival material. The processes of fact retrieval, that is, the making of historical narratives, are defined by the labour and judgement of archivists and also the institutional and national settings in which they are assembled. And such archives are never static; ethical guidelines change, time limits elapse, and responsible archivists reassess such protections. Even as they shape what is possible to know about the past, present-day notions of access, privacy, and protection are constantly being remade and adapted to our own understanding of history.

Ethical regimes also help to define what gets collected and made into historical sources. Joanna Radin (chapter 20) found herself in the position of working *with* one of her historical actors to help him to archive his papers, and her chapter reflexively considers her ethical responsibilities, as both a historian and a "human subjects researcher." She demonstrates that cultivating relationships with historical actors in a responsible way might involve honouring a scientist's desire for some things *not* to be kept, to prevent some materials from "becoming historical." Here we see how personal relationships of trust can significantly affect what materials are kept for as-yet-unknown futures.

HIDDEN LIVES

Radin engages with the creation of an archive that is ultimately biographical. The aggregation of papers around a single person's life is so common as to seem natural. Often initially collected together in the context of an institution, the biographical archive reflects the structural dispositions of a field and society and so often reinforces a focus on "great men" and their correspondence. On the one hand, this tendency creates a profusion of material on some individuals and gaps in relation to others. On the other hand, comprehensive archives focused on single individuals can also provide the only access to otherwise obscure people and their activities. Either way, this archival tendency also helps to reinforce biography as an "invisible motor of historical argument," (Jardine, chapter 22)—easy to reach for but undertheorized. Not only do archives relegate some people to be unnamed and unacknowledged (another of Trouillot's layers of silence), but even the lives of those who are present in the archive are always fragmented, such that the record is both evidence of who is worthy of being remembered and the choices of what were, or are, considered worthy of safe-keeping.[60] As Trouillot implied, biographers are restricted by the evidence they amass in the archive because what gets recorded is a result of choices made to prioritize some documents, voices, and traces of lived lives.

Lan A. Li's biographical account of Lu Gwei-djen (chapter 2) recovers a legacy eclipsed by her famous collaborator Joseph Needham—indeed, his archives are so profuse that, today, a dedicated institute houses them.[61] Li shows that becoming attuned to the "private epitext," in marginalia and informal records, including in Needham's papers, offers a route to reattribute intellectual credit that Lu herself had abdicated in favour of her male collaborator. Julia E. Rodriguez (chapter 1) narrates the work and contributions of Mexican American Zelia Nuttall, an anthropologist of the Americas. What Rodriguez provides are rich insights on *how* historical obfuscation occurs, asking why some scientists, despite having made tremendous contributions to a field, are forgotten. In the case of Nuttall, Rodriguez shows how race, gender, and nation came together to relegate Nuttall to the margins.

Omnia El Shakry (chapter 11) offers a radically different approach in her chapter on prisons as sites of inquiry and the politics of recuperating and amplifying the voices of incarcerated subjects. El Shakry rethinks the meanings and possibilities that result from re-analysing drawings made by incarcerated women as part of a late-1950s clinical study, "Prostitution in Cairo," at al-Qanatir prison. Rejecting a straightforward recovery project, El Shakry instead draws on psychoanalytical theory and experiences to offer ways of understanding the drawings as more than "transparent objects." She boldly asks what it would require to understand these women as *collaborators* in clinical inquiry and shows how their political and institutional positions contributed to their invisibility as knowledge makers. El Shakry thus provides a moving biography of women whose names remain unknown, in the form of a speculative "microhistory."[62]

As all three case studies show, biography and its techniques can share more than a person's story—it can also comment on and contextualize the science of a historical milieu. To historian of science Mary Terrall, the trouble with biography is that the "intimate, and potentially unhealthy, relationship linking biographer to subject seemed to risk falling into either some version of hagiography or its opposite, a critique of the flaws and mistakes of the hapless (and defenceless) subject."[63] In either case, choices made in the accretions of the archive and by the historian inevitably leave out some aspects of the life being "exhumed."[64]

The choice to reveal or obscure the identity of participants or historical actors is itself a sign of hierarchy and power in scientific institutions and practice.[65] Decisions to make visible forms of knowledge that have previously been obscured are not always rooted in emancipatory politics of representation. Sociologist Hannah Landecker makes this point in her analysis of stories about Henrietta Lacks, whose cells were extracted for medical diagnosis but were subsequently multiplied and distributed, without her consent, to become standard research material in human biology. Landecker shows that the history of revealing or hiding the identity of the donor of experimental material—in this case, HeLa cells—hinged on shifting systems of value.[66] During moments of celebration—for example, following the success of the polio vaccine in 1952—the revelation of Lacks' name was seen as a way of performing gratitude and recognition. By contrast, in 1966, when it was claimed that HeLa cells had "contaminated" cell cultures across several laboratories, Lacks' class and race were made the focus of public scrutiny. Exposing Lacks' identity as a working-class Black woman gave purchase to the racist language used to describe the movement of HeLa cells as "vigorous," "aggressive," "surreptitious," "a monster among the Pyrex," "indefatigable," "undeflatable," "renegade," "catastrophic," and "luxuriant."[67] Henrietta Lacks' initial invisibility and subsequent notoriety indicate how the process of making visible can be ameliorative and destructive depending on the power and politics of its enactment. The choice to reveal or obscure the identity of participants or historical actors is itself a sign of hierarchy and power.

Some biographies are created and consumed not through text, and not just using paper sources, but through exhibition displays of historical objects.[68] Museums are rich sites for analysing the processes and practices of selection, display, and concealment that scaffold the labour of making history. Grappling with the problem of how to represent the life of an individual associated with few surviving objects, Boris Jardine (chapter 22) reflects on the curatorial work behind an exhibition at London's Science Museum about computer pioneer Alan Turing. The exhibition's curator convened a series of group interviews with a group of over-fifties LGBTQ+ community advocates—interviews that are permanently embargoed and had no direct representation in the exhibition itself. Jardine reflects on what he and the curator gained from their private testimonies and how the experiences of listening and recording such testimonies related to the fragmented aspects of Turing's life that were allowed to emerge among the curated choices of display objects, placards, and exhibition design.

Margaret M. Bruchac (chapter 4) shows that beyond the meanings created by their relationships to people, objects themselves may have animacy, that is, the quality of being alive with literal power, "as conveyers of messages, participants in diplomacy, and speaking (or temporarily silent) beings that communicate across time and space."[69] The chapter analyses wampum belts, a particular class of objects made and remade by North American Indigenous artisans in the seventeenth and eighteenth centuries for adornment, political communication, and cultural memory. Bruchac draws attention to the physical and conceptual sensing and recognizing that constitutes *seeing* an object. As she puts it, objects that are "unseen" are not necessarily invisible but are subject to the limited senses and restricted conceptions of agency and animacy of custodians and audiences.

Biographies of people, places, and objects provide powerful opportunities to reveal or obscure their role in the production of scientific knowledge. By foregrounding some lives, roles or contributions over others, historians participate in the creation of hierarchies of power. Gaining a sense of a person's life behind the words on a page, or objects in an exhibition, helps one to comprehend the spirit or ethos of a work of science, even if that meaning lies in the background. Revivifying the past through biography prioritizes the experiences of some over others, and thus can be an opportunity to redress previous omissions.

UNSEEN AFFECTS

Scholarship that pays attention to the labour inherent in tacit and embodied practices, and care work specifically, is gaining traction.[70] While these forms of labour tend to be made invisible by systems of attribution, chapters in this collection take a variety of positions with regards to recognizing and revealing them. Xan Chacko (chapter 16) follows the curatorial labour that readies seeds for long-term storage in the frozen vaults of seed banks. She challenges the taken-for-grantedness of the endeavour of seed banking by describing the affective and embodied entanglements between the seed curators and their collections. Caitlin Donahue Wylie (chapter 19) productively displaces the simple project of reclaiming value for these invisible forms of labour in science, with the example of fossil preparators—the people who carefully extract and stabilize specimens for scientific study. Preparators are rarely included in published scientific reports, but Wylie suggests that their relative invisibility is not the outcome of any underhanded plot to silence. Rather, they choose invisibility. The infrastructural work in these examples is deliberately obscured, and both Chacko and Wylie show how this fits into a more general pattern of invisibility of care.[71]

Sociological analyses of labour and anthropology of medicine have produced theories of care that draw attention to human-human relations of support.[72] But another recent turn in STS has evidenced care and affect in relations *between* species, beyond the human.[73] This scholarship has clarified how power is organized in systems of care in the laboratory, field, and other scientific spaces, showing that care is neither morally neutral nor equally distributed across knowledge-making practices. Different from concepts like "moral value" and "concern," care indexes a deeply subjective and materially consequential entanglement of a person with their objects and/or subjects of enquiry. Care matters because it shapes the embodied practices, aspirations, and concerns of knowledge makers across the hierarchies of science.[74] Feminist scholarship dedicated to care in scientific labour has demonstrated how the "carers" and "cared for"—human or other-than-human—are involved in the co-creation of "care practice" that changes both the entities and the nature of the scientific enquiry.[75]

Alexandra Widmer (chapter 23) elaborates on the partially invisible care activities that surround and support the practice of fieldwork in the human sciences. Using personal experiences of travelling with her own children to Vanuatu, her ethnographic field site, Widmer theorizes the concept of hospitality with the uncompensated and unacknowledged caregiving and receiving that made it possible for her to gather data. Revealing these caring connections is more than a matter of recognition because the choice to stay hidden could be purposeful; by this we mean that it is not a simple task of unveiling a previously hidden mode of knowledge, people or practice, to gain a better *understanding*, but that *reality itself* is woven through with layers of the (un)known and (un)knowable.[76] Susannah Chapman (chapter 12) calls attention to the more-than-human dimensions of rice varietal innovation in The Gambia. Chapman argues that to deny the forces—God, *jinn* (a kind of spirit), and the rice plants themselves—that participate in breeding projects with farmers relegates the care practices of farmers and their rice fields to the realm of superstition. Rather than translate those practices into ones deemed legible through the lens of scientific knowledge, such as weeding or transplanting, Chapman suggests instead the paradigm of breeding could include actors and practices that have been hitherto invisible. This could radically alter not just the science but also the value-practices of intellectual property law.

Stuart McCook (chapter 13) also engages plant breeding but with coffee in Latin America, where the globalization of production rendered some aspects of farm labour invisible to transnational scientist-breeders. Breeders and farmers are divided in how they see and understand coffee production: both care for the crop, but they do so in ways that have fallen under each other's radar of perceptibility. McCook traces these invisibilities, by exploring how farmers resisted the adoption of novel coffee varieties owing to the labour and investment needed to "technify" their farms. He also shows how this resistance eventually caught the attention of scientists, who modified their breeding programs to cater to different farming needs. McCook argues that a technological unity based on their shared interests and values has the effect of boosting both groups' visibility.

Care is hard to record, tricky to classify, and so is often deliberately rendered invisible—even where it profoundly affects what can be known about the world. Moreover, the labours of care are unevenly distributed in ways that are racialized, gendered, and classed, making it all the more important to trace their invisibilities. This offers a route to mapping where and how care is valued in the production of knowledge and the world beyond.

The fields we consider here range from biomedicine to botany, forensic science to linguistics, and archaeology to psychiatry. While diverse in their materials and aims, we have knit them together to argue for the remarkable productivity of examining scientific knowledge production and representation with "invisible labour" as a guide. Each case study brings invisibility front and centre to explore how scientific activities are valued, and by whom, and reflect on the varied ways in which power, reward, coercion, motivation, and remuneration can operate. There are startling connections to be made regarding the people and activities that drive these case studies, and by staying with invisibility irrespective of the scale at which it operates, this collection responds to the call to pay attention to intermediary or "mesoscopic" levels in the histories of science.[77] Our coverage of "science" is by no means comprehensive, but we aim to provide a conceptual toolkit that emboldens the reader to engage areas of science and technology beyond our collection. As editors, we are extremely grateful to all who have contributed their stories and perspectives to the project.

NOTES

1. These words are uttered by the protagonist as he recalls the circumstances of his momentous scientific discovery: H. G. Wells, *The Invisible Man* (Project Gutenburg, 2004 [1897]), chapter 19.

2. Wells was retelling the story of Gyges in *Republic*, through which Plato argued that a person's ethical conduct is maintained through scrutiny by others: Philip Holt, "H. G. Wells and the Ring of Gyges," *Science Fiction Studies* 19, no. 2 (1992): 236–47.

3. Ralph Waldo Ellison, *Invisible Man* (New York: Random House, 1952), 1.

4. On "invisibility" as a pervasive literary form in Black American literature: Todd M. Lieber, "Ralph Ellison and the Metaphor of Invisibility in Black Literary Tradition," *American Quarterly* 24, no. 1 (1972): 86–100.

5. Thanks to Simon Schaffer for drawing out these themes in relation to the central questions of this book.

6. On labour processes in science: J. D. Bernal, *The Social Function of Science* (London: G. Routledge and Sons, 1939); Mike Hales, *Living Thinkwork: Where Do Labour Processes Come From?* (London: CSE Books, 1980); Mike Cooley, "The Taylorization of Intellectual Work," in *Science Technology and the Labour Process*, eds. Les Levidow and Bob Young, vol. 1 (London: CSE books, 1981), 46–65.

7. On "virtual witnessing" in science: Steven Shapin and Simon Schaffer, *Leviathan and the Air-Pump: Hobbes, Boyle and the Experimental Life* (Princeton, NJ: Princeton University Press, 1985); Steven Shapin, "The Invisible Technician," *American Scientist* 77 (1989): 554–63; for techniques of scientific representation: Lorraine Daston and Peter Galison, *Objectivity* (New York: Zone Books, 2007); Catelijne Coopmans, Janet Vertesi, Michael Lynch, and Steve Woolgar, eds. *Representation in Scientific Practice Revisited* (Cambridge, MA; London: The MIT Press, 2014); on the invisibilities created through scientific representation: Omar Nasim, "Making Invisible: The Other Side of Visualization in Science," in *Visualization: A Critical Survey of the Concept*, ed. Erna Fiorentini (Berlin: LIT Verlag, 2021).

8. Simon Schaffer, "Astronomers Mark Time: Discipline and the Personal Equation," *Science in Context* 2, no. 1 (1988): 115–45; Steven Shapin, *The Scientific Life: A Moral History of a Late Modern Vocation* (Chicago, IL; London: University of Chicago Press, 2008); on the even temperament of the "invisible college": Richard Yeo, *Notebooks, English Virtuosi, and Early Modern Science* (Chicago, IL: University of Chicago Press, 2014), xvi and 69–95; for a late-nineteenth-century view of the virtues of self-elimination in science: Karl Pearson, *The Grammar of Science*, second edition (London: Adam and Charles Black, 1900 [1892]), 6–7.

9. On invisibility as empowering (as well as marginalizing) in systems of work: Lucy Suchman, "Making Work Visible," *Communications of the ACM* 38, no. 9 (1995): 56–64; Susan Leigh Star and Anselm Strauss, "Layers of Silence, Arenas of Voice: The Ecology of Visible and Invisible Work," *Computer Supported Cooperative Work* 8 (1999): 9–30; for an expansive historical story of the power of the invisible in science: Philip Ball, *Invisible: The History of the Unseen from Plato to Particle Physics* (London: Vintage, 2015).

10. Alison Wylie, "Why Standpoint Matters," in *Science and Other Cultures*, eds. Robert Figueroa and Sandra G. Harding (London: Routledge, 2003), 26–48; Alison Wylie, "Feminist Philosophy of Science: Standpoint Matters," *Proceedings and Addresses of the American Philosophy Association* 86, no. 2 (2012): 47–76.

11. Maureen Burns, "Invisibility," in *Theorizing Visual Studies: Writing through the Discipline*, eds. J. Elkins et al. (London: Routledge, 2013): 146–49.

12. Some scholars in subaltern studies argue that a person in a marginalized position has a less biased and more accurate perspective than a person in power: Gayatri Chakravorty Spivak, "Can the Subaltern Speak?" in *Can the Subaltern Speak?: Reflections on the History of an Idea*, ed. Rosalind Morris (New York: Columbia University Press, 2010): 21–78; in addition, feminist scholars have argued that only by taking seriously the perspectives and consciousness of marginalized people will solutions in the name of justice not reiterate harmful biases and constraints: Audre Lorde, "The Master's Tools Will Never Dismantle the Master's House," in *Sister Outsider: Essays and Speeches*, ed. Audre Lorde (Trumansburg, NY: Crossing Press, 1984), 110–13.

13. For a wide-ranging and readable historiography of recent science: Lynn K. Nyhart, "Historiography of the History of Science," in *A Companion to the History of Science*, ed. Bernard Lightman (Boston, MA: Wiley Blackwell, 2016), 7–22.

14. Sharon Traweek, *Beamtimes and Lifetimes: The World of High Energy Physicists* (Cambridge, MA: Harvard University Press, 1988); Steven Shapin, "The Invisible Technician"; Bruno Latour and Steven Woolgar, *Laboratory Life: The Construction of Scientific Facts* (Beverly Hills; London: Sage Publications, 1979).

15. "Tacit knowledge" was a phrase originally proposed by physical chemist and philosopher Michael Polanyi in 1958, in his book *Personal Knowledge: Towards a Post-Critical Philosophy* (Chicago, IL: University of Chicago Press), and was elaborated on by sociologists of science in the 1970s and 1980s: for example, Harry Collins, "The TEA Set: Tacit Knowledge and Scientific Networks," *Science Studies* 4 (1974): 165–86; the Special Issue, edited by Harry Collins: "Knowledge and Controversy: Studies of Modern Natural Science," *Social Studies of Science* 11, no. 1 (1981).

16. For a review, see Léna Soler, Sjoerd Zwart, Michael Lynch, and Vincent Israel-Jost, eds., *Science after the Practice Turn in the Philosophy, History and Social Studies of Science* (New York; Oxford: Routledge, 2017).

17. Projit Bihar Mukharji, "Occulted Materialities," *History and Technology* 34 (2018): 31–40; Donna Haraway talks about the work of creating "a view from above, from nowhere, from simplicity": Haraway, "Situated Knowledges: The Science Question in Feminism and the Privilege of Partial Perspective," *Feminist Studies* 14, no. 3 (1988): 575–99, 589.

18. Roger Sanjek, "Anthropology's Hidden Colonialism: Assistants and Their Ethnographers," *Anthropology Today* 9 (1993): 13–18.

19. See Ann Laura Stoler, *Along the Archival Grain: Epistemic Anxieties and Colonial Common Sense* (Princeton, NJ: Princeton University Press, 2008); Warwick Anderson, "From Subjugated Knowledge to Conjugated Subjects: Science and Globalisation, or Postcolonial Studies of Science?," *Postcolonial Studies* 12, no. 4 (2009): 389–400; Suman Seth, "Putting Knowledge in Its Place: Science, Colonialism, and the Postcolonial," *Postcolonial Studies* 12 (2009): 373–88; Sandra Harding, *The Postcolonial Science and Technology Studies Reader* (Durham, NC: Duke University Press, 2011); Eve Tuck and K. Wayne Yang, "Decolonization Is Not a Metaphor," *Decolonization: Indigeneity, Education & Society* 1, no. 1 (2012): 1–40; the Special Issue, edited by Gabriela Soto Laveaga and Pablo F. Gomez: "Thinking with the World: Histories of Science and Technology from the 'Out There,'" *History and Technology* 34, no. 1 (2018); and the Forum, edited by Gabriela Soto Laveaga and Warwick Anderson: "Decolonizing Histories in Theory and Practice," *History and Theory* 59, no. 3 (2020).

20. Katherine McKittrick, ed., *Sylvia Wynter: Bring Human as Praxis* (Durham, NC: Duke University Press 2015); see also Donna Haraway, *Primate Visions: Gender, Race and Nature in the World of Modern Science* (New York: Routledge, 1989); Kim TallBear, *Native American DNA: Tribal Belonging and the False Promise of Genetic Science* (Minneapolis: University of Minnesota Press, 2013).

21. Anne Fausto-Sterling, "Gender, Race, and Nation: The Comparative Anatomy of 'Hottentot' Women in Europe, 1815–1817," in *Deviant Bodies: Critical Perspectives on Difference in Science and Popular Culture*, eds. Jennifer Terry and Jacqueline Urla (Bloomington: Indiana University Press, 1995), 19–48; Deirdre Cooper Owens, *Medical Bondage: Race, Gender, and the Origins of American Gynecology* (Athens: University of Georgia Press, 2017); for a new exploration of the invisible labour of subjugated research subjects: Rohan Deb Roy,

"Decolonise Mosquitoes," draft of keynote address at the Annual Summer Conference of the Animal History Group, July 2021.

22. Alondra Nelson, *The Social Life of DNA: Race, Reparations, and Reconciliation after the Genome* (Boston, MA: Beacon Press, 2016); Ruha Benjamin, *Race after Technology: Abolitionist Tools for the New Jim Code* (Cambridge, UK; Medford, MA: Polity, 2019); Katherine McKittrick, *Dear Science and Other Stories* (Durham, NC: Duke University Press, 2021).

23. Sandra Harding, *Sciences from Below: Feminisms, Postcolonialities and Modernities* (Durham, NC: Duke University Press, 2008); Wenda Bauchspies and María Puig de la Bellacasa, "Feminist Science and Technology Studies: A Patchwork of Moving Subjectivities. An Interview with Geoffrey Bowker, Sandra Harding, Anne Marie Mol, Susan Leigh Star and Banu Subramaniam," *Subjectivity* 28 (2009): 334–44; Gabriela Soto Laveaga, *Jungle Laboratories: Mexican Peasants, National Projects, and the Making of the Pill* (Durham, NC: Duke University Press, 2009).

24. Robert Proctor and Londa Schiebinger, eds., *Agnotology: The Making and Unmaking of Ignorance* (Stanford, CA: Stanford University Press, 2008); Koen Vermeir and Dániel Margócsy, eds., *British Journal of the History of Science*, "Special Issue: States of Secrecy," 45 (2012).

25. Donna Haraway, "Situated Knowledges." See also Alison Wylie, "Why Standpoint Matters"; John Berger, *Ways of Seeing* (New York: Viking Press, 1973).

26. Ellison, *Invisible Man*, 1.

27. Ludwik Fleck, "To Look, to See, to Know," in *Cognition and Fact: Materials on Ludwik Fleck*, eds. Robert S. Cohen and Thomas Schnelle (Dordrecht, Netherlands: Reidel Publishing, 1986), 129–51, 137.

28. On the Greek philosophical and Christian privileging of vision, see Martin Jay, "The Rise of Hermeneutics and the Crisis of Ocularcentrism," *Poetics Today* 9, no. 2 (1988): 307–26.

29. Carolyn Birdsall, "Talking History: Oral History between Archives, Historiographic Method, and Digital Humanities," presentation given at the workshop "Betwixt and Between: Sound in the Humanities and Social Sciences," MPIWG, February 16, 2018.

30. Oxford English Dictionary, s.v., "Laboratory," accessed September 30, 2021, https://www.oed.com/view/Entry/104723.

31. By "body" we take an expansive view that includes human, mechanical, virtual, animal, and botanical bodies that labour to create knowledge and value. Bodies are literally and figuratively shaped by the work they do: S. Lochlann Jain, *Injury: The Politics of Product Design and Safety Law in the United States* (Princeton, NJ: Princeton University Press, 2006); Michelle Murphy, *Sick Building Syndrome and the Problem of Uncertainty* (Durham, NC: Duke University Press, 2006).

32. The words "work" and "labour" have sometimes had distinctive meanings, although in this book we often use them interchangeably. Karl Marx was focused on the "productive labour" of capitalist industry: Karl Marx, *Capital* (London: J. M. Dent, 1933); Hannah Arendt distinguished between the "work" that resulted in the production of durable objects and the "labours" necessary for life: Hannah Arendt, *The Human Condition* (Chicago; London: University of Chicago Press, 1958), 79–93. On the consequences of understanding intellectual work as "labour": Hales, *Living Thinkwork*; Cooley, "Taylorization."

33. The phrase "bio-labour" is used by Laura Stark in chapter 6; on the gendered and racialized donation of tissues, surrogacy, and participation in clinical trials function in post-Fordist labour politics: Catherine Waldby and Melinda Cooper, *Clinical Labour: Tissue*

Donors and Research Subjects in the Global Bioeconomy (Durham, NC; London: Duke University Press, 2014).

34. Scholarship on the devaluing of women's labour and the continuing invisibility of housework spans half a century and is far from resolved. For example: Ruth Schwartz Cowan, *More Work for Mother: The Ironies of Household Technology* (New York: Basic Books, 1983); Arlene Kaplan Daniels, "Invisible Work," *Social Problems* 34, no. 5 (1987): 403–15; Kathi Weeks, *The Problem with Work: Feminism, Marxism, Antiwork Politics, and Postwork Imaginaries* (Durham, NC: Duke University Press, 2011); Erin Hatton, "Mechanisms of Invisibility: Rethinking the Concept of Invisible Work," *Work, Employment and Society* 31 (2017): 336–51; for a recent rich and wide-ranging collection that deals intersectionally with gender, race, and disability: Marion Crain, Winifred Poster, and Miriam Cherry, eds., *Invisible Labor: Hidden Work in the Contemporary World* (Oakland: University of California Press, 2016).

35. Mar Hicks, *Programmed Inequality: How Britain Discarded Women and Lost Its Edge in Computing* (Cambridge, MA: MIT Press, 2017); Margot Lee, Shetterly, *Hidden Figures: The American Dream and the Untold Story of the Black Women Mathematicians Who Helped Win the Space Race* (New York: William Morrow, 2016); for more on *Hidden Figures*, see Sabine Clarke's commentary.

36. Arlie Hochschild, *The Managed Heart: Commercialization of Human Feeling* (Berkeley: University of California Press, 1983); Pam Smith, *The Emotional Labour of Nursing: How Nurses Care* (Hampshire; London: Macmillan, 1992); for a new appraisal of this field: Agnes Arnold-Forster and Alison Moulds, "Introduction," in *Feelings and Work in Modern History: Emotional Labour and Emotions about Labour* (London; New York: Bloomsbury Academic, 2022).

37. Susan Leigh Star and Karen Ruhleder, "Steps toward an Ecology of Infrastructure: Design and Access for Large Information Spaces," *Information Systems Research* 7, no. 1 (1996): 111–34; Yujie Chen, "Invisible Labor for Data: Institutions, Infrastructure and Virtual Space" (PhD diss., University of Maryland, 2015); Megan Finn et al., "(Invisible) Internet Infrastructure Labor," *AoIR Selected Papers of Internet Research* 3 (2013); Gregory J. Downey, "Making Media Work: Time, Space, Identity, and Labor in the Analysis of Information and Communication Infrastructures," in *Media Technologies: Essays on Communication, Materiality and Society*, eds. Tarleton Gillespie et al. (Cambridge, MA: MIT Press, 2014), 141–65; Lilly C. Irani and M. Six Silberman, "Turkopticon: Interrupting Worker Invisibility in Amazon Mechanical Turk," *CHI 2013* (2013), 611–20.

38. Elizabeth A. Povinelli, "Do Rocks Listen? The Cultural Politics of Apprehending Australian Aboriginal Labor," *American Anthropologist* 97, no. 3 (1995): 505–18.

39. On the ways that institutions turn people into research subjects: Susan M. Reverby, *Examining Tuskegee: The Infamous Syphilis Study and Its Legacy* (Chapel Hill: University of North Carolina Press, 2009); Nathaniel Comfort, "The Prisoner as Model Organism: Malaria Research at Stateville Penitentiary," *Studies in History and Philosophy of Biological and Biomedical Sciences* 40, no. 3 (2009): 190–203; Theodore Porter, *Genetics in the Madhouse: The Unknown History of Human Heredity* (Princeton, NJ: Princeton University Press, 2018).

40. Hannah Atkinson et al., *Race, Ethnicity and Equality Report* (London: Royal Historical Society, 2018).

41. See also, Laura Stark (chapter 6); Burns, "Invisibility"; Susan C. Lawrence, *Privacy and the Past: Research, Law, Archives, Ethics* (New Brunswick, NJ: Rutgers University Press, 2016).

42. For an early statement about the problem of recovering historical anthropological expertise: Sanjek, "Anthropology's Hidden Colonialism"; at a similar moment,

anthropologists began paying close attention to the politics of their own text construction—who is represented, who does the representing, who gets excluded from citations—see Rayna Rapp, *Testing Women, Testing the Fetus: The Social Impact of Amniocentesis in America* (New York; London: Routledge, 2000), 17.

43. For a monumental account of how "go-betweens" "made and changed the contents and the paths of knowledge" in the long eighteenth century: Simon Schaffer, Lissa Roberts, Kapil Raj, and James Delbourgo, eds., *The Brokered World: Go-Betweens and Global Intelligence, 1770–1820* (Sagamore Beach, MA: Science History Publications, 2009); Kapil Raj, "When Human Travellers Become Instruments: The Indo-British Exploration of Central Asia in the Nineteenth Century," in *Relocating Modern Science: Circulation and the Construction of Knowledge in South Asia and Europe, 1650–1900* (Basingstoke: Palgrave Macmillan, 2007), 181–222.

44. Lyn Schumaker, *Africanizing Anthropology: Fieldwork, Networks, and the Making of Cultural Knowledge in Central Africa* (Durham, NC; London: Duke University Press, 2001).

45. Schumaker, *Africanizing Anthropology*, 12.

46. For more recent scholarship on the enculturation, training and guidance of Western researchers, this time in Egypt: Taylor Moore, "Living Room Magic: Ritualistic Ethnography and Esoteric Intimacies," *Superstitious Women: Race, Magic, and Medicine in Egypt (1875–1950)* (PhD diss., Rutgers, 2020).

47. The judgements that contribute to the making of racialized genetic field data are also explored elsewhere, for example, Michell Chresfield, "Genetics, Health and the Making of America's Triracial Isolates, 1950–1980," in *Companion to the Politics of American Health*, eds. Martin Halliwell and Sophie Jones (Edinburgh, UK: University of Edinburgh Press, forthcoming 2022).

48. Michael Gordin, *Scientific Babel: How Science Was Done Before and After Global English* (Chicago, IL: University of Chicago Press, 2015), 10–13; see also the 2016 special issue of *Annals of Science*, "Translating and Translation in the History of Science"; this topic was also taken up in an *Isis* focus section on "Linguistic Hegemony and the History of Science" in 2017.

49. Scott Montgomery, *Science in Translation: Movements of Knowledge through Cultures and Time* (Chicago, IL: University of Chicago Press, 2000); Patrick Manning and Abigail Owen, eds., *Knowledge in Translation: Global Patterns of Scientific Exchange, 1000–1800 CE* (Pittsburg, PA: University of Pittsburg Press, 2018).

50. For emphasis on translation in the history of medicine: Harold Cook, *Matters of Exchange: Commerce, Medicine, and Science in the Dutch Golden Age* (New Haven, CT: Yale University Press, 2007); Shinjini Das, *Vernacular Medicine in Colonial India: Family, Market and Homeopathy* (New York; Cambridge: Cambridge University Press, 2019).

51. Lydia Liu, ed., *Tokens of Exchange: The Problem of Translation in Global Circulations* (Durham, NC; London: Duke University Press, 1999), 1–12.

52. Liu, *Tokens of Exchange*, 13.

53. For a classic example of the large literature on this, see Walter J. Ong, "Before Textuality: Orality and Interpretation," *Oral Tradition* 3 (1988): 259–69.

54. Bruno J. Strasser, Jérôme Baudry, Dana Mahr, Gabriela Sanchez, and Elise Tancoigne, "'Citizen Science'? Rethinking Science and Public Participation," *Science & Technology Studies* 32, no. 2 (2019): 52–76; Elena Aronova, "Citizen Seismology, Stalinist Science, and Vladimir Mannar's Cold Wars," *Science, Technology and Human Values* 42, no. 2 (2017): 226–56.

55. In another vivid story of science as a route to political visibility, Gabriela Soto Laveaga describes how Mexican peasant farmers achieved visibility in president Luis Echeverria's populist political programme via their farming of the wild yam Barbasco for steroid-based pharmaceuticals: Gabriela Soto Laveaga, *Jungle Laboratories.*

56. See also Susan Lindee, "Invisible/Visible Radiation: Skin in the Game at Hiroshima and Fukushima," in *Invisibility and Labour in the Human Sciences*, MPIWG Preprint 484, 2016, 53–64; Susan Lindee, "Survivors and Scientists: Hiroshima, Fukushima, and the Radiation Effects Research Foundation, 1975–2014," *Social Studies of Science* 46, no. 2 (2016): 184–209.

57. Carolyn Steedman, *Dust: The Archive and Cultural History* (Manchester, UK: Manchester University Press, 2001).

58. Michel-Rolph Trouillot, *Silencing the Past: Power and the Production of History* (Boston, MA: Beacon Press, 1995), 26. Trouillot notes that he did not mean this to be a literal, sequential description of history-making but rather a framework for disaggregating where and how power and silence operate in archives. Trouillot's original quote used emphases that we have not included here.

59. Michelle Caswell, *Archiving the Unspeakable: Silence, Memory, and the Photographic Record in Cambodia* (Madison: University of Wisconsin Press, 2014).

60. For a critical approach to the selective nature of biographical archives: Joan L. Richards, "Introduction: Fragmented Lives," *Isis* 97, no. 2 (2006): 302–5.

61. For the Needham Institute, Cambridge: http://www.nri.cam.ac.uk./.

62. Gabriela Soto Laveaga, "*Largo Dislocare:* Connecting Microhistories to Remap and Recenter Histories of Science," *History and Technology* 34, no. 1 (2018): 21–30.

63. Mary Terrall, "Biography as Cultural History of Science," *Isis* 97, no. 2 (2006): 306–13, 308.

64. Jacques Derrida and Eric Prenowitz, "Archive Fever: A Freudian Impression," *Diacritics* 25, no. 2 (1995): 9–63, 61.

65. Burns, "Invisibility."

66. Hannah Landecker, *Culturing Life: How Cells Became Technologies* (Cambridge, MA: Harvard University Press, 2007).

67. Landecker, *Culturing Life*, 171.

68. There is extensive scholarship on the biographies of objects and their roles in producing identity and cultural value; for example the Special Issue, edited by Chris Gosden and Yvonne Marshall, "The Cultural Biography of Objects," *World Archaeology* 31, no. 2 (1999); Lorraine Daston, ed., *Biographies of Scientific Objects* (Chicago, IL: University of Chicago Press, 2000).

69. Bruchac, chapter 4, p. 72.

70. For the importance of mundane practices of care and repair see: Annemarie Mol, "One, Two, Three: Cutting, Counting, and Eating," *Common Knowledge* 17, no. 1 (2011): 111–16; Ruth Müller and Martha Kenney, "Agential Conversations: Interviewing Postdoctoral Life Scientists and the Politics of Mundane Research Practices," *Science as Culture* 23, no. 4 (2014): 537–59; Xan Sarah Chacko, "Creative Practices of Care: The Subjectivity, Agency, and Affective Labor of Preparing Seeds for Long-term Banking," *Culture, Agriculture, Food, and Environment* 41, no. 2 (2019): 97–106; Simon Werrett, *Thrifty Science: Making the Most of Materials in the History of Experiment* (Chicago, IL; London: University of Chicago Press, 2019).

71. Notwithstanding the continuing invisibility of infrastructural work, there are notable efforts being made to put the inner working of systems on display at sites like the Millennium

Seed Bank where the seed curators' workspaces are made visible to public visitors. The transparent glass walls that wrap around the seed laboratories seem to make the curators hypervisible. Susan Leigh Star theorizes this hypervisibility of infrastructure in "The Ethnography of Infrastructure," *American Behavioral Scientist* 43, no. 3 (1999): 377–91.

72. For studies of the sociology and philosophy of care see: Rutger Claassen. "The Commodification of Care." *Hypatia* 26, no. 1 (2011): 43–64; Arlie R. Hochschild, *The Commercialization of Intimate Life: Notes from Home and Work* (Berkeley: University of California Press, 2003); Eva Feder Kittay, "The Ethics of Care, Dependence, and Disability," *Ratio Juris* 24, no. 1 (2011): 49–58; Joan C. Tronto, *Moral Boundaries: A Political Argument for an Ethic of Care* (New York: Routledge, 1993).

73. For recent engagements with care for the non-human, see Carrie Friese, "Realizing Potential in Translational Medicine: The Uncanny Emergence of Care as Science," *Current Anthropology* 54 (S7) (2013): S129–38; Aryn Martin, Natasha Myers, and Ana Viseu, "The Politics of Care in Technoscience." *Social Studies of Science* 45 (2015): 625–41; Annemarie Mol, Ingunn Moser, and Jeannette Pols, eds., *Care in Practice: On Tinkering in Clinics, Homes and Farms* (Bielefeld, Germany: Transcript Verlag, 2015).

74. "Carers" and the "cared for" are co-constructed through the practice of care. This is analogous to Ian Hacking's argument that acts of representation are contingent on making interventions into the processes that one is attempting to capture. Ian Hacking, *Representing and Intervening: Introductory Topics in the Philosophy of Natural Science* (Cambridge, UK: Cambridge University Press, 1983).

75. For examples of feminist attunement to care in science, see Melissa Atkinson-Graham, Martha Kenney, Kelly Ladd, Cameron Michael Murray, and Emily Astra-Jean Simmonds, "Care in Context: Becoming an STS Researcher," *Social Studies of Science* 45, no. 5 (2015): 738–48; Annemarie Mol, *The Logic of Care: Health and the Problem of Patient Choice* (London: Routledge 2008); Michelle Murphy, "Unsettling Care: Troubling Transnational Itineraries of Care in Feminist Health Practices," *Social Studies of Science* 45, no. 5 (2015): 717–37; María Puig de la Bellacasa, "Matters of Care in Technoscience: Assembling Neglected Things." *Social Studies of Science* 41, no. 1 (2011): 85–106.

76. Donna Haraway, *Modest_Witness@Second_Millennium.FemaleMan©_Meets_Onco Mouse: Feminism and Technoscience* (New York: Routledge, 1997); Eduardo Viveiros de Castro, "The Crystal Forest: Notes on the Ontology of Amazonian Spirits," *Inner Asia* 9, no. 2 (2007): 153–72; Marisol de la Cadena, *Earth Beings: Ecologies of Practice across Andean Worlds* (Durham, NC: Duke University Press, 2015); Marisol de la Cadena and Mario Blaser, eds., *A World of Many Worlds* (Durham, NC: Duke University Press, 2018).

77. See Robert E. Kohler and Kathryn M. Olesko, "Introduction: Clio Meets Science," *Osiris* 27 (2012): 1–16.

I

PEOPLE

Commentary: People and the Processes of Erasure

Sabine Clarke

In the wake of the Black Lives Matter movement, a number of new Twitter accounts were created, united by a common purpose—to spotlight the presence of Black scientists working in a wide variety of scientific fields. Communities such as "Black in Microscopy," "Black in Physics," "Black in Chem," and "Black in Environ" often used a similar language to describe their mission—they wished to "illuminate" or "amplify" scientists of colour.

Raising the visibility of Black scientists through social media can be seen as a political act, and one that serves a number of functions. One is to build a community that supports its members (scientists of colour are visible to each other). Another is to expand the image of a professional scientist, maybe to inspire scientific achievement amongst young people (scientists of colour are visible to a wider public). A third is to fight racial injustice by drawing attention to processes of marginalization or erasure (scientists of colour are visible within their respective fields and organizations). It is clear that there is more than one arena in which a person can be rendered invisible, and concomitantly, there is more than one account of science. Science is recorded in the informal writings of the people who do the work; it has a more formal narrative that operates through academic publications, and it also has a public culture of honours and prizes that help create narratives of scientific achievement. The last is slightly adjacent to the second; public discourse works not just to celebrate scientific success but also functions as a discourse about national prestige, for example.[1] Science is not just described in textbooks, patents, and journals; ideas about the character of science, its norms and purpose, are also produced through processes such as commemoration, and in popular books and films. To be visible in one register is not necessarily to be visible in all.

The chapters in this section allow reflection upon the various types of people who can be invisible in science and *where* they are invisible. They focus on three main groups. The first includes people who have claim to be the authors of knowledge. This includes people whose work resulted in advances in understanding, maybe through discovery, innovation or interpretation but who have not necessarily gained a place in accounts of their respective field. The second group are the assistants, including people such as technicians, guides, and collectors. These are the people who do the jobs that allow science to function but by convention are not generally credited for their role in helping to generate new knowledge. This group is often absent from the more official and public narratives of science, only existing in informal or personal accounts of scientific work. Finally, the third group are the people who comprise the material of science: the research subjects enrolled in clinical trials, or the objects of ethnographic study. Members of this group are not erased by the same processes as technicians. They are not so much left out of formal scientific narratives but rendered invisible by processes in which individuals are reconfigured into information or data.

These chapters allow an exploration of the processes by which invisibility occurs, and how this varies according to the role that people play in science, or the narrative of science we are considering. They also explore the function that invisibility might play, through an exploration of those who seek anonymity. They point to a bigger question as well. What does recovery or illumination offer us? What is the bigger project, and can historians accomplish it?

HOW DOES (IN)VISIBILITY OCCUR?

Historians have considered the variety of people that are present in the notes, diaries, and correspondence of scientists. In these narratives we can see the colleagues, the technicians, the research subjects or the wives who accompanied their husbands on field trips.[2] Some of these people might not go on to appear in the more formal narrative of an expedition, research paper, or book. Or they might, but then not appear in the public narrative that is constructed through prize-giving, newspaper reports, exhibitions, or school textbooks and Wikipedia. Invisibility may occur at different levels and by different processes; some are part of doing science, some are part of the narratives used in public to talk about science.

Two of the chapters in this volume consider women—Julia Rodriguez describes the work of the anthropologist Zelia Nuttall (chapter 1), and Lan Li writes about the nutrition scientist and historian Lu Gwei-djen (chapter 2). While both Nuttall and Lu had a high degree of visibility in academic circles during their lifetimes, they have been marginalized in subsequent accounts of their respective fields.

It might seem that the project for the historian is to prove the size of the accomplishments of Nuttall and Lu in order to demonstrate the need for their restoration. This is fraught with problems, not least as it implies that we must first *weigh* our subjects. Women in the past faced barriers which restricted their full access to the things that mark achievement, such as institutional positions, or prizes, so how would we take them *out of time* and evaluate them? What the historian can do more confidently is shed light on the processes that produce obscuration. We can identify the sites of erasure, and show how certain narratives have functioned to diminish, or suppress, significance.

Nuttall and Lu did not necessarily face the same barriers to scientific achievement and recognition as women in earlier periods, such as limited access to formal education and opportunities to publish, or restricted participation in learned societies.[3] Nuttall was an important figure in Americanist archaeology and anthropology at the turn of the twentieth century. She was the author of over seventy papers and manuscripts, and the Mayan *Codex Nuttall* was named for her by the Peabody Museum in 1902. Lu was one of the first women to gain a doctorate in biochemistry at the University of Cambridge; she wrote a number of books and was an important contributor to the monumental *Science and Civilization in China.*

One possible reason for the relative obscurity of Nuttall and Lu nowadays is the nature of their legacy. While both were able to do many things that had previously been difficult for women, neither gained an academic position in her lifetime, led a research school, or trained students. The status and profile bestowed by institutional affiliation contribute to the visibility of a scholar. In practical terms, it can also allow an individual to shape their own legacy so that their perspectives and approaches are passed down to students and younger colleagues. An institutional position, particularly a high-ranking one, makes it likely that an individual can be part of a "family tree" of researchers. The chemist Norman Haworth, for example, taught students and led a research school at the University of Birmingham, United Kingdom, from the 1920s to 1940s. In doing so he was able to foster a network of chemists that reached across academia and industry as his students took positions in other universities and in the chemical firms of the Midlands and North of England. This network extended and perpetuated his research legacy and provided opportunities for future collaborations, including those that stretched across the industry/academia boundary.[4]

Scientists are also made visible, of course, through the practices of citation. Talking about how the claims of scientists become facts through incorporation into the literature of other researchers, Latour says that "to survive or to be turned into fact, a statement needs the *next generation* of papers."[5] The scientific paper is a historical one, in that the current project is situated by the author(s) in relation to previous work. We can observe that it is not just a claim that might disappear if it is ignored in subsequent papers, it is people too.

Some names may also be obscured through the organization of undergraduate and postgraduate teaching. There is a narrative of scientific progress inscribed in the curriculum; it is a history of science, in which students are told who made the major discoveries in their field and are pointed towards the "classic" papers. Invisibility within your field can come down to whether or not your name lives on in citations, or curricula, and this depends upon who teaches the students, or writes the next generation of papers. Nuttall's obscurity may, in part, have arisen as she did not create curricula or a research school. Rodriguez tells us however, that Nuttall's exclusion from official professional positions is not enough of an explanation for the neglect of her achievements since her death in 1933.[6]

Apart from the fact that they did not have academic positions, establish a research school or teach students, what else may be the source of the relative invisibility of Nuttall and Lu? We can note that they were both commemorated through the practices of honorary naming. A codex was named for Nuttall; Lu has a research fellowship and a prize named in her honour at the University of Cambridge. These types of practices are a way in which individuals who have made achievements in particular fields become known in wider culture.[7] Londa Schiebinger has written on the significance of botanical naming practices as a way of honouring individuals and creating a place for them as part of elite in natural history. She notes how few women had plants named for them in the eighteenth and nineteenth centuries, in comparison to men, and how European scientists were elevated over Indigenous people.[8]

In general, women are less likely to be the recipients of the major prizes in science.[9] Obituaries, honorary naming, and the practices of commemoration can move people from their place in the internal narratives of science into the public narrative about science. When we look at the public narrative, however, we can see how the achievements of Nuttall and Lu are framed in ways that act to diminish their significance.

Public descriptions of the lives of Nuttall and Lu subordinate their achievements to those of other people in their fields. Nuttall is painted as an individual who facilitated the work of her fellow (male) anthropologists. She was known "for her ability to find lost or forgotten manuscripts and bring them to the attention of scholars."[10] Nuttall set up home in Mexico around 1902, and her house was a hub for knowledge production in Mexican anthropology. Rodriguez tells us that she has been portrayed by other anthropologists "as a hostess with access to local knowledge and connections for those who did the 'real' research."[11] Conventional narratives of Lu's life and work have a tendency to focus on her personal relationship with Joseph Needham, whom she taught Chinese (less is said about her role as one of the authors of *Science and Civilization*). She is presented as Needham's muse.[12] Both Nuttall and Lu are constructed as individuals whose significance comes from their role as a conduit for the achievements of their male colleagues.

Naomi Oreskes has argued that if women are marginalized in science it is not because they are less objective in the way that they do their work, but because women are not easily accommodated by a narrative of science that promotes "the ideology of scientific heroism."[13] She tells the story of Eleanor Lamson, associate astronomer at the U.S. Naval Observatory, who was acknowledged privately for her work in producing new geophysical data from gravity measurements taken on board the U.S. submarine S-21, but who was absent from the public account and celebration of this expedition.[14] Popular accounts of the voyage constructed this scientific study as one requiring the qualities of strength, endurance, and intelligence, and they portrayed the male scientists involved as uniquely equipped to carry out this difficult work because of their single-minded determination and stoic disposition. In fact, Oreskes tells us that rather than packed with adventure, the defining characteristics of the expedition were probably boredom and routine. Nonetheless, the heroic story that emerged could not accommodate the figure of Lamson. A public discourse of science that relies upon such ideas as domination and conquest and that casts its practitioners as risk-taking, adventuring heroes tends to exclude women.

It is not the case that all women are as absent from the public narrative of science as Lamson was. Women can be present in public discourse, but the roles and attributes that are available to women—hostess, handmaiden, muse—strip them of agency and authority. Representations of Nuttall and Lu place them outside of the narrative of true academic achievement. The cliché is that genius manifests itself in solitude but that the scientific work of women involves care and relationships, maybe occurring within a domestic context.[15] On the issue of place, we might contemplate the fact that Charles Darwin spent the vast majority of his life working amongst his family at Down House in Kent, but the significance of this site for the production of his scientific knowledge is rather obscured by the focus on his more "heroic" voyage on HMS Beagle.[16]

Major achievement in science comes through overcoming adversity, and so the figure of the great scientist is characterized by strength, endurance, emotional detachment and a willingness to defy convention. Women's contributions are rated as less significant as they are represented, and indeed they sometimes represent themselves, in a way that does not fit this public narrative of scientific attributes. (It seems that women cannot win. Evelyn Fox Keller notes that the low visibility of the geneticist Barbara McClintock is said to have stemmed partly from the fact that she was *too* much of an individual; she lacked sufficient allies to gain supporters for her ideas.)[17]

SEEKING ANONYMITY

We can, then, identify the places in which erasure takes place, ranging from those narratives that might be said to be produced "inside" science to those that are generated more on the "outside." In considering how marginalization has occurred, for example, in analysing the nature of representation, we can be confronted, however, with the fact that the people who we hope to recover spoke of their lives and their achievements in ways that contributed to their invisibility. Recovering people can be seen as a form of justice, in which we hope to "right a wrong." So how do we approach a discussion of those people who seem to have found anonymity, or modesty, to have value?

María Fernanda Olarte-Sierra shows how anonymity functions for some forensic scientists working in Colombia (chapter 3). She speaks of three groups of scientists. Some experts involved in identifying victims of violence work largely behind the scenes, others act as expert witnesses in court, and the third group play a social role in working to return the remains of victims to their families. In the first case, anonymity functions at two levels. It confers credibility on laboratory results by conveying the idea that knowledge has arisen objectively, and impartially, without an author—Donna Haraway's "God trick."[18] Olarte-Sierra also shows us the importance of anonymity for laboratory investigators who might otherwise face retribution as their work involves interpreting the evidence and circumstances of violent crimes. Anonymity has an epistemic function but also a practical one. Science is a way of producing both certified knowledge and also a job. By exploring various aspects of the process of seeking justice for the families of victims in which visibility is key, Olarte-Sierra shows us the significance of considering context—national, institutional, and social—for the ways that people may negotiate between the power of using the spotlight and the value of invisibility. The visibility (in terms of tangible presence) of forensic experts is important when they are doing another job—providing care for families who are receiving the remains of their relatives. While on the one hand, emotional labour is often marginalized as economically unimportant women's work, in the context of forensic science in Colombia, it has a function which goes beyond the personal realm and makes a contribution to a national process of healing. Returning to the question of where people are rendered visible or invisible, we can see that alongside the idea that we should remember that there are different accounts of science (some recorded in academic literature, some played out in public) there is a wider political and social context that shapes the representations of people and their work, which is not the same everywhere, or at every moment in time. In the context of forensic science in Colombia, care becomes part of a public discourse about national reconciliation.

Anonymity may function in important ways for a different group of people involved in science. Historical scholarship has worked to recover the people who play subordinate roles in knowledge production, the most well-known intervention being Steven Shapin's "invisible technicians."[19] The focus of this scholarship is on individuals who are under the authority of another, without the power to produce knowledge autonomously, as servants to a master.[20] From this work we see how the process of producing scientific knowledge (or more accurately, the narrative produced in scientific papers that employs such devices as the passive tense) works to obscure the role of technicians in the same way that it plays down the role of machines. The aim is to contrive an impression that facts are produced through direct and unmediated access to nature. Assistants of various types are obscured as part of a narrative of scientific discovery in which the leader or leaders of a project or expedition are the only ones who see the bigger patterns or larger significance from their positions at the top of the hierarchy—less a view from "nowhere," as a view from above. Laboratory technicians, field guides, collectors or Indigenous people are reduced in the creation of a formal scientific discourse to mere instruments that allow elites to see nature.[21]

If historians have sought to recover the people that scientific texts have worked to render invisible, then what about objects? Can we read these in ways that make people in the past visible? Margaret Bruchac offers a way of recovering people and the meanings they forged in objects by looking at the traces left on artefacts and their constituent parts. Her focus is the wampum belts produced by North American Indigenous people and the ways in which the beads that are used in these objects were created, recycled, and repurposed across time (chapter 4). By studying beads and their trajectories as part of different objects at different points in the past, she suggests we have an opportunity to understand better the shifting meaning of things, and therefore the people who made them.

Alexandra Noi considers individuals who worked as guides and assistants in fieldwork in Siberia but who are not visible in formal scientific texts (chapter 5). As well as recovering their stories, she draws attention to the function of a form of sought anonymity for these individuals. Her focus is on ex-prisoners of the Siberian Gulags who were employed as labour for forestry research in the Taiga in the 1950s and 1960s. Noi notes the registers in which they appear and disappear, telling us that the traces of these people, referred to as *bichi*, can be found in the memoirs of scientists but not the formal products of research. We are shown both how these individuals had been marginalized by the state, in political and geographical terms, but also how the hiring practices of scientific fieldwork allowed people to maintain a silence about their previous lives while entering into a new set of social relations. Many of the papers here consider the processes that produce invisibility, suggesting of course that a person must be visible to somebody, somewhere, first, before erasure occurred. Noi studies a different trajectory from "nobody" to "somebody," while also noting how erasure of that new identity occurs later on in the production of a formal account of the scientific findings of an expedition.

The study that Noi gives us sheds light on the way that visibility and invisibility may mean something very different from the perspective of an individual than it does to historians. Historians have noted the erasure of people in the process of making science, who was erased (Indigenous guides, technicians), how they were erased (in producing the particular genre that is scientific writing), and why (in the service of manufacturing authority and objectivity). These understandings of the way that science works are not necessarily important for people for whom performing a role in science is a job. Oral historians have noted that while scholars are concerned with the social and epistemological practices that constitute science at the level of the laboratory, expedition or discipline, scientists might be more preoccupied with their pension plan, or their experiences of a terrible boss. Historian Thomas Lean has said that interviews with laboratory technicians, engineers, and scientists reveal that "they often don't see their activities in the same way as historians tend to. For its practitioners the activities of science are interesting, but for many it is ultimately just a job. In a curious sort of way interviews sometimes give the impression that we as historians of science have a tendency to treat science as something more special than its protagonists do."[22]

While we may explore the location and the processes of erasure, it does not follow that this is how a person experiences their role, or that this fact is key to their identity. You can be subordinate in terms of the bigger project, but it does not follow that you are powerless in your working life. Noi (chapter 5) tells us that employment for *bichi* depended upon word-of-mouth recommendations, suggesting that there was competition (even a hierarchy) amongst these workers. It is not necessarily the case that the bigger hierarchies of the scientific enterprise are of greater importance to an individual than the more immediate power relations that are concerned with their ability to gain employment and succeed in it. An individual may not of course be unaware of wider power structures; it is just that they are dealing with the ones that impinge upon them most obviously.

Seeing science as an array of different jobs (which, while interdependent involve categories of workers that each have their own sub-culture) suggests that it would in fact be possible to participate in science without believing, or supporting, any of the larger epistemological goals of the project. Laura Stark considers the way in which we should understand those people who volunteered to be the subjects of medical research in the mid-twentieth century (chapter 6). In research of this type, and also the anthropological work described by Elise K. Burton, individuality is erased as part of processes that turn people into information (chapter 7). These processes include measuring, classification, and the compilation of data and statistics.[23] Some historians have explored how invisibility can be the end-product of processes that are concerned with finding patterns at the level of the community or group, and in doing so, subsume the person.[24] Burton shows the complexity of the process of classifying people into groups when it involves local intermediaries or informants. Her work examines the intersection of two particular modes of erasure. One is the process in which the subjects of research can find their understandings of their identity cast aside in favour of new and simplistic collective categories under the gaze of anthropologists. The other concerned those who were doing the classifying—Burton explores the way that the visibility and invisibility of people involved in producing knowledge was informed both by their relationship to the communities who were conceived as research subjects, and also each other. In the study of different ethnic groups across Iran, Iranian scientists were valued for the knowledge they supposedly had of community characteristics, but the contribution they made was initially rendered invisible in the accounts produced by the French and American scientists on the project. Local informants were both powerful and visible in the field and invisible in the formal products of the study, apart from when their hypotheses were found to be wrong.

Stark's research is concerned with the paradox that the people who comprise research material, and whom we often see as exploited in some way, entered into such arrangements willingly, and even happily. She tells us that for some groups of religious volunteers, participation in clinical research was valorized through a discourse of sacrifice and "self-abnegation."[25] Stark offers us an answer to the question that we sometimes arrive at as we seek to illuminate the power relations that obscure particular people: why don't all the technicians or research subjects complain? Or the women fight more against their marginalization? Lan Li tells how Lu Gwei-djen supported "patriarchal hierarchies" even as those hierarchies worked to diminish her intellectual achievements (chapter 2). We have noted that people may not feel their own erasure, exploitation or suppression, if this occurs as part of a narrative that they may have little investment or interest in, such as the scientific paper. More importantly, there are narratives which are more readily available to them in which they are validated, lionized, or made visible in other ways. Or to put it another way, our identity is worked out in conversation with a wider set of cultural expectations, constructions, and norms than just the ones that describe work in science. This allows people who have a role in science to mobilize religious, military, and patriotic ideals as "sacrifice," "service," and "duty" in the formulation of identity.

While some tropes are widespread such as the general, cultural expectations of how men and women behave, some norms may exist more at the level of the institution or project. People are constituted in relation to the culture of organizations, institutions or groups, as well as a more general public culture, or the culture of science. Thomas Lean has used oral history to explore the nature of identity in the workforce of the British electricity industry.[26] He notes that the technical staff who maintained Britain's national grid are largely invisible in historical accounts and public perceptions of the industry. Interviews show these workers to have a clear sense of themselves as technicians engaged in a project of national and technical importance, however, and this is largely informed by the "industry's internal depictions of itself," rather than any public representation of their work. The internal organizational culture of the electricity industry during the period when it was nationalized promoted the values of engineering, public service, and masculinity. In their self-fashioning, workers were able to draw upon a narrative that deployed the notion of "heroism of practical knowledge."[27] Invisibility or suppression within a particular narrative cannot necessarily be equated to suppression of an individual. A person may have status, validation, and visibility and constitute their sense of self through reference to a number of discourses or constructions that are available to them, and which may exist outside of those traditionally deployed by scholars in describing science. People have a job in a place, or on a project, not in "science" as a field of knowledge-making practices, as historians and sociologists understand it.

Historian Penny Summerfield has written about the sense of self that we constitute in our personal narratives using the resources available to us in our culture. She refers to the process in which subjectivity is constructed through self-narrativization as being one that aims for "composure."[28] This refers both to composition, in the sense of telling a story about yourself, and also gaining a sense of composure by producing a coherent identity. The absence of public discourses or cultural representations that incorporate a person's experiences can make it difficult to gain composure or coherent subject identity. She suggests that women whose activities do not match the traditional expectations of feminine behaviour by, for example, taking on a role normally reserved for men, "may have difficulty finding words and concepts with which to compose their memories, whether in anecdotal snapshots or extended narratives."[29] Summerfield tells how interviewees could be silent when it came to important moments in their lives in which they had done things that might be considered unusual for their sex because no public narrative that included such experiences was available to them.

Alison Blunt has also explored the ways in which an individual may strive to produce a self-narrative in their writing that places them within the conventions or expectations of their time, despite the fact that they behaved in ways that were transgressive. Her study of the nineteenth-century explorer Mary Kingsley shows the way that in her travel writing Kingsley downplayed the dangers she sometimes faced in West Africa or the fact that she travelled alone, by emphasizing the femininity of her dress, or using humour to undercut the threats she faced.[30] We can reflect upon the fact that women scientists whose actions might potentially place them outside the limits of propriety may choose to present themselves in stereotypical ways. Transgressive actions can prompt a form of self-narrativization which deals in feminine ideals such as modesty, empathy, and care. Self-erasure may be part of the process of producing a coherent self for an individual—as it might have been for Lu who had left her home country, defied educational expectations and had an affair with a married man. What we may identify as invisibility or marginalization may be the outcome of a process in which an individual has sought composure. It comes from a mode of self-narrativization in which a person seeks to reconcile their choices and actions with the expectations of their place in the world.

WHAT IS THE BIGGER PROJECT?

What we do with our analysis of the various accounts that exist of individuals in science, both those that an individual might tell about themselves and the ones in which they figure, depends on what we believe to be our bigger project. Scholars have noted the problem facing the historian who may wish to recover or rehabilitate their previously invisible subject. Is the goal to add a person to the pantheon of heroes? Popular books for children and adults that are focused on the achievements of women such as Jane Goodall or Mary Anning appear to be doing just that.[31] The problem, if we follow the analysis of Oreskes, is that in doing this we are probably relying upon ideas about the way that science works that are informed by the ideology of heroism, a set of "masculine" ideals that doesn't accurately reflect the nature of science. Or, according to Londa Schiebinger, "retains the male norm as the measure of excellence."[32] In popular accounts, science is about discovery, often a flash of insight or an unexpected find, it is thrilling and characterized by exciting noise, smells, and visual phenomena, but most importantly, it is about individual endeavour, often in the face of adversity.

Arguably, the stereotype of the lone scientist is the most pervasive and problematic aspect of popular representations of science. It is striking how the covers of books about women scientists aimed at children and teenagers show a single figure on the cover against a backdrop that represents their field of work. Some of these show the lone figure dwarfed by enormous seas, skies, jungles and so on to convey how massive nature is, and therefore how great the accomplishments of these women are in the face of such vastness. There is the oceanographer Sylvia Earle as a small figure diving alone in the deep blue sea; the astronomer Henrietta Leavitt, in solitude, head tilted up to the night sky; Katherine Johnson thinking hard about maths, on her own (see figure PI.1).[33]

Katherine Johnson's fame grew considerably with the publication of Margot Lee's book, *Hidden Figures*, and the release of a film loosely based on it. *Hidden Figures* tells of the important computational, mathematical, and engineering work done by three Black women who worked at NASA—Johnson, Dorothy Vaughan, and Mary Jackson. It is an inspiring tale and an important corrective to the popular image of the Space Race as an arena dominated by the actions of white men. The film is

Figure PI.1. One of set of free, downloadable posters created by Amanda Phingbodhipakkiya to celebrate women and science. This poster shows the figure of Katherine Johnson seated at her desk working on a mathematical problem. Copyright Amanda Phingbodhipakkiya (CC BY 4.0).

also useful as an insight into the commonplace portrayal of previously invisible or marginalized individuals in science. The story of Johnson, Vaughan, and Jackson relies on the clichés of heroic endeavour, showing their fortitude, bravery, and even recklessness as they fight to overcome the sexism and racism that works to exclude them. Interestingly, while the story might seem to be a collective one with three women at the heart of the narrative, the scenes of scientific or technical triumph for Johnson, Vaughan, and Jackson that occur in the film are very much separate moments in which their individual genius wins the day. Each is afforded their own, discrete "eureka" moment in their respective fields of computing, engineering, and mathematics. That films and books employ narratives that are centred on the way that individuals defy convention and produce brilliant insights is unsurprising, but it is striking how inaccurate they are in portraying science as it really happens. In the world of book covers, for example, female scientists, mathematicians, astronauts, explorers and so forth have no colleagues, or technical support. The visibility afforded to "remarkable women," involves erasing other people in this popular format.

The alternative to recovering individuals is to focus on the interdependence of all the types of work necessary to generate and promote knowledge. The project then becomes rehabilitating not individuals but neglected modes of work in science including support, networking, managing, and so on. In thinking about the various types of work that science depends on in this way, we also then need to remember that heroic metaphors are not actually accurate descriptions of the work done by men. They are just as likely as women are to be doing a job in science that could be described as "care," such as a well-respected laboratory manager who ensures that his team are protected from wider institutional politics, for example, or who always secures the equipment that they need.

Historians and sociologists have emphasized for some time now that science is a collective enterprise. We know that nothing becomes a "fact" in science without endorsement from the wider community, and that dialogue and exchange are defining characteristics of the field. Large projects have made the correspondence of important figures such as Charles Darwin and Joseph Hooker widely available, and scholarship has emphasized the importance of networks, translation, and internationalism in the establishment of scientific ideas. In the related field of the history of technology, we see increasing interest in people who are the "maintainers" rather than just inventors.[34] The maintainers are the people who repair and maintain objects and systems, and without whom these things would not function.

The challenge for historians is not so much to reorganize our studies so we focus on the various and interdependent types of work on which science depends, as this has been key to the way that we pursue our work for a while, but to reconceptualize the way we engage with the public conversation about science. The problem, of course, is that much of that narrative is not actually concerned with providing a description of science, but rather it should be considered to be doing work of various kinds. The public discourse about science is made up of stories intended to inspire, entertain, teach, celebrate the nation or justify policy decisions. The fact that our cultural constructions of science have a function in this way is not perhaps in itself a problem. The issue is more when the purpose of this sort of narrative has been obscured. The sites in which a public narrative is produced that should most concern the professional historian are the ones that might be taken to be a true and validated account of how science actually works. By this reckoning, it is the textbooks, centenaries, encyclopaedia, and museums with their focus on individual achievement that might be considered as opportunities to emphasize the connected nature of science work and its practitioners.

NOTES

1. For a very good account of the politics of prizes, see Elisabeth Crawford, *Nationalism and Internationalism in Science, 1880–1939: Four Studies of the Nobel Population*, (Cambridge: Cambridge University Press, 2009). On commemoration, see George E. Haddad, "Changing Practices of Commemoration: Representing Robert Koch's Discovery of the *Tubercle Bacillus*," Osiris 14 (1999): 118–37, plus other chapters in this volume.

2. For work that has considered collaboration between couples in science, see Annette Lykknes, Donald L. Opitz, and Brigitte Van Tiggelen (eds.), *For Better or For Worse? Collaborative Couples in the Sciences* (Basel, Switzerland: Birkhauser, 2012).

3. Londa Schiebinger, *The Mind Has No Sex: Women in the Origins of Modern Science*, (Cambridge, MA: Harvard University Press, 1991); Anna B. Shteir, *Cultivating Women, Cultivating Science: Flora's Daughters and Botany in England, 1760–1860* (Baltimore, MD: Johns Hopkins University Press, 1996); Barbara T. Gates, *Kindred Nature: Victorian and Edwardian Women Embrace the Living World* (Chicago, IL: University of Chicago Press, 1998).

4. Sabine Clarke, *Science at the End of Empire: Experts and the Development of the British Caribbean, 1940–1962* (Manchester, UK: Manchester University Press, 2018), chapter 6.

5. Bruno Latour, *Science in Action: How to Follow Scientists and Engineers through Society* (Maidenhead, Berkshire, UK: Open University Press, 1987), 38; see also Latour and Steve Woolgar, *Laboratory Life: The Construction of Scientific Facts* (Princeton, NJ: Princeton University Press, 1986), chapter 2.

6. Rodriguez, chapter 1, 40.

7. Londa Schiebinger, *Plants and Empire: Colonial Bioprospecting in the Atlantic World* (Cambridge, MA: Harvard University Press, 2004); on commemoration, see *Osiris*, volume 14, no. 1 (1999), *Commemorative Practices in Science: Historical Perspectives on the Politics of Collective Memory*.

8. Schiebinger, *Plants and Empire*, "Linguistic Imperialism."

9. Per Lunnemann, Mogens H. Jensen, and Liselotte Jauffred, "Gender Bias in Nobel Prizes," *Palgrave Communications*, 5, article no. 46 (2019), https://www.nature.com/articles/s41599-019-0256-3.

10. "Zelia Nuttall," Wikipedia, last modified November 13, 2021. https://en.wikipedia.org/wiki/Zelia_Nuttall.

11. Rodriguez, chapter 1, 43.

12. "Loved China. Maybe Too Much. Also Loved Women," *The Globe and Mail*, posted May 31, 2008. https://www.theglobeandmail.com/arts/loved-china-maybe-too-much-also-loved-women/article720009/.

13. Naomi Oreskes, "Objectivity or Heroism? On the Invisibility of Women in Science," *Osiris* 11 (1996): 90.

14. Oreskes, "Objectivity or Heroism," 95.

15. On domestic contexts for science, see Donald L. Optiz, Staffan Bergwik, and Brigitte Van Tiggelen, *Domesticity in the Making of Modern Science* (London: Palgrave Macmillan, 2016).

16. On the significance of domesticity for Darwin, see, Paul White, "Darwin's Home of Science and the Nature of Domesticity," in *Domesticity in the Making of Modern Science* (London: Palgrave Macmillan, 2016), chapter 3.

17. Evelyn Fox Keller, *Feeling for the Organism: The Life and Work of Barbara McClintock* (New York: Holt Paperbacks, 1983).

18. Donna Haraway, "Situated Knowledges: The Science Question in Feminism and the Privilege of Partial Perspective," *Feminist Studies* 14, no. 3 (1988): 575–99.

19. Steven Shapin, "The Invisible Technician," *American Scientist* 77, no. 6 (1989): 554–63.

20. Shapin, "Invisible Technician."

21. Schiebinger, *Plants and Empire.*

22. Thomas Lean, "The Voices of Science Past," *Viewpoint: The Newsletter of the British Society for the History of Science* 95 (June 2011): 3.

23. For an interesting discussion of data production and invisible labour, see, Michael J. Scroggins and Irene V. Pasquetto, "Labor Out of Place: On the Varieties and Valences of (In) visible Labor in Data-Intensive Science," *Engaging Science, Technology and Society* 6 (2020): 111–32.

24. John Harley Warner, *The Therapeutic Perspective: Medical Practice, Knowledge, and Identity in America, 1820–1885* (Princeton, NJ: Princeton University Press, 1997); James C. Scott, *Seeing Like a State: How Certain Schemes to Improve the Human Condition Have Failed* (New Haven, CT: Yale University Press, 1999).

25. Stark, chapter 6, p. 95.

26. Thomas Lean, "The Life Electric: Oral History and Composure in the Electricity Supply Industry," *Oral History* 46, no. 1 (2018): 55–66.

27. Thomas Lean, "Life Electric," 58.

28. Penny Summerfield, "Culture and Composure: Creating Narratives of Gendered Self in Oral History Interviews," *Cultural and Social History* 1, no. 1 (2004), 65–93. Summerfield draws upon the idea of composure formulated by Graham Dawson, *Soldier Heroes: British Adventure, Empire and the Imagining of Masculinities* (London: Routledge, 1994).

29. Summerfield, "Culture and Composure."

30. Alison Blunt, *Travel, Gender and Imperialism: Mary Kingsley and West Africa* (New York; London: The Guildford Press, 1994).

31. A nice selection can be found at https://www.amightygirl.com/.

32. Schiebinger, *The Mind Has No Sex*, 6.

33. Claire A. Nivola, *Life in the Ocean: The Story of Oceanographer Sylvia Earle* (New York: Farrar, Strauss and Giroux 2012); Robert Burleigh, *Look Up! Henrietta Leavitt, Pioneering Woman Astronomer* (New York: Simon and Schuster, 2013); Thea Feldman, *Katherine Johnson* (New York: Simon Spotlight, 2017).

34. Lee Vinsel and Andrew L. Russell, *The Innovation Delusion: How Our Obsession with the New Has Disrupted the Work that Matters Most* (New York: Currency, 2020).

1

Under the Mexican Sun

Zelia Nuttall and Eclipses in Americanist Anthropology[1]

Julia E. Rodriguez

In early November 1909, from her home in Mexico City, Mexican-American archaeologist Zelia Nuttall responded to a letter from Franz Boas. He had written to request assistance in the foundation of an International School of American Archaeology and Ethnology in Nuttall's adopted country. The project had been underway for a number of years, with the dual goals of giving U.S. anthropologists a foothold in Mexico's rich and fertile terrain and expanding Mexican anthropological institutions. Boas had appealed to Nuttall as an instrumental figure in this effort: "Excuse me if I trouble you again with our affairs; but your interest in American archaeology is such, that I know I can always rely upon your help when matters pertaining to the advancement of science are concerned."[2] Nuttall wrote back immediately, "You can count on me for doing all I can to further the cause of our beloved science."[3]

Who was Zelia Nuttall? And why has she, and her significant body of work, been largely forgotten? Nuttall was not a minor figure in Americanist and Mexican anthropology but the intellectual equal of recognized giants in the discipline, including Boas and his Mexican student Manuel Gamio.[4] Her first publication, on terracotta artefacts she picked up in Teotihuacán as a young woman, appeared in 1886, and her last, on a ruin in Monte Albán appeared in 1932, a year before her death. In between, Nuttall published a book and over seventy papers and manuscripts in all the major journals of her day, including the *American Anthropologist* and *Science*. While she lacked the institutional and financial support enjoyed by her male counterparts, she was an authority in an astonishing array of subfields for close to five decades.

While scholars continue to expose the marginalized status of women in anthropology at the turn of the century, historians have largely overlooked Nuttall in the history of Americanist anthropology. Americanists like Boas and Gamio, as well as Harvard's Alfred Tozzer and the Mexican bureaucrat Leopoldo Batres, have gained central roles in the narrative of Americanism, while only traces of Nuttall's life and work remain. This cannot be attributed to her status outside of professional spaces. In fact, in her lifetime she was a major presence at Americanist events and institutions in Europe, Mexico, and the United States. Nuttall fought against the forces of concealment that conspired to marginalize her from the public spaces of anthropological and archaeological knowledge production. But after her death in 1933, her success in the field faded from view and her legacy was overshadowed. The eclipse of Nuttall as a major figure of Americanist anthropology can be attributed to multiple interlocking factors, including her gender and her location in Mexico, itself shaped by enduring geopolitical power dynamics that favor U.S.-centric histories of anthropology. Disregarding Nuttall's contributions has also hidden the interdependent nature of knowledge production in Americanist anthropology and archaeology.[5] Nuttall stood at the centre of a complex community of local and transnational actors, including not just visiting prominent scientists but also her family members, female friends, domestic staff, neighbours, craftspeople and people living and working near archaeological sites. Her status in this complex world reflected both the power and the limits of her privilege.[6]

This chapter expands on previous attempts to draw attention to Nuttall's contributions to anthropology by putting her gender and other biographical factors in the larger context of Americanist science.[7] An examination of Nuttall's life reveals how layered forces, including gender expectations, institutional exclusion, and the geopolitics of knowledge production not only shaped the lived experiences of marginalized peoples in science but continue to throw shadows on the process of historical memorialization. Nuttall's extraordinary career and her erasure from anthropology's central historical narrative are all the more extraordinary since her contributions to knowledge are visible in the published record.

I first encountered Nuttall's work in the course of my research on the formation of Americanism in the late nineteenth century, a new scientific project that was counterpart to the better-known Orientalist movement and precursor of twentieth-century anthropology and Latin American Studies.[8] As Nuttall began to emerge as a major protagonist of this transnational scientific network, I looked for her personal papers, visited her historic home, the Casa Alvarado, in the Coyoacán neighbourhood in Mexico City, searched archives in both Mexico and the United States and even attempted to contact her descendants—all in vain. It appears that after her death, Nuttall's possessions were dispersed, if not destroyed. She had spent her entire inheritance, dying penniless and in debt. Subsequently, her beloved home and most of its belongings had to be sold. The material realities of Nuttall's lifelong exclusion from secure and paid positions in anthropological institutions would shape her life course as well as the reception of her legacy. Nevertheless, Nuttall's nearly fifty years of work produced an immense body of publications and letters, which allow us to reconstruct her influence on Americanist anthropology.[9]

Nuttall's life story tells us not just about women in science and the gendered spaces of U.S. and Mexican anthropology. It also tells us something about the history of Americanism itself. Nuttall's experience provides a window on the complex dynamics generated as individuals, ideas, local conditions, and geopolitical forces coalesced to create the new field of Americanist anthropology. An intersectional approach—taking account of Nuttall's marginal statuses as a woman and as a transnational scholar living *on site* in Mexico—shines new light on the politics of scientific recognition in the history of the human sciences.[10]

Zelia Nuttall's unique biographical details explain much of her extraordinary scholarly accomplishments in early Americanist anthropology. The first detail was her Mexican heritage; the second her wealth; the third her transnational upbringing and multilingual abilities. Underlying these factors was her intelligence, her passion for discovery and intellectual life, and her boldness. She was entirely self-taught in her chosen fields. From young adulthood it was clear to all who encountered her that Nuttall was a bright and determined woman.

Zelia María Magdalena Nuttall was born on September 6, 1857 in San Francisco to a well-off Irish-American father, Dr. Robert Kennedy Nuttall, and the daughter of an elite Mexico City family, Magdalena Parrott Nuttall. The couple established their home in San Francisco but raised their children largely in Europe, where they learned multiple languages. Over the course of her career Nuttall would publish her scientific work in French, German, and Spanish in addition to English; her linguistic abilities were mentioned by most admirers.

In her early twenties, Nuttall was briefly married to French anthropologist Alphonse Louis Pinart, with whom she travelled through the West Indies, France, and Spain. After divorcing him in 1888, she fought for, and was granted, full custody of their only child, Nadine, as well the right to reclaim her maiden name—a remarkable series of events for that era. The marriage was brief and unhappy, but this period introduced Nuttall to anthropology and demonstrated her interest in, and ability for, both fieldwork and ethnology.[11]

After her divorce in 1888, and before her permanent transfer to Mexico City in 1904, Nuttall published what would be her longest work, a book titled *Fundamental Principles of Old and New World Civilizations* (1901); over twenty articles on Mexican artefacts, archaeology, and the Maya calendar; and what would become known as the *Codex Nuttall* (1902). She regularly attended scientific meetings such as the International Congress of Americanists (ICA) and the American Association for the Advancement of Science (AAAS), joining a tiny group of women attendees that sometimes included North American Alice Fletcher and Argentine Juliane Dilenius.

Nuttall moved to Mexico City between 1902 and 1904. Settled in her beautiful colonial house in the historic neighbourhood Coyoacán, she raised her daughter and worked. She was part of a cadre of European and North American anthropologists who spent decades living and working in Latin America at the turn of the century.

She explained the advantages of her move to Mexico in a letter to her mentor Fredrick Ward Putnam in 1902:

> I have been realizing that if I am to spend a part of each year here for several years to come, and am to work here, I *must* have my books at hand and a fixed place of residence to retire to, to work up my material. . . . The best thing for me to do is to establish my headquarters in one of the suburbs of the City of Mexico, where I can have . . . perfect tranquility, a large garden, and a spacious house. . . . I will be at the fountain-head and able to do lots of good work. . . . The climate here suits me splendidly.[12]

Nuttall's thirty-year residence in Coyocán, along with her Mexican heritage, likely account for her championing of Mexico in a way that few of her foreign-born colleagues did. Embedded in this rich field, Nuttall began work on diverse topics such as ethnobotany, Mexican art, architecture, colonial history, and folklore. While after her death Harvard's Alfred Tozzer stated that Nuttall's accomplishments "[rested] more firmly upon her ability to find lost or forgotten manuscripts and bring them to the attention of scholars," there is no doubt that her work went well beyond treasure hunting, as her significant body of work reveals.[13]

Nuttall's legacy is perhaps most evident in the Codex (pre-Columbian manuscript) that bears her name. Nuttall immediately recognized that the manuscript was a marvellous and untapped source, with a "wealth of detail and interest."[14] While the Codex's material qualities indicated mastery of Mixtec technical skill and cultural complexity, Nuttall also plumbed it for sociocultural knowledge, for example, noting its "wealth of fresh knowledge, especially concerning the dress, ceremonial observances, and position of women."[15]Another of her major preoccupations was the Aztec Calendar. In 1893 she wrote to Daniel Brinton, a prominent anthropologist at the University of Pennsylvania, that she was wrapping up her work as a judge at the Chicago Exhibition, and looking forward to "my next line of investigation [which] will be on the symbolism of the day names, their connection with deities and constellations or planets, and historical data. I have a mass of material collected already and shall make all efforts to publish my conclusions as soon as possible."[16]

Nuttall's peers were equally excited by her findings. Brinton wrote in 1895 that "the profound studies of the Mexican Calendar undertaken by Mrs. Zelia Nuttall have vindicated for it a truly surprising accuracy which could have come only from prolonged and accurately registered observations of the relative apparent motions of the celestial bodies."[17] In 1902, Eduard Seler called the study of Mexican glyphs "so intricate and difficult a subject"; the Nuttall Codex, he remarked, was "precious . . . [and] rightly bears the name of its discoverer."[18]

The trajectory of Nuttall's career took shape in the context of expanding transnational scientific networks of Americanism, anthropology, and archaeology. This time period—the 1870s to about 1930—saw the rise of professional anthropology and archaeology and a transition from European to U.S. dominance of these fields. Before 1900 there were very few college courses on Latin America in North America; similarly, only after 1910 was Spanish widely taught. That said, as the twentieth century approached, focus on Latin America intensified due to the economic interest in the region and a push for hemispheric diplomacy; these agendas required regional "experts" in a variety of areas, including culture.[19] Insofar as Nuttall moved in U.S. scientific circles, the rarity of her expertise on Mexico and her linguistic abilities put her at a great advantage. If scientific knowledge is the result of collaboration, Nuttall was a central node in a complex community of local, national, and transnational actors affected in different ways by the social structures in which they lived. Enjoying relative wealth and autonomy in Mexico City, Nuttall enjoyed privileges greater than most Mexican women—including paid domestic and professional labour (and including the Indigenous guides shown in figure 1.1). Her social position compared to visiting foreign male scientists, at times, gave her advantages (such as her familiarity with places and people) but also defined the limits of her social status.[20] Despite

Figure 1.1. Zelia Nuttall (second from right), enjoying an outing in Xochimilcho, the floating gardens of Mexico City with fellow anthropologists Alfred Tozzer (on her immediate right) and Franz Boas. The scientists are flanked by two unnamed Indigenous guides, one of whom propels the boat. The photograph was taken in 1909 when Tozzer and Boas visited Mexico and relied on Nuttall's local knowledge. The photograph serves as a reminder that all three were dependent on Indigenous expertise and labour, which they were able to command owing to their wealth and social status. Courtesy of the Peabody Museum of Archaeology and Ethnology, Harvard University, 2004.24.28554.

her coveted access to Mexican land and people, Nuttall's two male colleagues are more richly preserved in anthropology's archive.

In spite of her remarkable and visible productivity, however, other anthropologists frequently described Nuttall primarily as a hostess with access to local knowledge and connections for those who did the "real" research. But abundant evidence belies the limited role attributed to her. Over the course of her long career, Nuttall had decades-long affiliations with emerging anthropology institutions at Harvard and the University of Pennsylvania, as well as a shorter but significant relationship with the University of California. She also held an honorary post at Mexico's National Museum. Nuttall was one of few women elected to the AAAS and the American Philosophical Society and served on the ICA Council three times between 1894 and 1910.

Formal and informal networking was part of her strategy and probably accounted for her reputation as a conduit for male scientific accomplishments. On the personal level, Nuttall sought introductions, funding, and intellectual collaboration. She allied herself with influential men like Putnam at Harvard and Boas at Columbia. Powerful men in U.S. anthropology museums, such as Brinton and Charles Bowditch (at Harvard), identified Nuttall's talents early on and encouraged her to engage in the emerging field of Americanism. She had a particularly close relationship with Putnam, calling herself his "devoted god-daughter." In an 1886 letter she responded gratefully to his encouragement:

> My breath was fairly taken away by your words about reading a paper myself. You who have lived in the atmosphere of the emancipation of women can scarcely understand, I am sure, the effect my European bringing-up has still upon me. It seems to me almost over-bold to put myself forward in print and the idea of my standing up and endeavoring to convey information to persons savvier than I, fairly terrifies me.[21]

By 1892, Nuttall had gained more confidence in her work. She wrote to Putnam about her work on the Mexican Calendar: "It is the only piece of work I have ever made that I am *absolutely satisfied with, for it proves itself right—no one can doubt the system once having examined it.*"[22]

Another strategy available to Nuttall was sustained relationships with female benefactors like philanthropist Phoebe Hearst, who provided financial and moral support, and with other women anthropologists. Some, like Alice Fletcher, were allied with men in paid posts at universities, or worked on Smithsonian Institution projects. In 1886, Nuttall recounted the strong connection between these women, noting, "The pleasure I had in meeting Miss Fletcher was very great. . . . Besides being an ethnologist she is a philanthropist, she is a woman of head and heart, and as such, commands my warmest admiration."[23] Another collaborator was Adela Breton, a British-born friend and skilled artist who first accompanied Nuttall to Chichén Itzá in 1900. Breton was eventually in demand as copyist of pre-Columbian murals and artefacts, but according to her biographer, she was content to remain an auxiliary of male-dominated archaeology.[24] Not so for Nuttall. She was ambitious, but realistic about the limits to her welcome.

If women were excluded from the central posts of North American anthropology, the situation was more extreme in Nuttall's chosen land.[25] As a wealthy U.S. citizen with good connections, she was able to bypass many of the typical obstacles presented to ambitious women in patriarchal Mexico. Ruíz argues that Nuttall in fact created an alternate site for science in her house in Coyoacán – it was both a female-centred space *and* a site of Mexican anthropological production, "an archaeology practiced from *home*."[26] As important as Nuttall's home was, however, we cannot forget that she also succeeded in creating a presence in public institutions for over forty years. For a story with some striking resemblances in this regard, see Lan A. Li's (chapter 2) on Chinese biochemist Lu Gwei-djan.

In Mexico, Nuttall had to navigate around high status men like Gamio, Batres, and foreign anthropologists who spent significant time at Mexican sites, such as Boas, Saville, and Tozzer. She had both productive and conflictual relationships with these men. For example, soon after her arrival in Mexico, even before moving to the Casa Alvarado, she recounted that she was visited by the notoriously brash Batres; their relationship would deteriorate later.[27] Her relationship with prominent German anthropologist Eduard Seler was also tense. In May 1902, as the *Codex Nuttall* was about to be published, she wrote to Putnam that "Dr. Seler is quite wild with envy and disgust at me for not having cited him more in my book as an authority. He and his creature (Preuss) will seize every change to attack me and are doing so—but I do not care and am quietly preparing a defensive missile that will settle both at once. I am biding my time."[28]

Nuttall's relationship with U.S.-based Boas was more constructive, and yet behind her back he complained about her temperament. In a 1908 letter to Seler, Boas mentioned that "Mrs. Nuttall was here a few days ago. . . . From what I know in regard to the conditions in the City of Mexico I think she might hinder us a good deal if we have not her good will."[29] In reality, Boas and Nuttall collaborated intensely on the development of the International School in Mexico. He saw her as crucial to the project, given her connections mentioned in the 1909 letter cited before, in which he asked Nuttall to help him identify members of the organizational committee of the school, albeit without offering her an official position.[30]

This episode is at once emblematic of Nuttall's marginalization and a limited lens with which to understand her experiences. Nuttall and Boas' relationship was much more extensive than the planning of the International School. They interacted for decades at scientific conferences and worked together on the Chicago Exhibitions. He praised her work on its own merits and publicly recognized her contributions to the developing disciplines of archaeology and anthropology. He was only one example of powerful men with growing authority in anthropological institutions who selectively included Nuttall in the Americanist project. But she could only remain a member of the founding generation of anthropology as long as she was productive. By the end of her life, Nuttall was in financial ruin despite her long career, prolific output, collaborations, and recognition. The lack of a steady position ultimately made her vulnerable, and her exclusion from institutional posts meant that she had no students to build on her legacy. Nuttall had been her own best advocate, so when she died, her image began to fade.

Since the 1990s, a small number of scholars have studied Nuttall, and those who do express astonishment at the lack of attention to her. Ross Parmenter, a journalist who wrote a biography of Nuttall's former husband Pinart in 1966, spent decades researching and writing a comprehensive study of her life and work. Parmenter's efforts resulted in a (single-spaced) 3,000-page unpublished manuscript that he bequeathed to the Latin American Library at Tulane University (copies are at Harvard and the University of California as well). Aside from remaining sequestered in these university archives and therefore out of reach for a public readership, Parmenter's work is inaccessible in another way: the manuscript, rich with historical detail, lacks citations.[31]

The only serious scholarly analysis of Nuttall's work is Carmen (Apen) Ruíz's 2003 doctoral dissertation on gender and Mexican archaeology and her 2016 book, published in Spanish. Ruíz examined Nuttall through the frame of "insider/outsider," arguing that while she had more opportunities to do science than most women at the time, she was ultimately excluded from the professionalizing discipline due to the "strongly patriarchal character of Mexican archaeology" that prevented her from conducting fieldwork.[32] As the public record shows, however, these difficulties did not entirely prevent Nuttall from having a long and fruitful career recognized by her peers. No doubt, Nuttall enjoyed privileges compared to other women. But in the end, a lack of secure employment and institutional backing meant that at the end of her life and career she slipped into obscurity. The rising stars of her male colleagues cast a shadow on her place in the history of anthropology in two ways. First, in her lifetime, she was denied a central role in the Mexican institutions, which favoured men like Gamio and Batres. At the same time, histories of anthropology have centred the contributions of U.S. anthropologists and institutions, with Mexicans seen as passive sources of raw material for scholars in New York, Washington or Berkeley. Nuttall's biography upends both of these narratives. As she published in the major U.S. and European journals and collaborated with the most powerful figures in anthropology, she chose to immerse, centre, and ground herself in Mexico. At the same time, her embeddedness in the field, fluency in Spanish, and access to local knowledge ultimately did not pay off for Nuttall.

An examination of Zelia Nuttall's public and epistolary record brings her out of the shadows as an Americanist par excellence. Her oeuvre, and especially her final publications in the 1920s, should place her among the leading figures of movements like *indigenismo* and Pan Americanism, in addition to anthropology.[33] It is easy to understand the obstacles Nuttall faced in her lifetime; less so her fading from the historical record. Her story reveals the enduring imbalance of power in knowledge production beyond gender exclusion and the internal dynamics of a professionalizing discipline. It is also a reminder that historians have imagined the history of Americanism without much of America.

NOTES

1. Many thanks to Jackie Gilbert for research assistance. Funds for archival research were generously provided by the UNH History Department, the UNH Center for the Humanities and the National Science Foundation (Award # 0547125).

2. Boas to Nuttall, October 13, 1909, American Philosophical Society, Boas collection, Correspondence B, box 61. Nuttall, Zelia.

3. Nuttall to Boas, November 3, 1909. American Philosophical Society, Boas collection, Correspondence B, box 61. Nuttall, Zelia.

4. For histories of Mexican anthropology in the nineteenth century, see Mechthild Rutsch, *Entre el campo y el gabinete. Nacionales y extranjeros en la profesionalización de la antropología mexicana (1877–1920)* (Mexico City: INAH, 2008); Christina Bueno, *The Pursuit of Ruins: Archaeology, History, and the Making of Modern Mexico* (Albuquerque: University of New Mexico Press, 2016).

5. See Sabine Clarke's commentary for insightful discussion of interdependence and work modes.

6. See the volume Introduction for a discussion of how knowledge production is shaped by structures of power, such as politics, economy and labour.

7. Carmen Ruíz, "Insiders and Outsiders in Mexican Archaeology (1890–1930)," doctoral thesis, University of Texas, 2003; Apen Ruíz Martínez, *Genero, ciencia y política: Voces, vidas y miradas de la arqueología mexicana* (Ciudad de Mexico: INAH, 2015).

8. Karin Rosemblatt, "Mexican Anthropology and Inter-American Knowledge," *LARR* 53, no. 3 (2018); Also, Ricardo D. Salvatore, *Disciplinary Conquest: U.S. Scholars in South America, 1900–1945* (Durham, NC: Duke University Press, 2016).

9. Ross Parmenter, "Zelia Nuttall and the Recovery of Mexico's Past," manuscript [1900s], Latin American Library, Tulane University; examined at Tozzer Library, Harvard University, August 24, 2018.

10. In this sense, Nuttall's life story can be treated as a 'wormhole' revealing the multidirectional, multinodal process of knowledge production in the human sciences. See my essay "Beyond Prejudice and Pride: The Human Sciences in Nineteenth- and Twentieth-Century Latin America," *Isis* 104, no. 4 (2013); Gabriela Soto-Laveaga, "*Largo Dislocare*: Connecting Microhistories to Remap and Relocate Histories of Science," *History and Technology* 34, no. 1 (2028).

11. Before 1900, scientists often used the term "ethnology" to refer to cultural analysis of non-European peoples; at the time, "anthropology" usually referred to what we today understand as physical and forensic anthropology.

12. Nuttall to Putnam, May 27, 1902, Peabody Museum Archives (Harvard), Zelia Nuttall Papers 1886–1912, Correspondence, Box 2, folder 4. Emphasis in original.

13. Alfred M. Tozzer, "Zelia Nuttall" (obituary), *American Anthropologist* N.S., 35 (1933): 477.

14. Zelia Nuttall, "Introduction," in *The Codex Nuttall: Facsimilie of an Ancient Mexican Codex Belonging to the Lord Zouche of Harynworth* (Cambridge, MA: Peabody Museum, 1902), 4.

15. Nuttall, 'Introduction', *The Codex Nuttall,* 35.

16. Nuttall to Brinton, September 30, 1893, University of Pennsylvania Special Collections, mss. coll 690, folder 297, Zelia Nuttall correspondence 1893–1898.

17. Daniel G. Brinton, *A Primer of Mayan Hieroglyphics* (Boston, MA: Ginn & Co., 1895), 32–33.

18. Eduard Seler, "On the Present State of Our Knowledge of the Mexican and Central American Hieroglyphic Writing," *ICA* (1902): 157–59.

19. Helen Delpar, *Looking South: The Evolution of Latin Americanist Scholarship in the United States, 1850–1975* (Tuscaloosa: University of Alabama Press, 2008), 26; Salvatore, *Disciplinary Conquest.*

20. See Elise Burton (chapter 10) on Iranian physicists at the Pasteur Institute.

21. Nuttall to Putnam, June 26, 1886, Peabody Museum Archives (Harvard), Zelia Nuttall Papers 1886–1912, Correspondence, Box 2, folder 1.

22. Nuttall to Putnam, August 8, 1892, Peabody Museum Archives (Harvard), Zelia Nuttall Papers 1886–1912, Correspondence, Box 2, folder 2. Emphasis in original.

23. Nuttall to Putnam, January 31, 1886, Peabody Museum Archives (Harvard), Zelia Nuttall Papers 1886–1912, Correspondence, Box 2, folder 1.

24. Mary F. McVicker, *Adela Breton: A Victorian Artist Amid Mexico's Ruins* (Albuquerque: University of New Mexico Press, 2005), 1.

25. On obstacles in Mexico's patriarchal society, see Ruíz, "Insiders and Outsiders in Mexican Archaeology," 340.

26. Ruíz, "Insiders and Outsiders," 242–43. Emphasis in original.

27. Nuttall to Putnam, n.d. (probably 1902), Peabody Museum Archives (Harvard), Zelia Nuttall Papers 1886–1912, Correspondence, Box 2, folder 4.

28. Nuttall to Putnam, May 27, 1902, Peabody Museum Archives (Harvard), Zelia Nuttall Papers 1886–1912, Correspondence, Box 2, folder 4.

29. Boas to Seler, December 22, 1908, APS Boas collection, Correspondence B, box 61. Seler, Eduard, folder 6.

30. Boas to Nuttall, October 13, 1909, American Philosophical Society, Boas collection, Correspondence B, box 61. Nuttall, Zelia. On the question of International School, Ruíz has concluded (and I concur) that Nuttall's role was diminished by Boas and others who had a limited view of her as a source of local connections. Ruiz, "Insiders and Outsiders in Mexican Archaeology," 188–200.

31. Peter Diderich, "Assessing Ross Parmenter's Unpublished Biography about Zelia Nuttall and the Recovery of Mexico's Past," Newsletter of the SAA's History of Archaeology Interest Group 3, no. 3&4 (November 2013): 2. Treatments of Nuttall's life intended for general audiences include Amanda Adams, *Ladies of the Field: Early Women Archaeologists and Their Search for Adventure* (Vancouver: Greystone Books, 2010); Leila McNeill, "The Archaeologist Who Helped Mexico Find Glory in Its Indigenous Past," Smithsonian Magazine online, November 5, 2018, https://www.smithsonianmag.com/science-nature/archaeologist-who-helped-mexico-find-glory-its-past-180970700/, accessed 15 July 2019.

32. Ruíz, "Insiders and Outsiders," 340. Ruíz also discusses an even more ignored woman anthropologist, Isabel Ramírez Castañeda; see Ruíz, "Insiders and Outsiders," 243.

33. See, for example, Zelia Nuttall, "The Causes of the Physical Degeneracy of the Mexican Indians After the Spanish Conquest as Set Forth by Mexican Informants in 1580," Peabody Museum of American Archaeology and Ethnology (Harvard University) 11:2 (1926); and Zelia Nuttall, "The New Year of the Tropical Indigenes: The New Year Festival of the Ancient Inhabitants of Tropical America and its Revival," *Bulletin of the Pan American Union* 62 (1928).

2

Escaping Immortality

Science, Civilization, and Lu Gwei-djen

Lan A. Li

Malta was warm. The afternoon sun stretched across the island as Lu Gwei-djen walked ahead. She turned around and looked at the camera, her fashionably large glasses covering most of her face (figure 2.1). She clutched a cream-coloured purse, which matched her mid-heel loafers, which matched her blue and purple shirtdress. Lu had packed at least three floral shirts, including a formal blouse and a sleeveless turtleneck.[1] Having just turned seventy, Lu enjoyed a playful wardrobe. She explored Malta's rocky shoreline and climbed up limestone walls for a view of the Mediterranean Sea.

Lu Gwei-djen (1904–1991) was never seen this way beyond her close personal relationships.[2] In fact, she was rarely seen. Though Lu meticulously curated her wardrobe, she avoided publicity. She preferred to be hidden and overshadowed by her collaborators despite her own accomplishments. In 1937, Lu travelled from China to Britain to become one of the first Chinese women to receive a doctorate in biochemistry at Cambridge University. In 1956, Lu retired from science to become one of the founding contributors to the monumental series *Science and Civilization in China*. Lu herself was one of the most knowledgeable historians of medicine in China. Her own colleagues have claimed that she was fundamental to the entire field of the history of science in China.[3]

Yet, few people knew about her. This was in part Lu's own doing. Though she remained professionally active, Lu did not assign herself any credit. She avoided public interviews and imagined herself as the invisible "arch" that sustained the bridge connecting China to the English-speaking world. She annotated her collaborator's correspondences but never signed her name. When asked to share her biography, Lu said, "Let them work it out after I die."[4] Biographers have tried to write about Lu's early career as a scientist, focusing on her dissertation on blood pyruvate and her work on metabolism.[5] But rather than capturing the full scope of her intellectual life, many

Figure 2.1. Photograph of Lu Gwei-djen vacationing in Malta, dated 1973–1974. Photographer unknown. Reproduced with permission of the Needham Research Institute, Cambridge.

of Lu's obituaries described her as the woman who inspired scientist and Sinologist Joseph Needham (1900–1995) to take on *Science and Civilization in China*. Journalists such as Simon Winchester have crudely described Lu as Needham's "mistress and muse."[6] The only book-length biography of Lu Gwei-djen by Wang Guozhong centres on Lu's early life in China and her later relationship with Needham.[7]

This chapter bridges Lu Gwei-djen's otherwise invisible professional and personal lives. It explores how Lu served as a critical actor in translating and culturally mediating histories of medicine in China while negotiating the professional and political demands of her work.[8] As a scientist and historian, Lu navigated across a range of social identities that were national, political, social, professional, and gendered.[9] She had come from a liberal Nationalist family but sympathized with Communist ideals; she identified as Confucian but participated in Christian rituals.[10] Early on, Lu's feminist father encouraged an uncommon interest in the history of pharmacology, which inspired her career as a scientist and historian.[11] Yet, her ambition was fuelled by political persuasions that motivated her intellectual breadth and limited her range of expression.

This chapter highlights Lu Gwei-djen's work as a scholar, thereby de-centring Joseph Needham in her narrative. I explore Lu's activities by closely excavating her Cambridge papers, including private work notes and annotations, from roughly 1964 to 1983. In this period, she composed drafts on Daoist immortality, created a dictionary of technical terminology, revised correspondences related to *Science and Civilization in China*, interrogated physicians in China, published *Celestial Lancets*

(1980) and critiqued reports on acupuncture analgesia.[12] She also endured lung cancer and suffered a heart attack – experiences that created an ontological distance between her position on Chinese medical cosmology and her embodied sense of self. Though Lu remained publicly obscure, she existed in an extensive paper trail. Her evolving ideas lived in the marginalia, among an array of loose annotations that captured her research. In other words, Lu made her presence known through what literary theorist Gérard Genette describes as "private epitext."[13] These were the memos, summaries, and suggestions that Lu composed by hand. It is here where we can locate Lu and identify the complex identity politics that undergirded her life and work.

LU'S POLITICAL PERSUASIONS

In 1904, Lu Gwei-djen was born in Nanjing, a city that would have surrounded Lu with vibrant intellectual communities and offered alternative models of modernity. There, Chinese feminism facilitated radical ideologies, which theorized and disrupted Confucian hierarchies and long-standing and long-evolving legal, ritual, and social institutions.[14] In this rich political landscape, physicians attempted to promote the integration of native and foreign forms of knowledge.[15] But rather than establishing a unified body of medical theory, these projects instead intensified a growing plurality among physiologists and medical practitioners.[16]

Lu Gwei-djen also developed complex political worldviews within this intensified plurality of ideas. As a teenager, she actively participated in the 1919 May Fourth Movement to support reform initiatives.[17] But as Chinese politics changed, so did Lu's own politics. Despite her early avoidance of the English language, she would begin to learn and master it. In 1922, she enrolled in Ginling College, a Christian school affiliated with Smith College and the University of California. There, she majored in English and chemistry.[18] Like many Ginling College students, Lu held onto radical ideologies despite being supported by foreign interests.[19] After graduating from Ginling, Lu moved to Beijing and enrolled as a visiting scholar to study pathology and pharmacology at Peking Union Medical College. Lu then returned to Shanghai, a site of ongoing unrest as members of radical feminist groups protested against Japanese forces and were detained in German prisons.[20]

In 1937, Lu left Shanghai for the United Kingdom to begin her graduate studies at Cambridge. When Lu Gwei-djen left China, she took her politics with her. At Cambridge, she studied under the supervision of Dorothy Needham (1896–1987), whom Lu admired for her progressive politics.[21] She had found intellectual and political allies in Dorothy and her husband Joseph Needham.[22] In Dorothy Needham's lab, Lu continued her work on metabolism and analysed the patient data that she collected in Shanghai. In addition to her thesis on molecular biology, Lu also added a cultural survey of nutrition in China.[23] Lu quickly completed her doctoral research in 1939 and wanted to return to China. Yet, the events of World War II left her stranded in the United States.[24] For six years, she travelled from California to Alabama to New York. All the while, she maintained her correspondence with Joseph Needham, who had been discreetly recruited by the Chinese ambassador to England to help rebuild scientific programs in China.[25] When Lu finally made her way home in 1945, both her mother and brother had died. While grieving, she travelled from Shanghai to Chongqing, where Joseph Needham established the Sino-British Science Co-operation Office (SBSCO) at the British embassy. Lu then met political dignitaries like the future minister of culture Guo Moruo (1892–1978) and Communist leader Zhou Enlai (1898–1976), who urged Lu to stay abroad.[26] But she became critical to the Needhams, and more importantly, to the *Science and Civilization* series.[27] Lu returned to Cambridge and stayed there, becoming a British citizen in 1967.

ETERNAL LIFE

Lu Gwei-djen endeavoured to articulate Chinese medicine as a science. For Chinese medicine to belong to the Chinese state, Lu needed first to draw boundaries around its unique ontology. She then actively constructed narratives of Chinese medicine as a timeless artefact and as a modern practice.[28] Scholars of postcolonial science studies would have readily recognized the tensions embedded within the political narratives of modernity and establishing a national legacy. Yet, Lu Gwei-djen was not a postcolonial scholar. She had spent time in France, and she may have read Fanon. But rather than recoiling from white elites, she embraced them. She allied with the Needhams. She partnered with them and appealed to them. In her appeal, Lu aimed to establish "Chinese" medicine as an ethnically recognizable, politically stable, and theoretically relevant practice. In one case, she focused on articulating Daoist practices of immortality.

In 1971, Lu drafted "The Inner Elixir (*Nei Tan*); Chinese Physiological Alchemy," which introduced the inner elixir as a "Daoist" method of achieving longevity and immortality. In her manuscript, Lu translated *nei* as "inner" to describe physical interior spaces. She emphasized that the inner elixir, or *nei dan* (*nei tan* in Lu's draft), involved a corporeal immortality, since "no other was conceivable."[29] She insisted that Daoist inner alchemy was nothing like early modern European alchemy and warned that, "no greater mistake could be made than to analogize *nei tan* with the 'spiritual alchemy' of the West; it was physiological through and through."[30] In other words, Lu aimed to not only distinguish *nei dan* from other occult practices but even superior alchemical practice in Europe – superior in part because *nei dan* worked.[31]

For Lu, *nei dan* worked as a practice of immortality because it was rooted in a basic corporeal body, unlike the non-corporeal early modern European alchemical practices. The object of *nei* represented the interior space of a corporeal body that gave rise to other physiological structures like meridian paths. She even translated *nei* as "corporeal." In other words, *nei* not only *defined* the corporeal body but *became* the corporeal body. Other scholars were not as radical. For instance, when historian of science Nathan Sivin first encountered Lu's interpretation, he noticed that Lu and Needham had translated the title *Huangdi Neijing*, a classical medical text that dated to the first century, as "The Yellow Emperor's Manual of Corporeal [Medicine]."

Sivin found it odd that Lu translated *nei* as "corporeal." He argued that rather than translating *nei jing* as "corporeal medicine," a better translation would be "the inner canon." This way, *nei jing* as the "the inner canon" could directly reference its *wai jing* counterpart, which Sivin described as "the outer canon."

Lu replied, "Only not true!"[32] She insisted that "corporeal medicine" took into account the historical uses of *nei* (inner) and *wai* (outer), which corresponded to Yin and Yang areas of the body. These Yin and Yang areas of the body gave rise to Yin and Yang acupuncture-moxabustion paths. Visible materiality mattered in Lu's translations. Thus, *nei* not only explained how a distinctly Daoist/Chinese/corporeal body behaved, but it further characterized these material qualities as uniquely distinct. Lu further urged that Chinese medical theory and practice were distinct from other Asian systems. She pressed that unlike yoga, *nei dan* was less radical because it emphasized hygiene, or *weisheng*.[33] This form of hygiene simply "retrac[ed] one's steps along the road of bodily decay."[34] *Nei dan* as a practice seemed simple enough to explain. For instance, it delayed the atrophy of important bodily fluids that possessed *jing* 精, like semen and saliva.[35]

Despite her confidence in the materiality of Daoist inner elixir, Lu harboured some ambivalence in her drafts. For instance, Lu originally wrote that the early texts on physiological alchemy claimed that "*chhi* [*qi*] can preserve the invisible life," before removing "the invisible."[36] This suggested that Lu might have reasoned that if *qi* were material, it could not be invisible. Invisibility implied immateriality. In other cases, Lu had originally described the *nei dan* technique that expressed the "sanity" and "sobriety" of Daoist philosophers before replacing these two words with "empiricism" and "rationality."[37] It is possible that Lu made this change because "sane" and "sober" invoked attitudes rather than practices. In another revision, Lu wrote that the "unfamiliar" and "medieval" character of Chinese physiological alchemy "was in a certain sense akin to the insights of modern science," before modifying this final sentence to say that physiological alchemy "was in a *real* sense akin to the *optimistic and experimental* outlook of modern science, especially biochemistry, endocrinology, and geriatrics."[38]

For Lu, *nei dan* needed to be more modern to be taken seriously. In her revisions, Lu urged the reader to recognize how *nei dan* was physical, material, and physiological—it was real, embodied, and empirical. Unlike other Asian practices, *nei dan* was furthermore superior to early modern European practices. Lu could not afford to translate *nei* as a simple "inner" space. It was not a relative directional space but a material thing. Representing the deep corporeality of *nei dan* required a strategic selection of words that was shaped by the conceptual and political trappings of translation on the one hand and Lu's own embodied limitations on the other.

ILLNESS NARRATIVES

While Lu Gwei-djen emphasized the material reality of immortality practices, she did not embody these practices. On the morning of January 20, 1969, Lu admitted herself into a hospital after feeling pains in her chest. She recorded the morning's events in a private note, including the nurse who injected atropine into Lu's neck to keep her from salivating.[39] The nurses had lifted Lu onto a cot to wheel her down the hall. As Lu waited outside the operating room, a doctor asked to delay her appointment before vanishing without further instructions. The nurses then rolled Lu into a corner in another patient's room. Lu insisted on returning to her own ward. The nurses wheeled her back down the hall and lifted her into her cot.

Following the pneumonectomy, Lu dropped to ninety-five pounds. She often felt exhausted and short of breath, enduring a deep pain throbbing in her gut. Lu struggled to write, noticing that her grip on her pen had weakened, and her stamina was low. She sweated profusely after writing a few sentences. Undeterred, Lu still took many notes on medical history and corresponded regularly with Joseph Needham. She remained optimistic and carried on with her daily routine. "I feel very well," she reported, "can eat & sleep and walk a bit, that is about all."[40] Lu's condition gradually improved. After a few years, she would travel to China with Needham, tour hospitals and interview physicians. When she returned from her trip, she suffered a heart attack in 1973. Her doctor prescribed a daily dose of Valium to prevent muscle spasms along with thyroxine to steady her metabolism.[41] A few friends doubted that she would be able to recover fully.[42] But at seventy years old, Lu travelled to Tibet to meet Uighur physicians. And she vacationed in Malta two summers in a row.[43] There, she would wear her favourite clothes, scale rocky shorelines, and pose for pictures.

Lu reflected on her declining health. In another private note with Joseph Needham, they co-authored a fragmented poem, titled "Woes of Lovers as They Get Old." It read:

less intense emotional feeling
loss of urge or drive for s—mutually
"current" not flowing forcefully as in/with youth
lack of gracefulness, awkwardness
grace of movement
beauty almost dancing
insuff natural V_1 secretion (hence artificial aids)
"ankylosis," rheumatic pains off-putting, awkwardness
decay of body, wrinkles, slackness after thinning
unnecessary frustrations like towel under
less or no guilelessness, naturalness.[44]

Among the first four lines of the poem, Lu and Needham described aging as being characterized by "less," "loss," and "lack." They described a weaker flow of emotions. Their drive for sex had vanished. Their movements were now less beautiful, less graceful. Lu described a kind of physical atrophy and emotional erosion that further manifested in a decreasing internal "current." Their fluid body made itself apparent in the skin that slackened as liquids leaked from it. Lu and Needham compensated for a leaking, evaporating, and diminishing body. Lu used ointments to supplement for vaginal dryness and relied on towels to absorb involuntary urinary contractions and bowel movements. These additions contributed to an "awkwardness" that manifested in the accessories of age. In their poem, neither Lu nor Needham had translated *qi* as a kind of vital energy. They did not discuss the physiological manifestations of Yin and Yang. They did not invoke Daoist practices of immortality.

CONCEALING CONTRADICTIONS

Lu Gwei-djen would not live forever. Throughout her career, she had embedded herself within files, notes, translations, revenue records, reports, and drafts to collect, produce, and communicate knowledge of the body in spite of her own. Her politics was at stake in her writing and translation, motivated by a seemingly lifelong dedication to China. Though she wrote about the materiality of the Daoist inner elixir, a process that Lu articulated as thoroughly physiological, she did not embody the same reversal of tissues. She could potentially appreciate the alchemy of the body in her early training as a biochemist and nutrition scientist, yet this kind of biomedical alchemy did not neatly map onto Daoist body practices. Her intellectual work involved describing a material reality that she did not inhabit.

Lu's own suffering remained as invisible as her intellectual contribution. As a Chinese woman, Lu was cosmopolitan and highly mobile. But because she was a Chinese woman, she regularly confronted the limits of patriarchal hierarchies. When she travelled to China with Joseph Needham to meet with political dignitaries, *People's Daily* writers described her as Needham's "assistant."[45] She actively constructed this identity, announcing to colleagues that Joseph Needham was the bridge between China and the world, while she was the arch that sustained it. When Lu published *Celestial Lancets* in 1980, she insisted that Joseph Needham be listed as the first author, even when she had conducted most of the research. Upon her death, two obituaries memorializing Lu and her accomplishments appeared in Britain. She remained virtually unknown in China.[46]

Lu had executed a kind of self-annihilation, contributing to the narrative of white male heroes and white male saviours. She had allied with her white British colleagues, appealing to them and embedding herself among them by effectively silencing herself. Though she had actively engaged in professional correspondences, she kept herself concealed in her reading notes and annotations. She stood by her intellectual convictions without drawing attention to herself. Despite being radicalized at an early age, Lu's subjectivity transformed throughout her life and her work, attempting to provincialize European knowledge through a history of medicine that was uniquely Chinese and uniquely modern. All the while, Lu slipped between the intellectual, social, and discursive cracks of a subaltern positionality, having been trained in biochemistry before subverting biochemistry by explaining embodied practices that she did not inhabit.

NOTES

1. The remaining photos from her trip to Malta are found as uncatalogued archival material at the Needham Research Institute.

2. For a longer and more thorough examination of Lu Gwei-djen's work, see Lan A. Li, "Invisible Bodies: Lu Gwei-Djen and the Specter of Translation," *Asian Medicine* 13, no. 1–2 (September 10, 2018): 33–68, https://doi.org/10.1163/15734218-12341407.

3. Peng-Yoke Ho, "'Lu Guizhen Boshi Jianjie' [Introducing Dr. Lu Gwei-Djen]," *China Historical Materials of Science and Technology* 11, no. 4 (1990): 26.

4. See Jixing Pan, "Dr. Lu Gwei-Djen: An Outstanding Woman [杰出女性鲁桂珍博士]," *China Historical Materials of Science and Technology* 14, no. 4 (1993): 55. Lu had also refused the director of the Institute of Natural History at the Chinese Academy of Sciences when he requested an autobiography. This burden fell on her colleague Ho Peng-Yoke, recounted in Ho, "Lu Guizhen Boshi Jianjie," 1990, 25.

5. For an official biographical entry, see Bing Wang, "Lu Gwei-Djen," in *Biographical Dictionary of Chinese Women: The Qing Period, 1644–1911*, eds. Lily Xiao Hong Lee et al. (Armonk, NY: M. E. Sharpe, 1998), 382–83. Lu's resume was published in the 1980 edition of *Who's Who*. Additionally, see Robin Hesketh, "A Great Adventure: From Quantitative Metabolism to the Revelation of Chinese Science," *The Biochemical Journal* 2012 (January 1, 2012): 1–4, https://doi.org/10.1042/BJ20120049.

6. Simon Winchester, *The Man Who Loved China: The Fantastic Story of the Eccentric Scientist Who Unlocked the Mysteries of the Middle Kingdom*, First Edition (New York, NY: Harper, 2008), 35.

7. These include Boying Ma, "My Memory of Dr. Lu Gwei-djen [Wǒ yìnxiàng zhōng de lǔ guì zhēn bóshì]," *Wen Wei Po*, July 10, 1992; Gregory Blue, "A Passion for Science: Gwei-Djen Lu-Needham," *The Guardian*, December 12, 1992; Guozhong Wang, *Lu Guizhen yu Li Yuese [鲁桂珍与李约瑟]* (Guiyang: Guizhou Renmin Chubanshe 贵州人民出版社, 1999).

8. Numerous scholars have engaged with co-authorship, circulation and friction in translation processes in colonial and postcolonial contexts. See Lydia H. Liu, ed., *Tokens of Exchange: The Problem of Translation in Global Circulations* (Durham, NC: Duke University Press Books, 2000); Lydia Liu, Rebecca Karl, and Dorothy Ko, eds., *The Birth of Chinese Feminism: Essential Texts in Transnational Theory*, First Edition (New York: Columbia University Press, 2013).

9. Inderpal Grewal has described the multidimensionality of social identities as a kind of "transnational connectivity" or "co-formation." Inderpal Grewal, *Transnational America: Feminisms, Diasporas, Neoliberalisms* (Durham, NC: Duke University Press, 2006).

10. Author's conversation (via Skype) about Lu Gwei-djen with an interlocutor who prefers to remain anonymous, April 14, 2017.

11. Wang, *Lu Guizhen yu Li Yuese*, 1999, 2–5.

12. The word "acupuncture" is used to translate the word *zhenjiu*, a compound practice that involved needling (*zhen*) and applying high heat to the skin via burning dried mugwort (*jiu*). Acupuncture analgesia would encompass the broad range of therapeutics of both *zhen* and *jiu* (acupuncture-moxabustion). See Lan Li, "Pinpricks: Needling, Numbness, and Temporalities of Pain," in *Imagining the Brain: Episodes in the History of Brain Research*, First Edition, vol. 243 (Academic Press, 2018), 205–29.

13. In his seminal work *Paratexts* (1997), Gérard Genette describes public and private epitexts as inscriptions that circulate and inhabit a "virtually limitless physical and social space."

See Gerard Genette and Richard Macksey, *Paratexts: Thresholds of Interpretation*, trans. Jane E. Lewin (Cambridge; New York: Cambridge University Press, 1997).

14. See Liu, Karl, and Ko, *The Birth of Chinese Feminism*, 2013.

15. These include Sean Hsiang-lin Lei, *Neither Donkey Nor Horse: Medicine in the Struggle over China's Modernity* (Chicago, IL: University of Chicago Press, 2016); Bridie Andrews, *The Making of Modern Chinese Medicine, 1850–1960* (Honolulu: University of Hawai'i Press, 2015); Volker Scheid, *Chinese Medicine in Contemporary China: Plurality and Synthesis* (Durham, NC: Duke University Press Books, 2002); Kim Taylor, *Chinese Medicine in Early Communist China, 1945–1963: A Medicine of Revolution*, First edition (London; New York: Routledge, 2005).

16. Historians Sean Lei and Bridie Andrews have written extensively on the many physicians and reformers who transformed medical practice into a space for legitimating the modern Chinese state. Medicine in China at this point still ranged on a broad spectrum from itinerant healers to scholar-physicians. To learn more about the spectrum of Chinese healing practices, see Bridie Andrews, *The Making of Modern Chinese Medicine, 1850–1960* (Honolulu: University of Hawai'i Press, 2015), 25–50.

17. Winchester, *The Man Who Loved China*, 6.

18. Wang, *Lu Guizhen yu Li Yuese*, 232–35.

19. The first big student strike was in 1919, which led to the formation of Ginling's Student Union.

20. Ya-mei Yu, "Minguo Shiqi Nüxing Liuxue Yu Fazhan Dingwei" [Situation of the Female Studying Abroad and Its Developmental Location during the Republican Period], *Journal of Xuzhou Normal University* 32, no. 1 (n.d.): 14.

21. Both Dorothy Needham and her husband, Joseph Needham, served as delegates for the Association of Scientific Workers on the Cambridge Trade Council.

22. Specifically, she linked vitamin B_1 deficiency to a longer history of beriberi, which was often associated with foot *qi*—a small detail that indicated Lu's already interdisciplinary enterprise. Gwei-djen Lu, The First Half-Life of Joseph Needham, in *Explorations in the History of Science and Technology in China: Compiled in Honour of the Eightieth Birthday of Joseph Needham, FRS, FBA* (1–38: Shanghai, 1982), Shanghai Guji Chubanshe, 7.

23. This disease state and its diagnosis involved a range of symptoms and differing explanations. For a comprehensive history of the shifting history of foot *qi* in East Asia, see Hilary A. Smith, *Forgotten Disease: Illnesses Transformed in Chinese Medicine*, First edition (Stanford, CA: Stanford University Press, 2017).

24. Gwei-djen Lu, "Letter from Lu to Needham," April 8, 1945, C.29–31., Needham Files. Needham Research Institute.

25. Winchester, *The Man Who Loved China*, 53.

26. Gwei-djen Lu, 'The First Half-Life of Joseph Needham,' in *Explorations in the History of Science and Technology in China: Compiled in Honour of the Eightieth Birthday of Joseph Needham, FRS, FBA* (1–38: Shanghai: Shanghai Guji Chubanshe, 1982), 6.

27. Wang, *Lu Guizhen yu Li Yuese*, 73.

28. For more on the history of acupuncture analgesia, see Lan Li, "Pinpricks: Needling, Numbness, and Temporalities of Pain," in *Imagining the Brain: Episodes in the History of Brain Research*, First Edition, vol. 243 (Academic Press, 2018), 205–29.

29. Gwei-Djen Lu, "The Inner Elixir (Nei Tan); Chinese Physiological Alchemy" (1971), Needham Files, Needham/NRI2/SCC2/197/12, Needham Research Institute, 3.

30. Lu, "Inner Elixir," 5.

31. For a detailed history of late nineteenth-century encounters with the occult, see Alison Winter, *Mesmerized: Powers of Mind in Victorian Britain*, First Edition (Chicago, IL: University of Chicago Press, 2000); Emily Ogden, *Credulity: A Cultural History of US Mesmerism*, First Edition (Chicago, IL: University of Chicago Press, 2018).

32. Gwei-Djen Lu, Joseph Needham, and Nathan Sivin, "Correspondence with Nathan Sivin," November 29, 1972, Needham/NRI2/SCC2/320/2, Needham Research Institute.

33. Historically, the term *weisheng* represented aspects of Chinese cosmology that was involved with guarding life and conquering one hundred diseases. See Ruth Rogaski, *Hygienic Modernity: Meanings of Health and Disease in Treaty-Port China*, Reprint edition (Place of publication not identified: University of California Press, 2014).

34. Lu, "Inner Elixir," 5.

35. Lu, "Inner Elixir," 25.

36. Lu, "Inner Elixir," 8.

37. Lu, "Inner Elixir," 3.

38. Lu, "Inner Elixir," 25. Italics added to represent sections modified in Lu's draft.

39. Atropine is a drug used to decrease muscle and nerve activity.

40. Lu, 1970, "Letter from Lu to Needham," August 12, Needham Files, A.797. Cambridge University Library Archives.

41. Lu, 1974, "Letter from Lu to Needham," March 28, Needham Files, A.1025. Cambridge University Library Archives.

42. Peng Yoke Ho, *Reminiscence of a Roving Scholar: Science, Humanities and Joseph Needham* (Hackensack, NJ: World Scientific Publishing Company, 2005), 150.

43. These events can be traced in Lu's photos archived at the Needham Research Institute.

44. This poem was composed on a scrap of paper and identified as composed in Joseph Needham's handwriting with possible annotations from Lu Gwei-djen. Special thanks to Dr. John Moffett for granting permission to reproduce this poem and for confirming the handwriting. n.d., Joseph Needham Papers 1947–48, A.798. Cambridge University Library Archives.

45. Lu Gwei-djen was described as Needham's *zhushou*, or assistant, from reports of Needham's visit to China from the 1950s to the 1980s.

46. Obituaries for Lu appeared in *The Guardian*, *The Times*, and *The Independent*. *Lu Gwei-Djen: A Commemoration*, 6–14.

3

Producing and Delivering Truth

The (In)visibility of Forensic Scientists in Colombia

María Fernanda Olarte-Sierra[1]

Colombia is a country with a long-lasting internal armed conflict that started in the 1950s owing to political unrest. Since then, the Colombian population has endured steady violence from various and changing actors including drug traffickers, ideological and narco-guerrillas, paramilitary forces, the State, and the armed forces.[2] The various armed actors have affected both rural and urban populations through acts of violence that include killing, kidnapping, enforced disappearance, torture, gender-based violence, and rape, amongst others.[3] Nonetheless, over the years, there have been numerous attempts to achieve peace, the most recent being the agreement signed by the government and the FARC-EP guerrilla group in 2016.[4]

The history of violence in Colombia has launched forensic experts into a paramount role. In the context of the armed conflict, forensic experts help identify victims (work that is often about the reconstruction of events and identities from scarcely visible traces), so they and their families can be compensated, as well as provide judges with material that is considered evidence and that helps them make decisions in criminal cases against perpetrators.[5] Thus, forensic experts' work is central to the judicial system as a source of scientific evidence, testimony and authority, and is based on their assumed objectivity and non-prejudicial approach. As historian of forensics Simon Cole has pointed out, forensic knowledge belongs to a different epistemic culture to research science, in that the former produces new knowledge about the natural or social world, while forensic science seeks to recover the circumstances and events of a crime.[6] These differences are particularly striking in relation to the latter's research agenda (which is typically about finding better ways to further validate forensic knowledge for practical uses), rewards structure (which is not based on a prestige economy, but instead on efficiency and individual reputation), and the reproducibility of results (forensic claims are case-specific).[7] Although forensic knowledge is largely dependent on and at the service of the judicial system, forensic knowledge has also been put at the service of humanitarian actions in violent or war-ridden contexts.[8] In either position, forensic scientists are placed in a specific spotlight, because their claims have social consequences that are more evident than those of many conventional sciences. Otherwise put, these scientists produce knowledge of what Latour calls "matters of concern," since the facts that forensic experts make are of deep and widespread public and social consequence and since the systems that they operate are central to a country's governmental system.[9] Furthermore, forensic scientists are held in high regard as professionals who uncover the truth—that is, a very specific truth about violence and extreme suffering inflicted upon entire populations.[10] It is, above all, a truth that is scientifically made.

In Colombia, most of these cases regarding victim identification are under the jurisdiction of the Attorney General's Office (AGO), a state institution in charge of administering justice to civilians who commit crimes. However, because of the abovementioned peace agreement, two new groups of forensic scientists are now active in Colombia: one works in the Transitional Justice system and another at the humanitarian and extrajudicial Unit for the Search of the Disappeared (UBPD). In this text, I show how the scientists producing forensic science are sometimes visible and at other times invisible. At times, invisibility is desirable and necessary, and at other times forensic experts become (partially) visible. Such degrees of (in)visibility do not pose a threat to the scientific authority or social relevance of forensic experts. On the contrary, these instances of varying invisibility are precisely what provide forensic knowledge with its authoritative stance and the ability to produce knowledge that is considered testimony and evidence.

I base my analysis on approaches developed by feminist STS scholars. In Latin America, the dialogue between feminist studies and STS is growing but is still narrow.[11] Here, I address knowledge production practices from an ethos of care. I attend to the daily practices that sustain the worlds that forensic scientists help to create, that is, forensic worlds and their inhabitants. I note that care as knowledge is always in the making, is never fully stabilized, and is made through specific intentional practices.[12] This approach enables me to address the shifting focus of forensic experts' invisibility as sources of legitimacy and authority, and to address delivery practices as constitutive to the world-making effects of knowledge production.

(IN)VISIBILITY

One conspicuous feature of modern scientific knowledge production is that scientists tend to perform invisibility at specific stages of the research process to secure credibility and objectivity. As Haraway observes,

> self-invisibility is the specifically modern, European, masculine, scientist form of the virtue of modesty . . . that pays off its practitioners in the coin of epistemological and social

> power . . . This is the virtue that guarantees that the modest witness is the legitimate and authorized ventriloquist of the object world.[13]

Forensic science—as it is practiced in Colombia and and elsewhere—maintains this vision of the objective, anonymous scientist. Yet, forensic knowledge is produced by people through specific, daily and sustained practices, performed as interactions between humans and more-than-humans. We know that knowledge has world-making effects and that those who produce it are situated in specific moments in history, inhabiting particular geopolitical positions.[14] Therefore, knowledge—including forensic knowledge—is neither neutral nor innocent. Rather, it has political and material consequences for those about whom the knowledge is produced.[15] The manner in which such knowledge is delivered to the public has world-making effects too. Thus, to account for the wider social repercussions of forensic knowledge is to also pay attention to the performance through which forensic knowledge is delivered. Production and delivery practices are constitutive of scientific knowledge authority and credibility; in both, the visibility and invisibility of scientists play a central role.

What follows is an account of the processes and experiences that create the partial invisibility of three forensic scientists. Each story refers to an aspect of (in)visibility, ranging from complete anonymity to full disclosure of the forensic scientists' names and jobs. I explain how the invisibilities of forensic scientists figure in the perceived soundness of their work, and the effect these may have for forensic practice.

THREE DEGREES OF (IN)VISIBILITY

The following sections are based on ethnographic data gathered over the course of three research projects in which I have observed Colombian forensic scientists of the Attorney General's Office and at the National Institute of Legal Medicine and Forensic Sciences.[16] Forensic work on victims' identification performed by the AGO is organized in interdisciplinary teams composed of forensic anthropologists, dentists, and coroners. Upon examination of the body, they decide if forensic geneticists need to be involved. However, forensic DNA is increasingly becoming a standard procedure and key actor of forensic teams.[17]

One. Invisible by Choice and for Safety

Azucena is a forensic geneticist.[18] She works for the Attorney General's Office and the National Institute of Legal and Forensic Medicine. As a forensic geneticist, Azucena and her colleagues contribute to identifying Colombia's armed conflict victims, as well as work on civil or criminal cases, and on the adjudication of paternity when it is disputed or denied.[19] As such, she and the other forensic experts work on matters that are of widespread public interest and consequence. A few years back, my colleagues Tania Pérez-Bustos, Adriana Díaz del Castillo and I explored how forensic geneticists (mostly women) dealt with the fact that they were constantly made invisible in the media. As we came to understand, the news tends to portray forensic geneticists' work in terms of how forensic experts in government employ identify people through DNA testing.[20] We initially understood this absence of acknowledgement as disrespectful towards the forensic experts. But in the course of our research, we found that those scientists were not interested in being visible. When we asked Azucena about invisibility, she said:

> We don't like to be visible in the media. It has to be that way; we prefer it to be that way. We work with sensitive and high-profile cases, dangerous cases. We work with classified files and information, we cannot be public.[21]

For Azucena and her colleagues, being obscured by the media was one way of remaining safe. She felt that visibility would endanger her life, and it also felt secondary and irrelevant to her work. Indeed, not being visible to the media—and not being made visible by the media—also meant safeguarding her work processes and findings. Further along in our conversations, Azucena added:

> If we keep away from the spotlight, we can keep our objectivity and our impartiality. The media usually delivers misleading information and we prefer not to deal with that.[22]

Hence, to occupy a less-exposed position served a double purpose of being both safe and objective. It allowed her to produce disembodied knowledge from nowhere, to do a "god-trick" (to use Haraway's term) and at the same time to avoid risking her life.[23] For Azucena, invisibility meant being an objective, sound, and safe scientist.

Aside from the absolute need of securing a forensic expert's life, one cannot overlook the fact that forensic statements, which are understood as highly reliable and objective, gain authority when they are produced by the nameless forensic experts of the Attorney General's Office or the National Institute of Legal and Forensic Medicine.[24] The institutions themselves are highly visible, and highly authoritative, and lend that authority to the scientists' findings, when those scientists are kept anonymous. Thus, in this instance of mass consumption and mass media, the desired and very much deserved concealment of forensic experts' identity adds to the already well-established reputation of forensic sciences.

Two. When Duty Calls: Partially Visible

Given the role of forensic science to aid (principally) the judicial system, forensic scientists are often summoned as expert witnesses to contribute to legal truth-telling.[25] As Cole points out, forensic experts' claims are valued based on their reputation in case-solving and on the quality of the facts they produce.[26] This is true for Jaime, a seasoned forensic anthropologist who has worked for more than twenty years at the Attorney General's Office.[27] In one of our conversations I asked him about being an expert witness and how he dealt with the complete disclosure of his identity, particularly when testifying against the suspect of a crime. He said:

> When you go to a hearing, to the Court House, you are summoned as an expert. You are there because it was you who did the job; it is your name and your expertise that counts for that case.[28]

In the environment of the Court House, in front of judges and juries, the identity and reputation of an individual scientist matters to the credibility of their testimony. I pressed the matter further and asked about his safety. He answered:

> Well, in our job, we are at constant risk, we have been targeted as military objective by all the armed actors. We know that, but that cannot stop us from doing our job, and we trust that the Attorney General's Office takes care of us. I think we are more exposed when we go to the countryside to do exhumations; there we run the risk of getting caught in crossfire [between armed actors]. In Court, it's different. They summon *me* (or whomever they summon) because I am the only one who can explain, or talk or answer anything about the cases that I have worked. It is my expertise and my work that put me there, so of course, my name and my trajectory have to be visible.[29]

Thus, in court, nameless and invisible scientists, who produce knowledge that can be considered evidence in a given laboratory, are insufficient for the administration of justice. The validity and credibility of forensic knowledge are qualities created by an interplay between the invisibility that forensic scientists have when producing the facts, and the visibility they gain when presenting them as expert witnesses. Thus, when faced with the need for further contextualization or explanation of evidence, jurists, prosecutors or defence lawyers may summon forensic experts to account for their testimony. At that point, it is no longer possible or desirable to conceal the identity of the forensic scientist. The performance of the delivery of this knowledge requires that the scientist's evidence or testimony is attached to their name and reputation—making that knowledge both valid and valuable.

Three. Full Disclosure. To Touch and Be Touched

> We like to attend to the delivery ceremonies [of victims' remains to their relatives]. It is a moment in which we can bring some relief to the families. It is a moment of hope. We answer their questions and explain what they need explaining of [with regards to the remains and the causes of death].[30]

Beyond formal judicial scenarios, forensic experts inhabit other social settings scenarios and engage in other practices concerning peace and reconciliation.[31] They are often invited to attend "delivery ceremonies," that is, public and ritualized ceremonies in which victims' remains are returned to their families after an identification is established. In the context of these ceremonies, forensic experts serve as "translators" of the violence traceable in the bones. That is, they explain the marks on the bones as a way to explain the cause of death. This interaction and explanation are understood to contribute to healing by the families receiving the remains, because, as forensic dentist Angela once explained:

> Families want to be sure this is their relative; their son or daughter. So, they say, for instance, "when he was little, he broke his arm riding a bike. Do you see that here"? (in the bones). And then you explain: "yes, look, here is the fracture" [pointing to an imaginary bone]. And families can rest assured that this is their son. That it is the truth that we are telling.[32]

The law that oversees victims' compensation and reparation, includes the restitution of human remains of those who have died in the context of the Colombian armed conflict.[33] In the previous quotation, Angela, a long-time forensic dentist of the Attorney General's Office, shares her experience in such ceremonies. In the intimate experience of the ceremony, Angela expressed no worry about being identified and singled out as a forensic scientist. As I have described elsewhere, forensic work is, to a great extent, a careful and caring work.[34] Here, in Angela's example, we can recognize a particularly vivid dimension of care (Widmer, chapter 23). This personal and intimate encounter, in such a particular setting, permits forensic experts to deliver knowledge in complete command of their authority as scientists, while being visible as people; it also enables them to be in touch with and to be touched by those for whom they work.

Although the ostensive aim of forensic scientists in delivery ceremonies remains the same as in the previous two examples—to provide objective and unimpeachable facts—here, these facts are not for the media, the general public or the judicial system and experts. Rather, the facts are for victims' families and, in the longer term, to become part of the social accounts of the violent past. Their function, in this context is distinct. Here it is necessary that facts are conveyed face-to-face because they have a personal, even therapeutic dimension—one that has broader effects in individual and collective memory practices that make powerfully evident the wider social effect of forensic scientific knowledge.[35] Forensic knowledge is being made a constitutive part of our collective memories of violence and armed conflict.[36] This makes it evident that forensic experts do not need to remain invisible to hold authority and make powerful scientific, legal or social truth-telling claims.

FINAL REMARKS

The three scientists presented here articulate the circulation and presence of forensic knowledge in varied scenarios that range from mass communication, the administration of justice, individual encounters and active participation in memory practices. As such, they demonstrate that in forensic science, the visibility of scientists is carefully adjusted depending on how and for what purpose forensic knowledge is mobilized. To extend the visual metaphors that I have used throughout this chapter, the focus and zoom of the lens through which forensic scientists are viewed is adjusted in different judicial institutional settings. I have raised a range of possible effects that scientists' visibility or concealment can have on the credibility of the scientific knowledge that they make. Referring to three distinct occasions in which a forensic experts' identity is or is not disclosed, I argue that visibility, for this science, is a matter of degree. These varied degrees of (in)visibility do not necessarily threaten the reputation of forensic science. Rather, they reinforce it. The practice of refocusing (in)visibility is, then, a matter of the performance involved in producing and delivering forensic facts and truths. It can be through invisible laboratory work, partially visible expert witness accounts or fully visible engagement in delivery ceremonies.

Consequently, the work of forensic experts is much more than a testimonial. Due to the nature of their work, it can be said that in all instances and degrees of invisibility, forensic experts engage in caring and careful practices towards their work and towards those involved in their cases.[37] By understanding forensic knowledge and delivery practices as an ethos of care, I have shown how the truth they produce and deliver are relational to their spaces (i.e., laboratories, the courthouse), intentions (i.e., to act as an expert witness or as a "translator" of violence to relatives), and contexts (i.e., for scientists alone, judicial or wider social spaces). Hence, the tangible social effects of forensic experts' knowledge practices range from effects on individual lives (of the victims and their relatives, for instance) to social and community processes of collective memory practices and reconciliation, and the administration of (transitional) justice.

NOTES

1. Dr. María Fernanda Olarte-Sierra is a postdoctoral fellow in Medical Anthropology and the Global Health Institute for Cultural and Social Anthropology at the University of Vienna.

2. Fajardo, Dario, *Estudio Sobre Los Orígenes Del Conflicto Social Armado, Razones de Su Persistencia y Sus Efectos Más Profundos En La Sociedad Colombiana.* [Study on the Origins of Armed Social Conflict, Reasons for Its Persistence and Its Deepest Effects on Colombian Society] (Bogotá: Centro Nacional de Memoria Histórica, 2014); Centro Nacional de Memoria Histórica, *¡Basta YA! Colombia: memorias de guerra y dignidad* [Enough NOW! Colombia: Memories of War and Dignity] (Bogota: Centro Nacional de Memoria Histórica, 2013).

3. Tania Gabriela Rodríguez Morales, "Geografía Del Terrorismo En Colombia: Una Visión Retrospectiva" [Geography of Terrorism in Colombia: A Retrospective], *Revista de Paz y Conflictos* 9, no. 2 (2016): 179–98. See also, Dario Fajardo, *Estudio Sobre Los Orígenes Del Conflicto Social Armado*, 2014; Centro Nacional de Memoria Histórica, *¡Basta YA! Colombia*, 2013.

4. FARC-EP stands for Fuerzas Armadas Revolucionarias de Colombia—Ejército del Pueblo, The Revolutionary Armed Forces of Colombia—People's Arm. Jordi Palou-Loverdos, "Memoria y Justicia Transicional En Los Acuerdos de Paz de Colombia"[Memory and transitional justice in the Colombian peace accords], *Revista Especializada En Sociología Jurídica y Política* 12, no. 2 (August 2018), https://repository.ucatolica.edu.co/handle/10983/16338; Presidencia de la República de Colombia and FARC-EP, "Acuerdo Final Para La Terminación Del Conflicto y La Construcción de Una Paz Estable y Duradera," November 24, 2016.

5. María Fernanda Olarte-Sierra and Jaime Castro Bermúdez, "Notas Forenses: Conocimiento Que Materializa a Los Cuerpos Del Enemigo En Fosas Paramilitares y Falsos

Positivos" [Forensic Notes: Knowledge that Materializes the Enemy's Bodies in Paramilitary Graves and False Positives], *Antípoda* 34 (2019): 119–40, https://doi. org/10.7440/antipoda34.2019.06.

6. Simon A. Cole, "Forensic Culture as Epistemic Culture: The Sociology of Forensic Science," *Studies in History and Philosophy of Biological and Biomedical Sciences* 44 (2013): 36–46.

7. Although forensic science results do not and cannot be generalized, this does not mean that this knowledge is not based on standardizations and generalization of racialized and gendered traits; see Amade M'charek, "Beyond Fact or Fiction: On the Materiality of Race in Practice," *Cultural Anthropology* 28, no. 3 (2013): 420–42; Vivette García-Deister and Lindsay A. Smith, "Migrant Flows and Necro-Sovereignty: The Itineraries of Bodies, Samples, and Data across the US-Mexico Borderlands," *BioSocieties* 15, no. 3 (September 1, 2020): 420–37. https://doi.org/10.1057/s41292-019-00166-4).

8. Stephen Cordner and Morris Tidball-Binz, "Humanitarian Forensic Action—Its Origins and Future," *Forensic Science International* (2017), https://doi.org/10.1016/j.forsciint.2017.08.011

9. Bruno Latour, "Why Has Critique Run Out of Steam? From Matters of Fact to Matters of Concern," *Critical Inquiry* 30 (2004): 225–48; Francisco Ferrándiz, "Exhuming the Defeated: Civil War Mass Graves in 21st-Century Spain." *American Ethnologist* 40, no. 1 (2013): 38–54; Claire Moon, "Interpreters of the Dead: Forensic Knowledge, Human Remains and the Politics of the Past," *Social & Legal Studies* 22, no. 2 (2013): 149–69; Tania Perez-Bustos, María Fernanda Olarte-Sierra, and Adriana Diaz del Castillo, "Working with Care: Experiences of Invisible Women Scientists Practicing Forensic Genetics in Colombia," in *Beyond Imported Magic*, eds. Eden Medina, Ivan Marques, and Christina Holmes (Cambridge, MA: MIT Press, 2015), 67–86; Sévane Garibian, Elizabeth Anstett, and Jean-Marc Dreyfus, *Restos Humanos e Identificación: Violencia de Masa, Genocidio y El "Giro Forense"* [Human Remains and Identification: Mass Violence, Genocide and the "Forensic Turn"] (Buenos Aires: Miño y Dávila Editores, 2017); Amade M'charek and Sara Casartelli, "Identifying Dead Migrants: Forensic Care Work and Relational Citizenship," *Citizenship Studies* 23, no. 7 (October 3, 2019): 738–57, https://doi.org/10.1080/13621025.2019.1651102.

10. Adam Rosenblatt. *Digging for the Disappeared: Forensic Science after Atrocity* (Stanford, CA: Stanford University Press, 2015).

11. Perez-Bustos, Olarte-Sierra, and Diaz del Castillo, "Working with Care," 2015.

12. See María Fernanda Olarte-Sierra and Tania Pérez-Bustos, "Careful Speculations: Toward a Caring Science of Forensic Genetics in Colombia," *Feminist Studies* 46, no. 1 (2020): 158–77. https://doi.org/10.15767/feministstudies.46.1.0158; Perez-Bustos, Olarte-Sierra, and Diaz del Castillo 2015.

13. Donna Haraway, Testigo_Modesto@Segundo_Milenio.HombreHembra©_Conoce_Oncoratón®. Feminismo y tecnociencia. [Modest Witness@Second_Millenium.FemaleMan©_Meets_OncoMouse™: Feminism and Technoscience] (Barcelona: Editorial UOC, 2004), 23.

14. Donna Haraway, "Situated Knowledges: The Science Question in Feminism and the Privilege of Partial Perspective," *Feminist Studies* 14, no. 3 (1988): 575–99.

15. Annemarie Mol, "Ontological Politics. A Word and Some Questions," *The Sociological Review* 47, no. 1_suppl (May 1, 1999): 74–89. https://doi.org/10.1111/j.1467-954X.1999.tb03483.x.; Sheila Jasanoff, "Ordering Knowledge, Ordering Society," in *States of Knowledge. The Co-Production of Science and Social Order*, ed. Sheila Jasanoff, 13–45 (London and New York: Routledge, 2006); John Law and Ingunn Moser, "Contexts and Culling," *Science,*

Technology, & Human Values 37, no. 4 (December 15, 2011): 332–54. https://doi.org/10.1177/0162243911425055; Amade M'charek, "Beyond Fact or Fiction: On the Materiality of Race in Practice." *Cultural Anthropology* 28, no. 3 (2013): 420–42.

16. For information on these research projects, see Olarte-Sierra and Castro Bermúdez 2019; Perez-Bustos, Olarte-Sierra, and Diaz del Castillo, "Working with Care," 2015; Olarte-Sierra, Maria Fernanda Olarte, Diaz del Castillo, Adriana, Pulido Ronchaquira, Natalia, Cabrera Villota, Nathalia, and Suarez Montañez, Roberto, "Verdad e Incertidumbre En El Marco Del Conflicto En Colombia: Una Mirada a Los Sistemas de Información Como Prácticas de Memoria" [Truth and uncertainty in the context of the conflict in Colombia: a look at information systems as memory practices], *Universitas Humanística* 79 (2014): 233–54. https://doi.org//0.11144/Javeriana.UH79.vimc.

17. Olarte-Sierra, et al., "Verdad e Incertidumbre En El Marco Del Conflicto En Colombia," 2014.

18. Names are pseudonyms, except Jaime's.

19. Perez-Bustos, Olarte-Sierra, and Diaz del Castillo, "Working with Care," 2015.

20. Perez-Bustos, Olarte-Sierra, and Diaz del Castillo, "Working with Care," 2015.

21. Interview with Azucena, 2012.

22. Azucena, interview.

23. Donna Haraway, "Situated Knowledges: The Science Question in Feminism and the Privilege of Partial Perspective," *Feminist Studies* 14, no. 3 (1988): 575–99.

24. Corinna Kruse, "Legal Storytelling in Pre-Trial Investigations: Arguing for a Wider Perspective on Forensic Evidence," *New Genetics and Society* 31, no. 3 (September 1, 2012): 299–309. https://doi.org/10.1080/14636778.2012.687084.

25. Caroline Fournet, "Forensic Truth? Scientific Evidence in International Criminal Justice," (2017).

26. Cole, "Forensic Culture as Epistemic Culture," 2013.

27. I have known Jaime for six years and currently we are co-researchers that is why I refer to his quotes as conversations and not as interviews (see Olarte-Sierra and Castro Bermúdez, "Notas Forenses," 2019).

28. Conversation with Jaime, 2019

29. Conversation with Jaime, 2019.

30. Interview with Angela, 2012.

31. Moon, "Interpreters of the Dead," 2013; Garibian, Anstett, and Dreyfus, *Restos Humanos e Identificación*, 2017.

32. Interview with Angela, 2012. When forensic dentists attend delivery ceremonies, forensic anthropologists and coroners help to prepare them to present the bones to the families and to be able to answer questions like the ones Angela refers to.

33. Congreso de Colombia, «Law 1408/2010», 2010.

34. Olarte-Sierra and Pérez-Bustos, "Careful Speculations," 2020; Perez-Bustos, Olarte-Sierra, and Diaz del Castillo, "Working with Care," 2015.

35. Zoë Crossland, "Forensic Afterlives," *Signs and Society* 6, no. 3 (September 1, 2018): 622–47. https://doi.org/10.1086/699597; Ferrándiz, "Exhuming the Defeated," 2013; Moon, "Interpreters of the Dead," 2013; M'charek and Casartelli, "Identifying Dead Migrants," 2019.

36. Garibian, Anstett, and Dreyfus, *Restos Humanos e Identificación.*

37. Olarte-Sierra and Pérez-Bustos, "Careful Speculations."

4

Of Animacy and Afterlives

Material Memories in Indigenous Collections

Margaret M. Bruchac

Indigenous items in museums and archives have long been obscured, if not silenced, by theoretical and structural categories that limit conceptions of agency and animacy. An object that occupies space clearly has measurable dimensions, appearance, and substance, but how is it perceived? Objects are visible only in so much as they can be *seen*—by which, I mean physically and conceptually *sensed* and *recognized*. Then they can be classified—as iterations of other words, objects, categories, and things that are already known or knowable. Archivists and curators have long been trained to use selective filters to sort through these classifications, to see what is expected, to recognize a familiar form, to know what fits into the system. But if we look closer, if we consider subtle details in the construction of these object-beings, perhaps we can better understand how they reflect the intentions of those who brought them into being. Here, I reflect upon methods for identifying and interrogating material evidence that records the intentions of "invisible labourers" who might still speak to us through the traces that remain.

SPEAKING OF ANIMACY

Anthropological discussions of object agency have long circled around the notion that objects can only be imbued with meaning through the processes of handling and distribution by humans.[1] Indigenous knowledge-keepers, in contrast, assert that objects can have agency of their own, and that they can, therefore, choose to collaborate with humans in constructing and circulating meaning . . . or not. Traditionally, Indigenous peoples recognize the coexistence of "other-than-human" beings (animals, shape-shifters, ancestral spirits, natural forces, etc.), perceived as "persons" with the capacity to manifest, transform, and direct power.[2] In Algonquian languages and cultural practices, other-than-humans and seemingly inanimate objects may all be construed as agents (or relatives) with potential power.[3] The concept of "animacy" offers a means to understand the nuances of lived relationships among humans and others, past and present. As Robin Wall Kimmerer notes, in the Potawatomi language, "Beings that are imbued with spirit, our sacred medicines, our songs, drums, and even stories, all are animate. . . . The language reminds us, in every sentence, of our kinship with all of the animate world."[4] It is helpful, therefore, to consider who (and what) contributed to the material assemblages that make up these things we call "objects," and to consider how they participate in social relations across time and space, even when trapped in a museum. What ecosystems did they emerge from, who were they related to, how were they designed to engage with other objects and other persons?

Despite decades of silence, some things in museums may still be seen as kin to the people who brought them into being, potentially animate ritual partners waiting for an invitation to be re-awakened. For example, where a curator might see a stone pipe as an inanimate object or a practical implement, an Indigenous user might know the source of the stone to be a sacred place, and might recognize this same object as a powerful pipe-being, an agent with transformative powers that could be activated through prayer and medicine. As such, the pipe could serve as both an animate conveyer of messages and an active participant in ceremony. If that pipe happens to be attached to a hatchet, there are at least two potential (and equally potent) messages: war and peace.[5]

So, in a world where objects can be construed as kin, and where power can be conceptually invested in both spoken words and material objects, disparate forms may be woven together into assemblages that eloquently symbolize and materialize inter-tribal, inter-cultural, or trans-national understandings. Seemingly inanimate objects may wield, not just *imagined* meaning or *distributed* agency, but *literal* power as conveyers of messages, participants in diplomacy, and speaking (or temporarily silent) beings that communicate across time and space. Indigenous objects, even when severed from their origins, may still represent multiple beings and multiple "processes despite their seemingly fixed position in the museum."[6] The museums that house these objects are thus increasingly being called upon to reconsider, in the words of curator Joshua Bell, "What it is that objects *do* in the world, how they work as social actors, and [what is] their influence" (my emphasis), even when they are not in direct contact with the communities that created them.[7] Animacy may be present, whether or not it is perceptible to a curator.

THE LIVES OF WAMPUM BELTS

The complexity of Indigenous messaging is especially visible in the material traces of plant fibre, animal hide, and shell beads that come together to form the assemblages called "wampum belts." These objects, when woven into figurative or symbolic patterns, have long been utilized by North American Indigenous peoples and nations for adornment, political communication, and cultural records. The generic term *wampum*, borrowed from the Algonquian word *wampumpeag* for "white shell beads," refers to both white and purple cylindrical shell beads carved from marine whelk (either *Busycon canaliculatum* or *Busycon carica*) and quahog (*Mercenaria mercenaria*), harvested from the coastal waters along present-day Long Island Sound.[8] These beads were woven together using brain-tanned deerskin leather, sinew, and/or plant fibres derived from dogbane (*Apocynum androsaemifolium*), black Indian hemp (*Apocynum cannabinum*), hairy milkweed/white Indian hemp (*Asclepias pulchra*), swamp milkweed (*Asclepias incarnate*), or the inner fibres of slippery elm bark (*Ulmus fulva*).[9]

As part of my "Wampum Trail" survey, I have been examining wampum-making materials and belts in museum collections by carefully recording and comparing the evidence of origins, construction, reconstruction, repair, and re-purposing.[10] In historical wampum belts, there are multiple layers of relative (in)visibility, from the labours of the other-than-human marine, floral, and faunal beings whose bodies provided the raw materials, to the labours of the artisans who wove wampum, to the subsequent labours of traditional keepers and interpreters. The foundational layers of faunal and floral weaving materials that hold these assemblages together—from the sinews of deer to strands of plant fibre—embody flexible yet durable tensile strength. The individual beads retain visible traces of their origins in marine shell-beings that inhabit the liminal environments of fresh and salt water where rivers and oceans meet. Each white bead (best taken when the shell-beings are young) has a natural pearl-like sheen, having been carved from the inner whorls of whelk. Each purple bead (best taken when these shell-beings are decades old), carved from the outer edges of quahog, has a deep purple sheen; the oldest beads are so dark as to be nearly black.[11] The dramatic difference in colour density enables these beads to effectively participate in a binary signalling system; in general, white signals ease, harmony, and peace, and purple signals difficulty, complexity, and conflict. The colours also reflect the ecosystems in which these beings lived, since the cleanest waters (before colonial settler contact) produced the healthiest mollusks and the densest colours.[12]

Then there is the process of discovering, or more appropriately, *recognizing*, the intentions in the minds and hands of the human artisans who co-inhabited these ecosystems. They knew exactly when and how to harvest, carve, and drill whelk and quahog to create sound beads of relatively uniform size and shape. They knew which plant materials to harvest to weave into fibres for weft strands, and how best to hunt and process animal hides to create the leather strands used for warp. They utilized a square-weave pattern that included distinctive crosses and twists to secure the weft to the edges of the belt; slight alterations in these techniques reflect, not just practical choices, but cultural and regional traditions that bespeak particular artisans. All of those knowledges and all of those liminal spaces—between ocean and land, forest and shore, flora and fauna, dark and light beads, one and another weaving pattern, and loose and assembled components—come into play in the making of wampum.

Haudenosaunee and Algonkian traditionalists view wampum beads as animate partners in diplomacy; once these beads have been woven into a collar or belt, they are collectively transformed into speaking objects that can preserve and communicate negotiated understandings. The messages in these wampum belts are most clearly *seen* (to return to my sensory metaphor) when they are employed in acts of cultural performance, draped across the shoulders of a tribal dignitary, held up during a council meeting, or otherwise ceremoniously displayed with words spoken over them.[13] When trapped in stasis inside a glass case or a storage drawer, however, they are less visible, and less animate.

I began this research by working from a database that Tuscarora traditionalist Richard W. Hill, Sr. spent decades compiling; this included a list, year by year, of every reference in a colonial document to a wampum belt. It appeared, from this record, that there had been thousands of wampum belts, but only a few hundred still survived in tribal and museum collections. Where did they all go? After close material analyses of more than two hundred historic wampum belts and consultations with Hill and other scholars, I can state, based on material evidence, that many of the surviving belts incorporate beads and component parts taken from other belts.[14] Although some wampum belts were intended to be fixed in form and function to communicate a single, unchanging message, other belts show clear evidence of having been taken apart, recombined, rewoven, repaired, or repurposed.

Some of these historic belts contain beads that vary dramatically in size, density of colour, and condition (including very old desiccated beads), suggesting they were gathered together from multiple sources. Some beads were marked to amplify their intent or to signal their re-purposing. In the Huron-Wendat "Great War Belt," for example, red colouring (typically made from a mixture of red ochre, bear grease and vermillion) was vigorously applied around the white beads composing the image of a hatchet woven into the belt.[15] A white peace belt could be transformed into a war belt by applying red, but there is also evidence, in some cases, of red having been washed away so a belt could be re-presented in a more peaceful context.

Yet, some influential non-Native historians imagined that wampum belts (and other traditional ritual agents) must be fixed and unaltered to be considered legitimate records of cultural continuity. So, for example, in 1971, William Fenton insisted that the Hiawatha wampum belt (recording the founding relationships of the Haudenosaunee Confederacy) was neither *original* nor *authentic* because it contained a single blue glass bead of European origin.[16] Taking that bead as diagnostic, he pinned it to a particular moment in the 1700s and declared that the belt could be no older than the bead, inferring that oral traditions associated with the object were thereby false.[17] Yet, when Haudenosaunee leadership allowed me to examine the Hiawatha belt, I found that although there *does* appear to be a single glass bead dating to the early 1700s, the other components of the belt—warp, weft, and thousands of whelk and quahog shell beads—show clear signs of heavy use, repair, and reweaving. There is no oral or written record of these repairs, and it is impossible to authoritatively date the shell beads without destructive analysis, but they are consistent in size, shape, and density with beads and weaving materials in other belts from the 1600s and 1700s (if not earlier). This belt, like the oral history it records and reflects, has a life that precedes and transcends colonial events and scholarly arguments. The glass bead, therefore, is not diagnostic of the *origins* of this object; it only marks a particular moment of repair.

To date, out of the more than two hundred historic wampum belts examined, I have seen more than twenty shell bead wampum belts, woven in similar fashion, that contain one or two glass beads (figure 4.1). Some of these belts also include tubular glass, steatite, or clay beads that mimic shell wampum beads. A few have bone beads painted black on both ends, strung into the original weave and secured with red thread. Whose bones, we might ask? The question warrants further investigation. Given the specificity of these individual beads, and the care it took to select them and weave them into the assemblage, it is highly doubtful that these inclusions were accidental. In these same belts, the warp strands show pinch marks that do not strictly align with the current weft arrangement; this typically indicates that a belt was made from re-purposed materials (taken from other belts) and/or re-constructed. Although museum records rarely mention anomalous beads or evidence of repair, all of these details matter.[18]

Taken together, the material details in historical belts reflect the choices and care of multiple artisans and wampum keepers—those who made the original, those who made the repairs, those who inserted any of the unusual beads, and those who curate the belt today—over multiple generations. Since some of the eighteenth-century belts containing glass beads are associated with members of the Seven Nations Alliance, it is possible that the anomalous beads represent signatures of repair, or contributions from particular allies.[19] If these belts were re-purposed or re-woven more than once, they might retain memories of other agreements, with other people, at other times.

Figure 4.1. Detail of eighteenth century northeastern wampum belt, showing inclusion of a single blue glass bead (sixth bead in the third row from the top) alongside white and purple shell beads of varying sizes. Item #M1909, gift of David Ross McCord, origin unknown. Courtesy of the McCord Museum. Photograph by author.

THE AFTERLIVES OF WAMPUM BELTS

During the colonial era, the exchange of wampum, whether between Native nations or with European nations, was accompanied by spoken words and/or ritual performances that signified acceptance of whatever message the belt was intended to convey. A potential diplomatic partner would accept a belt to preserve a record of an agreement. In general, to dismantle a belt is to dismantle the agreement.[20] This provokes a question: does the evidence of earlier lives disappear when objects are lost, broken, or cast in new performative roles? Do the beads that move out of a belt and into another object carry intentions with them?

Evidence of the afterlives of wampum belts and beads may be found in a small selection of northeastern Native wooden clubs and wooden bowls, mostly dated to the seventeenth century, embedded with wampum beads. The obvious evidence of drilled holes and traces of fibre weft indicate that these beads had been part of a woven assemblage before being repurposed on a wooden object. The placement of these beads—for example, inside a burl bowl, or along the spine of a wooden war club—clearly signals more than merely decorative purposes.

One such example is a Mohegan bowl that appears, at first glance, to be simply utilitarian. Made to hold cornmeal mush or *succotash*, it appears similar to many other seventeenth-century examples from the American northeast. One handle of the bowl includes two rows of white wampum beads in the shape of the letter "L" (figure 4.2).[21] Each tubular bead is drilled all the way through lengthwise, and traces of the original plant weft fibre remain inside, indicating that these beads were re-purposed from an earlier woven object, likely a collar or belt. All of the beads in this bowl were made from the central columns of whelk shells, but in this case, they appear to be locally sourced, since their size and shape match others archaeologically recovered from a seventeenth-century Mohegan wampum-making site at Fort Shantok.

Figure 4.2. Mohegan burl bowl, showing inclusion of re-purposed white wampum beads in one handle. Courtesy of the Mohegan Tribe. Photograph by author.

How can we see the agency in this bowl, and in these beads? Mohegan oral tradition tells us that burl bowls were more than mere containers; bowls like this were ritually used for serving Native foods like pounded cornmeal mush or succotash, and they were passed down through the generations. This particular bowl has a long history, having been entrusted to the care of Lucy Occum Tantaquidgeon (1733–1830), an elder and knowledge-keeper for the Mohegan nation. In the language of wampum, white signifies harmony and healing, when compared to the dark purple beads that communicate complexity and difficulty. By bringing together two categories of ritually significant objects—burl bowls and white wampum beads—this assemblage becomes a container for peace. When it held communally ground and prepared cornmeal, this object-being encouraged tribal members (and allies) to, quite literally, consume the message of peaceful relations.[22]

There are other surviving objects with wampum inclusions in wood. One of the earliest examples is a maple wood musket stock recovered from a seventeenth-century shipwreck, inset with an array of white wampum beads in the shape of four crosses, two on each side of the stock, in a manner that might be considered both decorative and protective.[23] A more dramatic example can be found in a Wampanoag weapon dubbed "Metacomet's War Club," which lives in a mirrored display case at the Fruitlands Museum (figure 4.3). Said to have been collected in 1676 after the death of the Wampanoag sachem Metacom (King Philip), this club once held at least 88 purple and white wampum beads, set in rows of 44 along each side of the spine (most are now missing).[24] Other war clubs with wampum inclusions reside in French museums: one (#1943–128, apparently Mohawk in origin) at the Bibliothèque Sainte

Figure 4.3. Ball-club, known as "Metacomet's War Club," Wampanoag, Rhode Island, c. 1670, maple, quahog shell, whelk shell, horn. Previously known as King Philip's War Club. FM.1.0095.002. Courtesy of Fruitlands Museum, The Trustees of Reservations. Photograph by author.

Geneviève, and another (#71.197.3.14 D, presumed to be Wyandot) at the Musée du Quai Branly. The most dramatic example of a war club with wampum inclusions is a sabre-shaped club with an underwater panther carved into the handle, housed at the Fenimore Art Museum (#T0794); wampum beads are arrayed along the spine, and two beads (now missing) were inset to serve as the panther's eyes.[25]

On each of these objects, the wampum beads show visible evidence of previous use and wear—drilled holes, traces of fibre weft, and handling marks (chips, grease, traces of red ochre)—indicating that they were taken out of a woven wampum object. Heretofore, curators and scholars have interpreted wampum inclusions on wooden objects as being primarily ornamental or expedient, indicative of artisans utilizing available material for decorative purposes. Yet the meticulous placement of the beads in these clubs suggests both enhanced power and redirected intentions. In each case, a communally held wampum belt was dismantled, and its beads were used to adorn a personally held tool of warfare. When combined with other figurative and utilitarian elements on these weapons, these beads convey a dramatic message of diplomacy gone wrong.

War clubs with wampum inclusions thus provide a compelling example of broken belts physically transformed to convey broken messages. Imagine the origins of a wampum belt woven of purple and white shell beads, meticulously gathered from the liminal waters at ocean's edge and carefully crafted to record a crucial intercultural agreement. At the close of a bloody conflict, the making of that wampum belt would have enabled opposing nations to live together peacefully. But imagine the moment when that belt was forcibly broken into pieces. The beads, torn out of their original configuration, were then meticulously glued into carved notches that nestled along the backbone of a menacing war club, now spotted with blood. That conveys a very different message.

In closing, let it be said that differing understandings of relative levels of animacy do not just constitute differing belief systems; they influence our relations with, and our understandings of, objects. Material details can illuminate records of the ecosystems, artistic traditions, beliefs and experiences from which these object-beings emerged; they can also retain memories of the artisans who created them. And so, if we (and the museum custodians of such objects) lean into Indigenous languages and ontologies, we might also recognize the possibility that meanings can be reawakened when otherwise mysterious objects make themselves visible, through the intentions woven into them, through the traces of the places they have travelled, through the ways in which they visually "speak." If these object-beings have the possibility of being or becoming animate, they can be a source of potential engagement or potential power. One must know, at the very least, how to recognize them.

NOTES

1. Igor Kopytoff, "The Cultural Biography of Things: Commoditization as Process," in *The Social Life of Things*, edited by Arjun Appadurai (Cambridge: Cambridge University Press 1986), 66–68; Eduardo Kohn, "Anthropology of Ontologies," *Annual Reviews in*

Anthropology 44 (2015): 311–27. For a critique of the "new materialism" that draws upon Indigenous epistemologies, see Jennifer Clary-Lemon, "Gifts, Ancestors, and Relations: Notes toward an Indigenous New Materialism," *Enculturation* (November 12, 2019), http://enculturation.net/gifts_ancestors_and_relations.

2. A. Irving Hallowell, "Ojibwa Ontology, Behavior and World View," in *Culture in History: Essays in Honor of Paul Radin*, ed. Stanley Diamond (New York: Columbia University Press, 1960), 19–52. For embodied discussion of these relations, see Robin Wall Kimmerer, *Braiding Sweetgrass: Indigenous Wisdom, Scientific Knowledge, and the Teachings of Plants* (Minneapolis, MN: Milkweed, 2013).

3. In Algonquian and some other Indigenous languages, nouns are routinely classed and gendered, not as "masculine" or "feminine," but as "animate" or "inanimate." See, for example, Mary B. Black-Rogers, "Algonquian Gender Revisited: Animate Nouns and Ojibwa 'Power'—An Impasse?" *Research on Language & Social Interaction* 15, no. 1 (1982): 59–76.

4. Kimmerer, *Braiding Sweetgrass*, 55–56.

5. For fuller discussion of Indigenous meanings in wielding and depicting tomahawks, see Scott Manning Stevens, "Tomahawk," *Early American Literature* 53, no. 2 (2018): 475–511.

6. Joshua A. Bell, "A Bundle of Relations: Collections, Collecting, and Communities," *Annual Reviews in Anthropology* 46 (2017): 245.

7. Joshua A. Bell, "Museums as Relational Entities: The Politics and Poetics of Heritage," *Reviews in Anthropology* 41 (2012): 73.

8. On the definition of wampum, see James Russell Trumbull, *Natick Dictionary*, Bulletin 25, Smithsonian Bureau of Ethnology (Washington, DC: Government Printing Office, 1903), 340–41. For an overview of historical wampum use, see William M. Beauchamp, "Wampum and Shell Articles Used by the New York Indians," *New York State Museum Bulletin* 41 (1901): 321–480. Also see Lynn Ceci, "The Value of Wampum among the New York Iroquois: A Case Study in Artifact Analysis," *Journal of Anthropological Research* 38, no. 1 (1982): 97–107.

9. These materials can often be detected through non-invasive visual analysis, when amplified by close photography or the use of a microscope. Some examples of weaving techniques and fibres are listed in Frank G. Speck, *The Penn Wampum Belts* (New York: De Vinne, 1925), and in A. C. Whitford, *Textile Fibers Used in Eastern Aboriginal North America* (Washington, DC: Anthropological Papers of the American Museum of Natural History, Volume 38, Part 1, 1941). Also see William C. Orchard, *Beads and Beadwork of the American Indians: A Study Based on Specimens in the Museum of the American Indian* (New York, NY: Heye Foundation, Volume XI, 1929).

10. See Margaret M. Bruchac research blog, "On the Wampum Trail," online at: https://wampumtrail.wordpress.com/

11. See Margaret M. Bruchac, "Wampum Matters: Notes on the Technology and Materiality of Historical Wampum Beads and Belts." Unpublished research report, University of Pennsylvania, July 2017.

12. For discussion of colour signalling in wampum and other objects, see George R. Hamell, "Strawberries, Floating Islands, and Rabbit Captains: Mythical Realities and European Contact in the Northeast during the Sixteenth and Seventeenth Centuries," *Journal of Canadian Studies/Revue d'études canadiennes* 21, no. 4 (Winter 1986–1987): 72–94.

13. For an historic illustration and discussion of a distinctive wampum belt in use—the "Great War Belt" of the Wendat—see Jonathan Lainey and Anne Whitelaw, "The Wampum and the Print: Objects Tied to Nicolas Vincent Tsawenhohi's London Visit, 1824–1825," in *Object Lives and Global Histories in Northern North America: Material Culture in Motion,*

c. 1780–1980, eds. Beverly Lemire, Laura Peers, and Anne Whitelaw (Montreal & Kingston: McGill Queen's University Press, 2021), 179–80.

14. As part of this survey, I have examined historic wampum belts housed in the collections of the British Museum, Canadian Museum of History, McCord Museum, National Museum of the American Indian, New York State Museum, Peabody Museum of Archaeology and Ethnology at Harvard University, Pitt Rivers Museum, Rock Foundation, and Royal Ontario Museum, among many others. See Bruchac, "Wampum Matters."

15. See item number M20401 in "Our People, Our Stories," McCord Museum, Montreal, Quebec: http://collections.musee-mccord.qc.ca/en/collection/artifacts/M20401.

16. For a summary account of the founding of the Haudenosaunee Confederacy, see Paul Williams, *Kayanerenkó:wa: The Great Law of Peace* (Winnipeg: University of Manitoba Press, 2018).

17. William N. Fenton, "The New York State Wampum Collection: The Case for the Integrity of Cultural Treasures." *Proceedings of the American Philosophical Society* 115, no. 6 (1971): 437–61.

18. For instance, in the McCord Museum database, the "Great War Belt" of the Wendat is simply described as composed of "Shell beads, hide, plant fibre," with no mention of the bone bead included in the weave.

19. The Seven Nations Confederacy (*Tsiatak Nihononwentsiake)* was an eighteenth-century inter-tribal alliance of First Nations Catholic converts from Mohawk, Abenaki, and Huron-Wendat nations, including Akwesasne, Kahnawake, Kanesatake, Odanak, Oswegatchie, Wendake, and Wolinak. See David Blanchard, "The Seven Nations of Canada: An Alliance and a Treaty," *American Indian Culture and Research Journal* 7, no. 2 (1983): 3–23. Also see Alain Beaulieu and Jean-Pierre Sawaya, "L'importance strategique des Sept-Nations du Canada," *Bulletin d'Histoire Politique* 8, nos. 2–3 (2000): 87–107.

20. For general discussions of wampum use in the colonial era, see Beauchamp, "Wampum and Shell Articles Used by the New York Indians."

21. Charles C. Willoughby, "Wooden Bowls of the Algonquian Indians," *American Anthropologist* N.S. 10, no. 3 (1908): 423–34.

22. Adapted from the interpretive notes by Mohegan Medicine Woman Melissa Tantaquidgeon Zobel.

23. The musket stock, catalogued as 57M14N2–37, was recovered from what is believed to be the 1690 wreck of the *Elizabeth and Mary*, found in a cove at l'Anse aux Bouleaux, Quebec, in 1994. See Pointe-à-Callière, *1609, The Siege of Quebec . . . The Story of a Sunken Ship* (Montreal, QC: Pointe-à-Callière, Montreal Museum of Archaeology and History, 2000). See Charles Bradley and Karlis Karklins, "A Wampum-Inlaid Musket from the 1690 Phips' Shipwreck," *BEADS: Journal of the Society of Bead Researchers* 24 (2012): 91–97.

24. Bert Salwen, "Indians of Southern New England and Long Island: Early Period," in *Handbook of North American Indians, Northeast*, ed. Bruce G. Trigger, Vol. 15 (Washington, DC: Smithsonian Institution 1978), 171.

25. Edmund Carpenter, *Two Essays: Chief & Greed* (North Andover: Persimmon Press, 1968): 78–83. For further discussion, see Margaret M. Bruchac, "Visions of War: A 17th Century War Club Embedded with Wampum," *Historic Deerfield* Vol. 18 (Autumn 2020): 40–45.

5

Ex-prisoners of the Gulag in the Siberian Expeditions

Alexandra Noi

In this chapter, I examine the role of ex-prisoners of the Soviet Gulag in the scientific exploration of the Siberian taiga (a vast forest in northern Russia) in 1950s–1960s. The Gulag—an extensive network of forced labour camps that was established in the aftermath of the Bolshevik revolution and expanded under Joseph Stalin—was not only a place of incarceration and exclusion but was also the backbone of Soviet economy. The Gulag camps were, for the most part, located in remote and sparsely populated areas of the Soviet Union. By the time of Stalin's death in March 1953, more than 2.5 million prisoners were deployed as disposable labourers on grand industrial projects aimed at the exploration and extraction of the natural resources.[1] After Stalin's death, more than one million people, mostly charged with criminal offenses, were released as a result of a mass amnesty.[2] The former convicts came to be known by the derogatory nickname *bichi*, reflecting their marginalized status in Soviet society.[3] Unable to integrate into society, *bichi* were often hired as seasonal workers on scientific expeditions. Although the scientific exploration of remote lands with often-extreme environments is a well-established topic in the history of science, the contribution of the ex-prisoners has not been examined.[4]

This essay argues that the invisibility of ex-prisoners as the labourers of Soviet forest science was a complex phenomenon, which included not only their marginalization enforced by the state but also their own agency in making certain life choices and decisions as they tried to rebuild their lives in a meaningful way. This agency was manifested in their participation in the scientific expeditions in the Siberian taiga. *Bichi* were not featured in contemporary scientific research publications on forestry and were hardly registered in the expeditions' accounting documents as official team members. One can find the testimonies of their presence, however, in scientists' diaries and memoirs, and can retrieve their stories through oral histories. The *unseenness* of ex-prisoners (to invoke James Scott's powerful metaphor) has erased them from the public space; I argue that, having been excluded from the public space, they turned their unseen-ness into an asset and a strategy to make a transition from prison to society.[5]

This chapter stands at the intersection of environmental history, specifically of Soviet forestry and forestry science, and the history of the Gulag prison system. These two fields have only recently been put into conversation with each other. Historians turned their attention to how the Gulag inmates, along with other marginalized populations—the Indigenous people of Siberia, migrant labourers, and forcibly relocated peasants—shaped Russia's northern environment. These communities are understood as active actors that played an important role in exploring, reclaiming, and transforming the ecosystems of the Soviet remote regions.[6]

The mutual integration of Soviet science and the Gulag has been explored in the context of histories of secret Gulag laboratories, or *sharashki*. There, a number of prominent scientists, engineers, and technicians who had been imprisoned in the Gulag during the Stalinist purges of 1930s–1940s and persecuted as political enemies of the Soviet regime were working on various applied projects.[7] However, the Gulag and post-Gulag experiences of non-political convicts who served terms for criminal offenses had not been examined in existing literature, partly because they left many fewer historical traces than high-profile prisoners.[8] In this chapter, I shift the focus from professional scientists and political inmates to ordinary prisoners and examine their lives after their release and their underrecognized role in the scientific expeditions.[9] To do so, I approach Soviet science and the Gulag as archives.

It is well established that writing history is conditional on the (colonial) archives as sites of knowledge production and of exclusion and silence.[10] Both the Gulag prison system and Soviet science functioned as forms of archival institutions which collected and organized information about people. Invoking Bruno Latour, one may say that scientific and punitive systems have one thing in common: they both produce and keep highly organized literary inscriptions.[11] As the Gulag bureaucracy utilized prisoners as a labour force, it turned them into numbers in production plans. Historians have successfully used the Gulag records to reconstruct the histories of the involuntary labourers. But once prisoners were released into society, their traces largely disappeared from the state archives, since there were no state-run prisoner reintegration programs in the 1950s–1960s. However, those former convicts who were hired into scientific expeditions entered the domain of the "scientific archive," leaving behind valuable glimpses of knowledge and experience. My larger project, of which this paper is part, seeks to expand the Latourian metaphor of science as producer of literary inscriptions and considers together the "mania for inscription" in both prison and science bureaucracies.[12]

The principal sources for this case study are the collection of essays written by a scientist, Vladimir Sedykh.[13] At the time of writing, Sedykh is a professor of biology and forest scientist at the Sukachev Institute of Forest of the Russian Academy of Sciences in Krasnoyarsk.[14] Before engaging in academic research in early 1970s, he had worked as an engineer in forestry and geodetic expeditions in Siberia, the Far East, and Central Asia. In 1959–1961 he was a member of the first Siberian

Expedition organized by the Siberian branch of the Trust of Forest Aviation to extensively explore the forests of Siberia and the Far East.[15]

According to Sedykh, the goal of the Siberian Expedition was to bring taiga to the limelight, or literally "to the condition of being known" (*privesti taigu v izvestnost*)—that is to say, to make the rich, yet understudied, natural resources of Siberia "visible" and accessible for exploitation and extraction.[16] The principal element of this endeavour consisted of forest surveying, which, historically, had been performed in Russia by specialists called *taksatory* (from Latin *taxatio*—evaluation).[17] This profession had been introduced to Russia during the eighteenth-century reforms of Peter the Great, who used the German cameralist science as a model.[18] The study and evaluation of forest vegetation continued throughout the imperial period and after the Bolshevik revolution, yet it was under Stalin and his successors Khrushchev and Brezhnev that scientists embarked on a full-fledged plan to identify, survey, measure, and describe the forest vegetation resources of Siberia.[19] In 1960, the Siberian expedition merged with the Western Siberian Aerial Photography and Forest Management Trust. The forests of industrial potential were first subjected to aerovisual observations, and then the aerophotographs (on a scale of 1:15,000–1:25,000) were given to specialists like Sedykh who used them on the ground to survey the forests on foot.[20]

The labour of *bichi* was indispensable for the work of *taksatory* such as Sedykh. Tempered under the harsh conditions of the Gulag, the *bichi* did not shy away from hard manual labour and could adapt to difficult circumstances of expeditionary life. They were part of larger teams that consisted of scientists and engineers, technical personnel, a cook (usually a female) and, sometimes, interns (recent university graduates)—altogether seven to nine people. Forests were delineated into parcels and assigned to these teams. The teams, along with all necessary tools, equipment, and victuals (estimated for months-long season in taiga), had to reach their respective parcels located as far as 100 to 300 kilometres from the nearest regional centre. Sometimes they used a boat, sometimes a helicopter, and sometimes a car, usually reaching the final destination by horse or reindeer. After a team arrived and settled in, a camping ground (*tabor*) would be set up and a short training session focused on safety rules would officially open the field season. The group was divided into smaller units with one *taksator* and two to three *bichi*. A unit would make an "entry" (*zakhod*) deep into the taiga interior for up to a week.[21] This is when the aerial images came in handy: *taksatory* would use them to orient themselves in the taiga and to cut a grid of forest quarters, in which the borders were denoted by poles.[22] *Bichi* were responsible for carving forest glades (*proseki*), cutting survey lines (*viziry*), felling trees, and mounting the poles.[23] The scientists counted the trees in the quarters; measured their diameters at the height of 1.3 meters; determined their height, age, and increment; examined the natural regeneration of the forest; evaluated the composition and amount of grass-fruticulose stories; and described the soil.[24]

In a typical season, a unit would make several such "entries" into the taiga. The work lasted from May to late October (the warmest months in Siberia), during which an expedition team studied an area of 30–40,000 hectares. On average, each member of the unit would walk more than 1,000 kilometres through the wilderness.[25] After the expedition season was completed, the scientists processed the data and produced comprehensive reports characterizing forest vegetation according to a number of specific parameters that included the composition and the age of the forest stand, the productivity of the forest, an average diameter and height, and so forth.[26] These data were then used to create comprehensive maps of the specific parts of the taiga and to compile volumes of forest inventory data. The information collected by different expeditions, including the geological ones, was ultimately applied to identify the forest resource bases and to evaluate their productivity and the extent to which they could be useful for industrial purposes.

The remote, barely accessible, and sparsely populated Siberian wilderness provided a space of invisibility. For the ex-prisoners, the seasonal jobs in taiga offered a sense of protection. As Sedykh's memoirs attest, *bichi* were employed semi-officially: often lacking passports or any identification papers, they were hired without many inquiries about their Gulag prehistories. In these remote locations, everyone knew everyone else, and the word of mouth could give more credence than government-issued IDs. Sedykh noted that the so-called expeditionary tact precluded scientists from bringing up issues that *bichi* tried to forget: "It seemed improper [*neprilichno*] to ask *bichi* questions" about their difficult past.[27] Indeed, in a party-state that sought to know and oversee everything about its subjects, at least on paper, this strategic invisibility provided the ex-prisoners with their own control over their lives. By participating in scientific expeditions, they found their "little corner of freedom" in Siberia, to quote the historian of science Douglas Weiner, and they exploited this opportunity to begin their post-incarceration lives with a clean slate.[28]

Bichi's strategic invisibility was made possible by the support of the scientists who hired them. Therefore, they cultivated mutually beneficial relations with the heads of scientific expeditions. In his memoirs, Sedykh argues that *bichi* were hard, honest and highly valued workers:

> Bichi . . . were former convicts who served all kinds of prison terms. Released from the camps and had been roaming for a while, some . . . set out to various geological, geophysical, geodetic, and forest mapping expeditions. . . . [These expeditions] demanded not sentimental daydreamers, but labourers accustomed to hard labour and hiking inconveniences, those who did not need to be enticed, persuaded, pressganged. The ex-prisoners were hired as these workers, with the release certificates and with no regard for their biographies, for the absence of passports. . . . Engineers and technicians in the forest expeditions treated them as equals, sharing with them bread, tea, sugar, [and] eating from one pot. They were neither insulted, nor humiliated or short-changed, and these workers went with professional engineers literally through fire and water.[29]

The fact that the scientists, an elite group in Soviet society, were not afraid to work, live, and sleep in tents side by side with people previously charged with various crimes suggests a high level of trust, as well as shrewd calculation on the part of both parties. Participants of expeditions depended on each other for survival in the Siberian wilderness: without mutual help, trust, and fair treatment it was impossible to endure the harsh natural conditions.

To be sure, the elitist nature of scientists' treatment of workers is implicit between the lines of Sedykh's own narrative, suggesting that the relationship between ex-prisoners and scientists was complicated and shaped by class and power hierarchies. A conspicuous example of uneven relationship comes from Sedykh's recollection of his first internship in a scientific expedition, which allowed him to gain his initial experience of working alongside ex-prisoners. The head of the expedition instructed him: "As you see, workers are hardened. They have been released [from the Gulag], but their habits haven't changed. When 10–40 of them gather in one place and have no work to do, expect card games, home brew, brawls and fights. In order to prevent this, I try to keep them busy with work in the taiga."[30] This quote reveals the team leader's patronizing and condescending attitude towards *bichi* and demonstrates a degree of alienation between the heads of expeditions and the workers, rooted in their different life experiences. *Bichi* were understood to come from the world of crime, persecution, interrogations, imprisonment, arduous labour and survival in a prison camp, followed by many post-release challenges. *Bichi* largely identified with their Gulag experience, which the scientists were not likely to comprehend or relate to.

In fact, a degrading attitude of scientists towards *bichi* was explicit in the hiring process. Describing this process, Sedykh, perhaps unconsciously, draws *bichi* and equipment under a common denominator—all require good quality: "Before the start of the fieldwork, every *taksator* has to solve two crucial tasks: first, to obtain solid equipment (if not new, then thoroughly repaired)—tents, sleeping bags, tilts, pots, work clothes, tools, and groceries; second, to hire trustworthy and healthy bichi with work experience in taiga and without addiction to chifir" (an exceptionally strong black tea that was popular among Soviet prisoners for its intoxicant qualities).[31] The drug addiction was apparently determined by a sneak peek at *bichi*'s teeth: dark, stained teeth signified excessive use of *chifir*, and heads of expeditions literally "looked a gift horse in the mouth," as the saying goes.[32] Furthermore, workers were classified into "seasoned" or "leather" *bichi*, the latter having no or minimum experience of working in taiga expeditions.[33] "Seasoned" *bichi* were valued higher: they had good reputation among scientists and could get a new job assignment easier and more quickly than their "leather" fellows.

The co-existence and co-labour of scientists and *bichi* were not easy. Nonetheless, it worked, and worked well. They teamed up to produce scientific knowledge each performing an important role. In Sedykh's account, the *bichi* are portrayed as significant part of the expedition on an equal footing with scientists and engineers. Their contribution, Sedykh emphasized, was crucial: *taksatory* could not have done the work of assessing and describing taiga vegetation resources without the background labour—manual, semi-skilled, and arduous—performed for them by the workers.

Forestry scientific expeditions brought together many different actors, from various walks of life—scientists, students, engineers and technicians, former prisoners, as well as the Indigenous people of Siberia, who all contributed to the production of scientific knowledge in their own distinctive ways.[34] Writing the histories of scientific labour has been conditioned by the scientific authority of the actors and the degrees of their archival visibility. The presence of non-elite, non-professional, amateur, and generally marginalized labourers is limited in the scientific archive and hence in historical scholarship. This chapter recovers the invisible labour of the Gulag ex-prisoners whose physical efforts and manual skills were decisive in making the scientific exploration of Siberian forests possible. Instead of seeing the Gulag survivors as merely victims of the Soviet legal system, I emphasize their own agency and argue that they used this invisibility in their own favour—to facilitate their post-Gulag re-entry into society and to create a new identity for themselves, that of a hard and decent worker.

NOTES

1. V. Naumov and Iu. Sigachev, eds., *Lavrentii Beriia, 1953: Stenogramma iiul'skogo plenuma TsK KPSS i drugie dokumenty* (Moscow, 1999), 19. This source indicates an exact number of 2,526,402 people.

2. V. N. Zemskov, "Situatsiia s zakliuchennymi v pervie poslestalinskie gody" [Situation with Prisoners in the Early Post-Stalin Years], *Izvestiia Samarskogo nauchnogo tsentra RAN*, no. 16 (3) (2014): 130–36, here 130.

3. In Russian, *bich* (single form, plural—*bichi*) denotes a vagrant or seasonal worker, and also stands for an acronym for "formerly intelligent person" whose life had been ruined by years of incarceration in the Gulag (L. Belovinskii, *Entsiklopedicheskii slovar' istorii sovetskoi povsednevnoi zhizni* [Encyclopaedic Dictionary of the History of Soviet Everyday Life] (Novoe literaturnoe obozrenie, 2015), 78).

4. For the histories of scientific explorations in extreme environments, see Michael Robinson, *The Coldest Crucible: Arctic Exploration and American Culture* (University of Chicago Press, 2006); Julia Lajus, "Colonization of the Russian North: a Frozen Frontier," in *Cultivating the Colony: Colonial States and their Environmental Legacies* (Ohio University Press, 2011): 164–90; Paul R. Josephson, *The Conquest of the Russian Arctic* (Harvard University Press, 2014); Andy Bruno, *The Nature of Soviet Power: An Arctic Environmental History* (Cambridge University Press, 2016); Bathsheba Demuth, *Floating Coast: An Environmental History of the Bering Strait* (W. W. Norton & Company, 2019). See an excellent review by Julia Lajus: "Russian Environmental History: A Historiographical Review," in S. Ravi Rajan et al., *The Great Convergence: Environmental Histories of BRICS* (Oxford University Press, 2018): 245–73. On Soviet amateur "citizen science", see Elena Aronova (chapter 14) and her article "Citizen Seismology, Stalinist Science, and Vladimir Mannar's Cold Wars," *Science, Technology, & Human Values* 42, no. 2 (2017): 226–56.

5. James C. Scott, *Seeing Like a State: How Certain Schemes to Improve the Human Condition Have Failed* (New Haven, CT: Yale University Press, 1998); on invisibility as empowering, see Wylie, chapter 19 in this volume.

6. Paul R. Josephson, *An Environmental History of Russia* (Cambridge: Cambridge University Press, 2013); Josephson, *The Conquest of the Russian Arctic*; Alan Barenberg, *Gulag Town, Company Town: Forced Labor and Its Legacy in Vorkuta* (New Haven, CT: Yale University Press, 2014); Bruno, *The Nature of Soviet Power*.

7. See Asif Siddiqi, "Scientists and Specialists in the Gulag: Life and Death in Stalin's *Sharashka*," *Kritika: Explorations in Russian and Eurasian History* 16, no. 3 (Summer 2015): 557–88; Josephson, *The Conquest of the Russian Arctic*, 115–69.

8. The scholarship on the Gulag releasees has been mostly concerned with exploring the fates of political prisoners, despite the fact that the majority of incarcerated people were common criminals. See Nanci Adler, *The Gulag Survivor: Beyond the Soviet System* (Transaction Publishers, 2002); Stephen Cohen, *The Victims Return: Survivors of the Gulag after Stalin* (I. B. Tauris, 2011); Nanci Adler, *Keeping Faith with the Party: Communist Believers Return from the Gulag* (Bloomington: Indiana University Press, 2012). On common criminals in the Gulag, see Miriam Dobson, *Khrushchev's Cold Summer: Gulag Returnees, Crime, and the Fate of Reform after Stalin* (Ithaca, NY: Cornell University Press, 2009); Barenberg, *Gulag Town, Company Town*. See Omnia El Shakry's approach in chapter 11 of this volume of seeing both political and non-political prisoners as knowledge creators.

9. *Bichi* were discussed as migrant workers, but no mention of their Gulag identities was made: Lewis H. Siegelbaum, *Broad Is My Native Land: Repertoires and Regimes of Migration in Russia's Twentieth Century* (Ithaca, NY: Cornell University Press, 2014); Juliane Furst, *Stalin's Last Generation: Soviet Post-War Youth and the Emergence of Mature Socialism* (Oxford University Press, 2010); Svetlana Stephenson, *Crossing the Line: Vagrancy, Homelessness, and Social Displacement in Russia* (Aldershot, UK: Ashgate, 2006). On Gulag inmates employed in Soviet geology expeditions, see Alla Bolotova, "Colonization of Nature in the Soviet Union: State Ideology, Public Discourse, and the Experience of Geologists," *Historical Social Research* 29, no. 3 (2004): 104–23.

10. See Carolyn Hamilton, ed., *Refiguring the Archive* (Dordrecht, Netherlands: Kluwer Academic, 2002).

11. Bruno Latour, "Drawing Things Together," in *Representation in Scientific Practice*, ed. Michael Lynch and Steve Woolgar (Cambridge, MA: MIT Press, 1990), 19–68.

12. Bruno Latour and Steve Woolgar, *Laboratory Life: The Social Construction of Scientific Facts* (Sage Publications, 1979), 48. See also Sebastian Felten and Christine von Oertzen, eds., "Histories of Bureaucratic Knowledge," special issue, *Journal for the History of Knowledge* 1, no. 1 (2020).

13. Vladimir Sedykh, *Taksatory i bichi: pervootkryvateli Sibirskoy taigi* [Forest Mappers and Bichi: Pioneers of the Siberian Taiga] (Novosibirsk: Nauka, 2012); Vladimir Sedykh, *Taezhnye budni* [Everyday Life in Taiga] (Novosibirsk: Sibirskie Ogni, 2017).

14. The Institute was established in 1944 by a prominent Russian-Soviet geobotanist Vladimir Sukachev and assigned to the Siberian Division of the Academy of Sciences of the USSR in Krasnoyarsk in 1959.

15. Sedykh, *Taksatory i bichi*, 303.

16. Ibid., 11.

17. In this chapter, I describe these practices in the Soviet context with words such as *evaluation*, *surveying*, *measurement*, or *mapping*, and call the practitioners *taksatory*. Related practices have been described in the Western context as mensuration. See John A. Kershaw, et al., *Forest Mensuration* (Hoboken, NJ: Wiley Blackwell, 2017).

18. Sedykh, *Taksatory i bichi*, 9. On the origins of Russian forest management, see Brain, *Song of the Forest*, 11–28.

19. Josephson, *The Conquest of the Russian Arctic*, 168–69. On the history of studying Siberian forests within the Academy of Sciences of the USSR, see L. K. Pozdniakov, "Iz istorii izucheniia lesov Sibiri" [On the History of Studies of Siberian Forests], in A. P. Okladnikov, *Akademiia Nauk i Sibir' (1917–1957)* [Academy of Sciences and Siberia] (Novosibirsk: Nauka, 1977), 122–29; E. G. Vodichev, et al., *Rossiiskaia Akademiia Nauk. Sibirskoe otdelenie: Istoricheskii ocherk* [Russian Academy of Sciences. Siberian Division: A History]

(Novosibirsk: Nauka, 2007); Josephson, *An Environmental History of Russia*, 155–62; Brain, *Song of the Forest.*

20. Sedykh, *Taksatory i bichi*, 13.

21. Sedykh, *Taksatory i bichi*, 54, 65, 77–78, 92, 120, 129–130, 155; Sedykh, *Taezhnye budni*, 11–20, 85–86; Leah Bendavid-Val, *Siberia: In the Eyes of Russian Photographers* (Munich: Prestel, 2013), 95.

22. Sedykh, *Taezhnye budni*, 85.

23. Sedykh, *Taksatory i bichi*, 300–301.

24. Ibid., 77.

25. Ibid., 54.

26. Ibid., 77.

27. Author's interview with Vladimir Sedykh, via phone, November 7, 2020.

28. Douglas Weiner, *A Little Corner of Freedom: Russian Nature Protection from Stalin to Gorbachev* (Berkeley: University of California Press, 1999).

29. Sedykh, *Taksatory i bichi*, 11–12.

30. Ibid., 65.

31. Ibid., 32.

32. Sedykh, *Taksatory i bichi*, 44; Sedykh, *Taezhnye budni*, 214.

33. Sedykh, *Taksatory i bichi*, 300–301; Sedykh, *Taezhnye budni*, 150–52, 156.

34. The topic of how the scientific expeditions used the labour of native peoples of Siberia and capitalized on their vernacular knowledge of those lands warrants more research. See Yuri Slezkine, *Arctic Mirrors: Russia and the Small Peoples of the North* (Ithaca, NY: Cornell University Press, 1994) and Bruno, *The Nature of Soviet Power* on how Soviet industrialization and collectivization policies impacted the native peoples of the Russian North.

6

The Bureaucratic Ethic and the Spirit of Bio-capitalism

Laura Stark

BUREAUCRACY AND THE PRODUCTION OF "INVISIBILITY"

One of the organizing questions of the history of science since the social turn of the 1980s has been, how is knowledge made? The premise of the field was that matters of fact and scientific truths were not pre-existing entities, nor the sum total of all possible certified truths. Rather, formal knowledge was an achievement. The task of scholars has been to study science as a practice—as an everyday activity—and to trace the political, material, institutional, and conceptual conditions through which knowledge is stabilized and disrupted.

The early organizing question of the field begat a second organizing question of a more recent vintage, namely, the question of "justice," which explores the assumptions built into scientific products and their patterned consequences.[1] Whereas "ethics" marks an interest in understanding and, possibly, adjudicating historical actors' proper treatment of people and things in the past often in the frame of the settler state, "justice" designates a focus on the ultimate shape that science takes—given that the content of science could take a variety of forms—and its systemic effects. Among others questions, justice-oriented studies ask: Whose values and desires have been embedded in knowledge-making enterprises? How? And with what political effects?[2]

I work on the history of morality, that is, how certain knowledge practices come into being as right or wrong, acceptable or unacceptable politically. For example, I have studied the modern market for "human subjects" of medical experiment, which came into being in its current form sometime after World War II and in that specific form embedded a white supremacist logic (Burton, chapter 7). This market form was unprecedented in that it was an anonymous, large-scale system in which the U.S. government gave money to private citizens in exchange for living bodies for experiments.[3] It stood in contrast to the smaller, intimate transactions in which scientists experimented on family members, their students or themselves, and also in contrast to large-scale markets that used not civilians but people with an obligation to the state, such as soldiers and prisoners. Since 2010, I have collected oral histories of Anabaptist "voluntary service" workers and other people who were "normal controls" at the Clinical Center of the U.S. National Institutes of Health (NIH), as well as oral histories of former program administrators and NIH scientists who used the Normals in their research. I have been comparing these vernacular accounts with NIH's archived record of its Normal Volunteer Patient Program.[4]

My aim has not been to find scandals, though the records hold plenty of unflattering information about NIH and other organizations. My aim, instead, has been to explore political questions through the history of science: How was clinical knowledge made in practice, given that (some of) scientists' research materials were rights-bearing civilians who had thoughts, interests, and inclinations of their own? And what were the structural consequences of these practices? The first step, though, was to create the archive needed to answer these questions—a vernacular archive of former "human subjects" and the scientists who used them, which would complement existing historical collections and, in a small way, disrupt the political bias of traditional archives (Bruchac, chapter 4). In creating this archive, these human subjects in the past became my own human subjects in the present.

It goes without saying that it was hard to find the people I aspired to include in the vernacular archive, but this difficulty pointed at an answer to my historical question: How was knowledge made? The answer: Through practices of erasure made possible within bureaucracies. These erasures—in the name of protection, privacy, and consent—in turn made clinical knowledge possible and, at the same time, made my historical ambitions nearly hopeless.

In creating the vernacular archive, I had anticipated that I would find records of force and (soft) violence, and though I certainly found some cases (prisoners and laid-off mine workers were "normal controls" too), I also found willing, enthusiastic civilians who participated for no financial reward, save a token stipend. In finding this politically uncomfortable evidence, I also located the source of the paradox that concerned me: that there were willing human subjects of medical experimentation enthusiastically participating in a system that itself extended from and created forms of discrimination and violence that they opposed. This paradox is a legacy, in part, of the historical record and the emphasis on archiving evidence of classic research abuses (Li, chapter 2). The Tuskegee Syphilis Experiments and Nazi doctors' abuses were horrifically real and remain important to study. But it is also a legacy of how previous scholars have interpreted the historical record. The history of human experimentation tends towards a scandal-driven approach so its scholars forget to marvel at its endurance and smooth functioning.

My aim in this chapter is to consider the strengths and weaknesses of two theoretical approaches to studying the experiences of people *within systems of exchange* in light of the structural consequences *of these systems of exchange*—scholarship in the Marxist tradition and work in the Weberian tradition. Each of these approaches offers a way to align seemingly contradictory evidence yielded by my own research: on one hand, situational evidence of (some) people's happy experiences as clinical materials; and on the other hand, structural evidence that the market for human subjects sustained scientific racism, state-sponsored violence, and economic inequality that were either counter to their interests or that they actively opposed. Work in the Marxist tradition aligns situational and structural evidence with recourse to an unsatisfying psychological concept, "false consciousness." By contrast, work in the Weberian tradition shows how the mechanism through which modern science operates—namely bureaucracy—naturally produces the paradox I found: happy human subjects of medical experimentation within a broader system that itself yields forms of discrimination and violence that they were actively trying to redress.

MARXIST APPROACHES

Histories of human experimentation tend to figure science as an agonistic field in which scientists are pitted against other, less powerful actors, like human subjects. This literature commonly draws on Marxist traditions in which actions are evidence of power struggles between workers and capital—embodied in various human dyads, including in the relations between a scientist and a subject. The punch line in this body of work is that knowledge is made through exploitation. At a structural level, this claim is persuasive. There are striking and unjust patterns of political and economic inequality between people who are, and those who are not, used as human subjects.[5]

The Marxist tradition might seem to offer a sensible framework for analysing structural inequality, and yet this framework is weaker for thinking through situational evidence of people's experiences in the modern economy, including their experiences of science in the age of capitalism. As a result, the literature on the history of human experiment tends to deal with counterevidence in an unsatisfying way. The historical record is replete with examples of cheerful labourers and, in the context of science, human subjects who themselves claimed to know precisely what they were doing. The happy, clear-headed worker under a seemingly exploitative capitalist arrangement confounds the Marxist tradition. To explain how such people could exist, Marx and followers attributed their perspectives to false consciousness, which is the inability of (Marx's version of) people to appreciate that they are participating in their own exploitation.[6] The concept implies a universal human nature and a shared experience of capitalism; and it undergirds scholarship created in Marx's debt.

Marx was cut from the same cloth and working in the same historical moment as Freud. Among late nineteenth-century German-speaking intellectuals, Marx was "like a fish in water," to use Foucault's felicitous phrase.[7] Marx's concept of false consciousness, which was the mental veil that hid people's own best interests from themselves, had great affinity with Freud's unconscious, which was the puppet master working behind the scenes in all of us and pulling our strings. A century later, the task of the expert was to show how, as Foucault wrote, "The whole of modern thought is imbued with the necessity of thinking the unthought . . . of making explicit the horizon that provides experience with its background of immediate and disarmed proof, of lifting the veil of the Unconscious, of becoming absorbed in its silence, or of straining to catch its endless murmur."[8] Foucault's damning description of the liberal enlightenment political project of the human sciences is also a harsh criticism of historians who purport to know people better than people knew themselves. In his theory, Marx treated with condescension the same people he wanted to rescue through his anticapitalist politics. Historians who follow in lockstep risk doing the same and adopting the liberal enlightenment project that Foucault warns against.[9]

My point is not to suggest there is or was a world without exploitation. Instead, my point is that false consciousness yields an unsatisfying analysis—politically and epistemologically—of historical evidence of action.[10] Still, evidence of happy human subjects, which appeared in my own work, is an awkward fit with twenty-first century political and moral sensibilities. Politically, it would be easier to rely on the concept of false consciousness than to accept responsibility for explaining uncomfortable evidence. Yet, devotees of Marxist approaches recreate the very paternalism that they claim to want to subvert.

How can scholars align structural findings about markets (they are exploitative) with situational findings (people often experience markets as satisfying)? The field of economic sociology has been developing ways of explaining this seeming contradiction between people's situational experience of markets and the structural patterns of inequality that markets create. The key has been to appreciate that markets trade simultaneously in symbolic values and in financial value.

Through the process of "valuation," markets—including markets for human subjects—*come to appear* as objects outside of and prior to human activity.[11] In fact, markets are social products that must be created and then sustained through ongoing work. The key insight of economic sociology is that symbolic practices—the ways in which people make meaning—affect market processes. When evidence of the injustices created by markets at a structural level butt up against evidence of the satisfaction of people involved in market transactions at an experiential level, they can be brought into smooth alignment with the help of another social theorist, namely Max Weber.

WEBERIAN APPROACHES

Weber has helped me to explore how bureaucratic arrangements, including those endemic to modern science, intentionally embed practices of erasure. Weber is useful for understanding the processes through which people are made invisible from particular vantage points—from the position of scientific authority in the past and from the position of historian in the present day. At this reflexive level, Weber's approach also resonates with Donna Haraway's concept of "situated knowledges."[12] The concept reminds scholars that all knowledge comes from specific *situations* (biographical bodies, geographical locations). But more radically for its time, the concept also described how the bits of insight that are authorized as formal knowledge are necessarily incomplete (a logical extension of the truism that people-in-contexts make facts) and that the bits of insight allowed to count as authorized formal knowledge reflect collective priorities about what *kinds* of insights can count as scientific. Haraway, like Weber, worries about the limits of Marxism. The Weberian approach to bureaucracy takes as its premise that invisibility is an historical achievement and cautions that scholars who posit "invisibility" rather than ask how invisibility is accomplished implicitly adopt a vantage point—an unacknowledged structural and moral position themselves.

For Weber, one defining feature of the modern Western world is that all facets of life are arranged within bureaucracies. Briefly put, in *The Protestant Ethic and the Spirit of Capitalism* (Weber, 2008 [1930 trans]), Weber showed that saving (in the spiritual sense) morphed into saving (in the financial sense), which begat the modern Western variant of capitalism that, for Weber, resulted in the organization of the world into bureaucracies. This final move—the organization of everything into bureaucracy—is Weber's signature observation. He accepted the motivating claims of Marxist theory, then played them out, and persuasively undermined them. The (successful) revolution that Marx forecasted in conditions of exploitation did not come in Weber's lifetime because, he argued, moderns were, ultimately, happy living, and making knowledge, in the chains of bureaucracy. People come to absorb the value structures of the bureaucracies they are captured within (employers, governments, churches), values that operate at a personal level (e.g., "work hard") and are advantageous to the bureaucracy itself.

As a set of historical claims, this account has problems. As a set of theoretical claims, however, this is a keen insight into the conditions of modern knowledge production.[13] Weber invites historians of science to consider the conditions under which bureaucracies obscure individuals. By exploring the case of NIH's market for human subjects, I aim to suggest not that Weber is right but that he is uncommonly helpful.

THE BUREAUCRATIC ETHIC AND THE MARKET FOR "HUMAN SUBJECTS"

In 1953, the U.S. federal government opened a research hospital at the new location of its main campus in Bethesda, Maryland. For scientist–administrators, one of the first orders of business was to sign contracts with suppliers of materials needed for clinical research. One kind of material they needed was human subjects for medical experiments. Their supply of sick people was secure: physicians throughout the United States and the world notified Clinical Center researchers of interesting patients and sent seemingly curious cases to the Clinical Center:

schizophrenic quadruplets, Native American children, and people with "dwarfism," for example.[14]

The supply of healthy people was a different matter. Prior to the middle of the twentieth century, medical researchers commonly ran experiments related to immediate wartime needs and ran them on specific kinds of people. (I am setting aside psychological research here.) The research tended to be intentional infection studies or studies of wartime prophylactics (e.g., pesticides and materials tests). Most important, the studies were done on people who had a debt to the state, such as soldiers, prisoners, or orphans.[15]

At the Clinical Center in 1953, scientist-administrators reactivated an arrangement they found quite amenable during World War II that fit this model of "procuring" subjects. NIH signed a contract, negotiated with the help of the Selective Service Administration, to use Religious Objectors as healthy experimental subjects. (The Korean War was a boon for the scientists, who needed an active draft and conscripts to justify the contract.) As soon as they set the legal-organizational infrastructure, NIH science-administrators opened the program to any member of the Historic Peace Churches, not only to draftees. The Quakers said no to the contract (for reasons that are unclear), but leaders of the two other peace churches, both Anabaptist (Mennonite Church and the Church of the Brethren) saw the contract as an opportunity to solve problems of their own bureaucratic organization.[16] The parishioners were channelled through formal programs already put in place by these churches' national organizations for missionary and voluntary service. It was 1954; NIH needed female bodies; and the churches needed service placements for Anabaptist women.

This was the start of the NIH's Normal Volunteer Patient Program. Its unprecedented character can be easy to miss: healthy human subjects of medical experiments no longer had a debt to the state, but could be everyday citizens. In 1959, NIH science-administrators signed the agency's first contract with a college, which served as a source organization, and then, in the 1960s, with labour unions, civic organizations, and (more traditionally) the Federal Bureau of Prisons—all the while renewing its original contracts with the churches.

At the origin of the Normal Volunteer Patient Program were young people who volunteered through their Anabaptist churches, which had a tradition of encouraging members to witness the teachings of Jesus through voluntary service. Here is the important Weberian point: The idea of getting credit (much less compensation) for their voluntary service undermined the spirit of humility and the aim of self-sacrifice that voluntary service was supposed to accomplish. One article in a church newsletter written by a Normal explained:

> Progress in medical research requires many things from many people. At the Clinical Center of the National Institutes of Health in Bethesda, Maryland, Brethren Volunteer Service sponsors a group of people who wish to do something positive to help bring the blessing of good health to others. The participants literally give themselves to science, and this sharing of themselves to further research is their way of [relieving] human need.[17]

The church, as much as the hospital, was the source of the institutional logic that organized their experience of the Clinical Center. I take seriously the institutional logic of "voluntary service," which celebrated self-abnegation—or at least promoted a lack of concern with credit, profit and worldly rewards.

The conscience of these Normals became a collective conscience through the bureaucratic organization of the Anabaptist churches. The Mennonite Central Committee and the Church of the Brethren were formal organizations empowered to sign legal contracts and had bank accounts for financial transactions. The creation of this first formal exchange of civilians for money was predicated on the churches' bureaucratic organization. This is not to say that their system was necessarily just, but simply that people recognized—and in many cases, seemed to appreciate and actively uphold—a system of authority and submission. In short, their masters were freely chosen. To think otherwise is to commit the Marxist infraction of imposing present-day political sensibilities on circumstances in the past in the form of presumed exploitation and false consciousness.

For its part, the NIH was part of the U.S. bureaucratic state. The tools of the state were designed to promote what today would be called transparency and privacy. In doing so, the NIH's bureaucracy intentionally concealed information, which simultaneously served the function of erasing personal accountability and individual credit from the formal record. Consider the words of the Clinical Center's long-time director, Jack Masur. "As part of the research team," Masur told hospital staff, "the patient is contributing so much he should not be asked or encouraged to allow his picture to be taken or his individual story to be told." Masur believed that sick and healthy individuals were indispensable to research, and by the same logic they should get no credit for it. "If the study on that patient yielded information that helped others, it could be told in an impersonal scientific article or in a medical conference setting," one colleague observed. "In vain, writers for the public media would protest that big stories are told in terms of little people," this NIH insider continued. "Dr. Masur had a profound conviction that a patient has a moral right to privacy."[18] Lo, the face of privacy hides the tail of secrecy.

By the 1960s, many of the human subjects of post-war medical experiment were healthy white Christians living in relative privilege, and to insist on their status as victims like any other deflects questions about the causes of systematic differences among the subjects of medical experiment and about the organizational arrangements that made service as a Normal possible, even necessary. To obscure these intricacies is to risk diminishing the true horror of instances when obedience was not a choice, but was the only possible response, short of death or displacement, to a command enforced through violence. At the same moment that Anabaptists were under study at the NIH Clinical Center, Public Health Service scientists in Alabama were in the second decade of the horrific Tuskegee Study of Untreated Syphilis in the Negro Male, even though the mass manufacture of penicillin, an easy treatment, had been a landmark achievement of American wartime medicine. At the Clinical Center, it went without saying that "normal controls"—people who functioned as a biomedical standard—were exclusively white middle-class people. In part, these circumstances could remain implicit because practices of racial exclusion were built into the medical models of disease.[19] Moreover, and more important than abstract models and more consequential than distant debates, legal segregation in the post-war period was apparent in the immediate space of American hospitals, as well as in people's daily lives around Washington, DC. The racial identities of Normals at the Clinical Center could be left unspoken—not because discrimination was absent but because the place was predicated on racist systems. This scientific racism and its fogginess in the formal historical record are both functions of the bureaucracy, which shaped clinical knowledge production in the past and the possibilities of historical knowledge in the present day.

BIO-LABOUR AND BUREAUCRACY: AN AGENDA

Capitalism is successful in large part because it is flexible.[20] As a product of capitalism, bureaucracy shares this same trait. As a result, science within bureaucracies simultaneously creates and obscures the patterned and broad-scale consequences of formal knowledge—both good and ill.[21] Taken together, studies of scientific practice and justice mark out a terrain of study of the ways in which political privilege is built into the practice of science, the mechanisms through which this privilege operates, and its consequences beyond the laboratory. Using Weber's theory of bureaucracy towards liberatory ends for the case of the NIH exchange in human subjects, I argue that clinical knowledge under capitalism has been made possible by the ability of bureaucracies to obscure individual involvement. This observation is a complement, not a rebuttal, to scholars' efforts to rehabilitate historically and recognize politically the labour of people made invisible from the dominant yet partial perspective of the state and science. Weber's approach offers a compelling description of knowledge-making in late capitalism, a useful way to account for a wide range of unintuitive actions and experiences in market transactions and, importantly, a warrant for historians to read historical evidence that might otherwise appear to be contradictory as entirely compatible.

Where Marx was a revolutionary, Weber was a reformer and as such the politics that Weber's theory implies for the historian can feel like apathy. Yet Weber's interest in process and positionality suggests how to improvise a way to greater justice; to create formal knowledge in order to point out the instability and incompleteness of formal knowledge itself.[22]

Weber's theory of bureaucracy raises important questions about how to interpret—and to create—the historical record:

- How can bureaucracy be used to better understand the process of bio-labour; and how can bio-labour refine theories of bureaucracy?
- What are the circumstances under which people derive power or lose power through processes to create invisibility that are endemic to bureaucracies?
- What frameworks are emerging to (re)interpret the meanings of actions?

- What are the political stakes in scholarly claims that actors participated in the erasure of their individual contributions to science? How do these processes contradict or substantiate current scholarship on credit and credibility, for example, in the history of authorship?
- In what ways do historians' theories make legible particular conceptions of labour, while obscuring others?

There has been an insistence on violence, force, coercion, and exploitation as the exclusive action of capitalism. These are central functions of capitalism, no doubt, but they are not its exclusive mode. The defining feature of modern political economy is bureaucracy, and attention to its slow grinding gears can accommodate a fuller, intricate account that makes sense of seeming contradictions, like the happy human subjects I studied as historical actors and as objects of my own research, the apparently willing victims I worked not to find.

NOTES

1. Benjamin K. Sovacool and David J. Hess, "Ordering Theories: Typologies and Conceptual Frameworks for Sociotechnical Change," *Social Studies of Science*, 2017.

2. Laura Mamo and Jennifer R. Fishman, "Why Justice?," *Science, Technology & Human Values* 38 (2013): 159–75; Michelle Murphy, *The Economization of Life* (Durham, NC: Duke University Press, 2017), 1; María Puig de la Bellacasa, *Matters of Care: Speculative Ethics in More than Human Worlds* (Minneapolis: University of Minnesota Press, 2017); Jenny Reardon et al., "Science & Justice: The Trouble and the Promise," *Catalyst: Feminism, Theory, Technoscience* 1 (2015); Kim TallBear, *Native American DNA: Tribal Belonging and the False Promise of Genetic Science* (Minneapolis: University of Minnesota Press, 2013); Liboiron, Max. *Pollution Is Colonialism* (Durham, NC: Duke University Press Books, 2021).

3. Laura Stark, "Contracting Health: Procurement Contracts, Total Institutions, and Problem of Virtuous Suffering in Post-War Human Experiment," *Social History of Medicine* 31 (2018): 818–46.

4. Laura Stark, "Vernacular Archive of Normal Volunteers," Harvard Dataverse. Collection H MS c464, Countway Library of Medicine, Boston, MA, 2017.

5. Jill A. Fisher, *Adverse Events* (New York: NYU Press, 2020); Susan Reverby, *Examining Tuskegee: The Infamous Syphilis Study and Its Legacy* (Chapel Hill: University of North Carolina Press, 2009); Susan M. Reverby, "'Normal Exposure' and Inoculation Syphilis: A PHS 'Tuskegee' Doctor in Guatemala, 1946–1948," *Journal of Policy History* 23 (2011): 6–28; TallBear, *Native American DNA*.

6. Roberto Abadie, *The Professional Guinea Pig* (Durham, NC: Duke University Press, 2010); Dwyer, Ellen, "Neurological Patients as Experimental Subjects," in *The Neurological Patient in History*, eds. Jacyna and Casper (Rochester, NY: University of Rochester Press, 2012), 44–60; Nathaniel Comfort, "The Prisoner as Model Organism," *Studies in History and Philosophy of Biological and Biomedical Sciences* 40 (2009): 190–203.

7. Michel Foucault, *The Order of Things: An Archaeology of the Human Sciences* (Vintage, 1994), p2. 62.

8. Foucault, *The Order of Things*, 372.

9. Laura Stark, "Out of Their Depths: 'Moral Kinds' and the Interpretation of Evidence in Foucault's Modern Episteme," *History and Theory* 55, no. 4 (2016): 131–47.

10. Janet Halley, "Rape in Berlin: Reconsidering the Criminalisation of Rape in the International Law of Armed Conflict," *Melbourne Journal of International Law* 9, no. 1 (2008): 78–124.

11. Michèle Lamont, "Toward a Comparative Sociology of Valuation and Evaluation," *Annual Review of Sociology* 38 (2012): 201–21; Viviana A. Zelizer, *The Purchase of Intimacy*, (Princeton, NJ: Princeton University Press, 2009); Marion Fourcade and Kieran Healy, "Moral Views of Market Society," *Annual Review of Sociology* 33 (2007): 285–311.

12. Donna Haraway, "Situated Knowledges," *Feminist Studies* 14 (1988): 575–99.

13. For example, Steven Shapin, *The Scientific Life* (Chicago, IL: University of Chicago Press, 2010). Laura Stark, "The New Bureaucracy of Everyday Life," in *Handbook of Science, Technology and Society*, eds. Kelly Moore and Daniel Kleinman (New York: Routledge, 2014).

14. Laura Stark, *The Normals: A People's History* (Chicago, IL: University of Chicago Press; under advance contract).

15. Allen M. Hornblum, *Acres of Skin: Human Experiments at Holmesburg Prison* (Routledge, 1999); Alison Bateman-House, "Men of Peace." *Public Health Reports* 124, no. 4 (2009): 594–602; Jordan Goodman, Anthony McElligott, and Lara Marks, eds., *Useful Bodies* (Baltimore, MD: Johns Hopkins University Press, 2003); Jonathan D. Moreno, *Undue Risk* (New York: Routledge, 2001); Eileen Welsome, *The Plutonium Files* (New York: Dial Press, 1999).

16. Stark, "Contracting Health."

17. Brenda Schnepp, "BVSers Serve in the Interest of Science," *Gospel Messenger*. April 14, 1962.

18. Anon. "Dr. Jack Masur, a Giant in Size Gentle in Spirit," *NIH Record*. March 18, 1968, vol XXI, no. 6, edition: 7.

19. Lundy Braun, *Breathing Race into the Machine* (Minneapolis: University of Minnesota Press, 2014); Christopher Crenner, "Race and Laboratory Norms," *Isis* 105, no. 3 (2014): 477–507; James H. Jones, *Bad Blood* (Free Press, 1993); Robert N. Proctor, *The Nazi War on Cancer* (Princeton, NJ: Princeton University Press, 2000); Londa Schiebinger, "Human Experimentation in Eighteenth Century," in *The Moral Authority of Nature*, eds. Lorraine Daston and Vidal Fernando (Chicago, IL: University of Chicago Press, 2004), 384–408.

20. Kaushik Sunder Rajan, *Biocapital: The Constitution of Postgenomic Life* (Durham, NC: Duke University Press, 2006).

21. Melinda Cooper and Catherine Waldby, *Clinical Labor* (Durham: Duke University Press Books, 2014); Epstein, *Impure Science*; Gabrielle Hecht, *Being Nuclear* (Cambridge, MA: MIT Press, 2012); Carol A. Heimer, "Inert Facts and the Illusion of Knowledge," *Economy and Society* 41 (2012): 17–41. Kowal, Radin, and Reardon, "Indigenous Body Parts, Mutating Temporalities, and the Half-Lives of Postcolonial Technoscience."

22. la paperson, *A Third University Is Possible* (Minneapolis: University of Minnesota Press, 2017).

7

"They Say They Are Kurds"

Informants and Identity Work at the Iranian Pasteur Institute

Elise K. Burton

In any kind of human subjects research, the value of the data heavily depends on the accurate identification of the subjects. Controlled clinical trials and other statistically mediated forms of medical, anthropological, and social research have made it a high-stakes enterprise to correctly sort individual people into generalizable identity categories like age, family status, socioeconomic status, gender and sexual orientation, and race or ethnicity. Who is it that performs and oversees this work of identity sorting? In many cases researchers rely on self-reporting: the subjects themselves provide the necessary information to be categorized for study. In other cases the self-representation of research subjects is partially or completely ignored by scientific actors, who instead turn to the judgements of other criteria or people to classify the study population (see also Surdu, chapter 8; Vimieiro, chapter 9). Under both types of scenarios, the scientific work of identity classification is frequently rendered invisible in official publications and reports.

This invisibility has special implications within cross-cultural or transnational research projects when scientists from the West or Global North studying communities from the non-West or Global South delegate identity work to local informants, which might include any combination of high-status community members, government liaisons, local guides and interpreters, or local professional scientists and scholars. Because the credibility of such informants is taken for granted, their identity work often becomes visible only in Western researchers' publications when it involves disturbances to "science as usual," such as uncooperative research subjects, infrastructural problems, or unexpected results. In the case described below, two French medical researchers published studies on different communities in Iran with clear hypotheses that genetic difference would map onto contemporary social divides of ethnicity and religion. However, their genetic data dramatically contradicted their expectations. Why were their predictions and the results so disparate? Much of the answer can be excavated from the publications' brief acknowledgements of three Iranian scientists who assisted in the fieldwork. This chapter explores the substantial gulf of socioeconomic status between the Iranian scientists and the Iranian research subjects, demonstrating how Iranian social politics informed the expectations and field practices of the French researchers. In thinking about the way these Iranian scientists were made invisible, my primary concern is the elision not only of their real contributions to scientific knowledge but also of the different scales of socio-political inequality embedded in transnational human subjects research.

In the following case study, I explore the invisible labour of determining ethnic identities in Iran for the purpose of population genetics, which until the 1980s relied principally on detecting the presence or absence of certain proteins in human blood serum, especially the ABO blood group antigens. Human population geneticists try to reconstruct the origins of human racial, ethnic, and national groups; trace their ancient or prehistoric migrations; and explain how different populations are related to one another. These reconstructions require individual humans with complex ancestries and social identities to be assigned along binary scales as belonging or not belonging to particular communities defined by religion, language, race or ethnicity, caste, tribal lineage or clan genealogy.[1] This diagnostic prerequisite meant that "non-biological knowledge entered the research design," which empowered not only community members or leaders but also "linguists, ethnographers, historians, sociologists, and others," including local scientists working alongside foreign researchers, to certify the authentic identity and ancestry of research subjects.[2] In other words, masked under a population sample size is the invisible labour of multiple people, both inside and outside of a community, to determine that community's legitimate boundaries. It is important to make this labour visible because genetics has served, historically and presently, as an allegedly objective arbiter of human group differences that various social actors have used to erase or disavow the material effects of colonialism, racism, and industrial environmental degradation.[3] Far from providing objective measurements of biological difference, human genetics has instead created archives of the embodied effects of socioeconomic discrimination and political repression against different communities.

In the 1950s, a pair of French physicians, working with Iranian employees of the Pasteur Institute in Tehran, decided to survey the ABO blood type frequencies of the entire country, searching for genetic differences between ethnic groups. They travelled to provincial cities and remote mountain and desert areas, hoping to collect blood from peoples who represented a single "pure" ethnic ancestry. However, the scientists struggled both to accurately categorize their research subjects and to interpret data that contradicted Iranian social expectations of which groups should be alike or different. I analyse scientific publications, the French scientists' oral history testimony, and an American colleague's field journal to address three interrelated questions: first, when subjects are not trusted to represent their own identity, who takes over this work? Second, what scales of power relations are involved in the work of determining identity? Finally, under what circumstances does this identity work (and its power relations) become visible in scientific analysis? In this case study, French and American researchers tasked Iranian scientists, whom they considered to be reliable local informants and interpreters, with the responsibility for identity work. Through their fulfilment of this responsibility, the Iranians reified the complex social and political structures that empowered them over rural and marginalized research subjects, while also subordinating them to the goals and priorities of foreign researchers.

GEOPOLITICAL AND SOCIAL CONTEXT: THE IRANIAN PASTEUR INSTITUTE

Between the 1950s and the 1970s, Iran under the Pahlavi regime played a strategic role in fending off communism in the Middle East. To support Iran's military and economy as a key First World ally, hundreds of American and European experts passed through the country to advise on infrastructural and technological development. Western scientists visiting Iran enjoyed a privileged position in the country's semi-colonial medical infrastructure, able to access the assets of the Iranian government, foreign aid organizations, and private medical institutions. However, in order to study human populations, they needed not only financial and technical support but also extensive assistance from Iranian researchers. One important organization for such assistance was the Tehran branch of the Pasteur Institute. The Iranian Pasteur Institute was founded in 1921, but after World War II, it took on a renewed significance as a centre for medical research in Iran. Although the Institute was funded by the Iranian government, a French director, Marcel Baltazard, managed the staff of Iranian scientists and technicians. From 1946 to 1961, Baltazard encouraged visiting foreign scientists to conduct their own research at the Institute's laboratory facilities and placed Iranian staff at their disposal for fieldwork assistance.[4] Under Baltazard's direction, the Pasteur Institute in the 1950s became a crucial node connecting the growing international network of epidemiologists and serologists to Iranian physicians and research subjects.

In 1951, Baltazard invited French physicians André and Joelle Boué to start research projects at the Institute. The Boués had initially travelled to Tehran to work as anaesthetists and were interested in strategies for resuscitating patients after anaesthesia. Considering blood transfusion as a possible solution, they accordingly began to pursue blood group research in collaboration with the transfusion services of Iran's version of the Red Cross (the Red Lion and Sun Society).[5] The Boués noticed that blood group frequencies within the diverse Iranian population did not match those of French patients. This piqued their interest in a broader serological survey, and André Boué began traveling around Iran to collect blood samples from different regions, shipping them to Joelle in Tehran for laboratory tests to determine an "average [ABO and Rh] distribution for the Iranian population."[6] Most of André's field trips took place during 1954 and 1955. The Boués' samples were augmented by the fieldwork of American researcher and future Nobel Prize winner

Daniel Carleton Gajdusek, a visiting researcher at the Tehran Pasteur Institute in 1954. Gajdusek tracked the incidence of certain infectious diseases across Iran and Afghanistan by collecting blood samples from rural villagers. When conducting his own research in places André had not visited, he collected extra samples to share with the Boués.

The Boués wanted to gather data from "primitive" tribal groups in the Iranian provinces, especially those living in near-complete geographic isolation from urban populations. However, conducting fieldwork in these areas involved enormous logistical challenges, along with cultural and linguistic barriers. For example, it took weeks to secure travel permits from government officials, and the Pasteur Institute had to buy vehicles capable of handling rural roads. More importantly, neither the Boués nor Gajdusek could speak the local languages, and they knew very little about Iran's history or ethnic diversity. Accordingly, they relied heavily on three Iranian scientists—Mahmud Bahmanyar, Rasoul Pournaki, and Ali Mashoun—to serve as interpreters, guides, and informants on religious and social customs. Yet the Boués say nothing at all about these important colleagues in their oral history accounts of their time in Iran. Likewise, the terse acknowledgements to Bahmanyar, Pournaki, and Mashoun that appear in their French publications do not clarify the specific contributions of the Iranian scientists. Gajdusek, although not always a reliable narrator, kept a field journal which provides detailed accounts of the Iranians' backgrounds, social attitudes, and working conditions.[7]

The status of Iranian employees at the Pasteur Institute reflected a delicate political balancing act between Iranian desires for national sovereignty and equal professional recognition, and French expectations of deference to their scientific management. For example, Baltazard referred to all of the Iranian scientists on his staff as his "collaborators," downplaying the fact that he effectively controlled the overall trajectory of their careers as well as their day-to-day work. Although the Tehran branch was not technically a French colonial institution, Baltazard deliberately used his directorship as an instrument of French influence in Iran. Meanwhile, Gajdusek noted Bahmanyar's complaints that the Iranian Pasteurians were not sufficiently acknowledged for their achievements.[8]

Regardless of Baltazard's domineering position, the Iranian physicians working at the Pasteur Institute did not have disadvantaged backgrounds. They belonged to the middle to upper-middle classes, were educated in foreign-run schools, and had travelled to Europe and North America. They held diverse political and social attitudes that defined their interactions with French staff and foreign visitors to the Pasteur Institute. Within the group, Mahmud Bahmanyar had a particular reputation as "a hopeless nationalist," an outspoken critic of the Anglo-American overthrow of Prime Minister Mohammad Mosaddegh in 1953.[9] However, as the son of a well-off Tabriz bazaar merchant and a graduate of an American missionary school, Bahmanyar's national pride had limits: he often became defensive or evasive when foreign colleagues encountered the poor living conditions of the lower classes, or witnessed local customs that he considered backward.[10] Pournaki and Mashoun, on the other hand, tended to indulge their Western colleagues' taste for the exotic.[11]

Rasoul Pournaki, like Bahmanyar, was born to a middle-class Azeri family, from the provincial city of Khoi. During a field trip to Mazandaran province, Gajdusek complained that Pournaki referred to poor rural villagers as "third-class people." Reflecting on the behaviour of Pournaki and Bahmanyar, he observed, "Few of my educated Irani friends are startled by the concept of "third-class people"—they see these members of another species all about them."[12] In contrast, Ali Mashoun, a cosmopolitan Tehran native, had a more charitable sensibility and regularly visited orphanages to provide immunizations for children.[13] Nevertheless, Mashoun's family was of the landowning class in Mazandaran province (figure 7.1). Gajdusek was disturbed to hear Mashoun describing his personal landholdings with "despotic" phrases of ownership, for example, "'my village' and 'my villagers.'"[14] Gajdusek recorded:

> Almost all educated Persians show, to me, a distasteful paternalistic attitude toward their rural countrymen without formal education. Although they may criticize their wealthy landowners and feudal lords, they themselves, by virtue of their education, expect to be served, and catered to and held in respected awe as though they were lords. The peasant is a "simple" person to them and never accepted as a brother.[15]

Despite Gajdusek's critiques, the interests of the Iranian and foreign scientists closely aligned when it came to research. Both groups were equally invested in the success of the fieldwork, and there is no record of any disagreement over the procedures and tactics (sometimes unethical) used together by the Iranians, Gajdusek, and the Boués to obtain blood samples. In terms of socioeconomic status and

Figure 7.1. Ali Mashoun stands beside a Jeep during a trip to Mazandaran province with D. Carleton Gajdusek in 1954. *Source*: Gajdusek, A Year in the Middle East (1991). Public domain.

professional aspirations, these men had much more in common with their French and American scientific colleagues than with their economically and culturally marginalized fellow citizens—that is, the research subjects from whom they helped to collect blood. Put in the position of mediating between these two groups, they made themselves indispensable to foreign researchers by acting as informants on Iranian society and history. Meanwhile, they performed their professional, ethnocultural, and socioeconomic superiority over subaltern research subjects by exerting control over interpretations of group identities. This involved highlighting religious, linguistic, and cultural differences between and within ethnic and geographical categories of people, shaping the Boués' expectations of where they would find genetic differences. They also helped to arbitrate what kinds of research subjects counted as "pure" representatives of these identity categories, expressing particular scepticism about the ancestry of certain Kurdish villagers.

THE EXPEDITIONS AND THE RESULTS

In August 1954, Pournaki accompanied André Boué to the Kurdestan region to gather samples from villagers and nomadic tribes. According to Boué, they were some of the first physicians rural Kurds had ever met. In a later oral history interview, he explained their process: "We arrived, we went to see the village chief. We obtained the right to go see women and children. We brought a lot of medications, especially antibiotics, of course. And so we could do all these studies on large numbers." Joelle clarified, "The important thing was getting blood from the chief of the tribe. We had to talk a lot, but after [this] everyone would be in agreement; especially if the chief's mother was in agreement!"[16] By "we" the Boués meant Pournaki, upon whom fell the burden of negotiation. Although the Boués recounted only moments of success in gathering large numbers of samples, not all villages were so cooperative, and the long "talks" would have required Pournaki to overcome a high degree of local suspicion and confusion about the use of extracted blood. Gajdusek noted that the Kurds of Mahabad had been resistant "to letting more than an mL or two of blood from their bodies," leaving the researchers with barely usable serum specimens.[17] Similarly, while Boué and Pournaki worked in Kurdestan, Gajdusek and Bahmanyar struggled to collect blood from Azeri villagers living on the border between Iranian Azerbaijan and the Soviet Union. Despite Bahmanyar's Azeri ethnicity and his fluency in the local language, they encountered significant challenges: "At village after village we met suspicion, resistance to letting us collect blood, and political-religious-political difficulties."[18] Pournaki and Bahmanyar, therefore, were not simple conduits of communication, neutrally translating between foreign scientist and rural villager. They represented a class of Persianized professionals aligned with the Iranian central government, with whom the tribes traditionally had hostile relations.

Despite these enactments of difference between Iranian scientists and research subjects, the Boués were utterly dependent on their Iranian colleagues to interpret the ancestral identities of the sampled communities. The influence of Pournaki and Bahmanyar appears indirectly in the Boués' data analysis, which described each locality according to the perceived ethnic "purity" of its inhabitants. Here, the Boués sometimes questioned or completely disregarded the inhabitants' own self-identification. For example, they considered the Kurds of Mahabad to be "the most pure" due to their nomadic lifestyle. Meanwhile, in the Sanandaj region, they hedged local accounts of ethnic intermixing against their own observations of geographic isolation:

> *According to information obtained on site, but that could not be verified,* Kurds in [the Sanandaj region] have undergone some admixture, partly due to the arrival of Lur tribes from the south, partly due to the settlement of refugee Sunni Turkish elements. *We report this opinion which seems valid for the city of Sanandaj itself, but that seems questionable for the tested villages,* isolated in the mountains and cut off from outside contact for three to five months of the year.

At the same time, they openly doubted the self-professed Kurdishness of the Sahneh villagers:

> Finally, a third series of tests was performed on blood samples from the village of Sahneh, at the southern boundary of Kurdistan. . . . *Although they say they are Kurds, we could not affirm the membership of these people in this group* because this village, situated on the Baghdad-Tehran route, was always a major axis of communication, a likely source of many mixtures.[19]

Accordingly, the Boués excluded Sahneh samples from most of their data tables. But the careful social-historical distinctions between different "types" of Kurds turned out to be a red herring. The data from all three localities satisfied a statistical homogeneity test, meaning that there was no reason to treat the Sahneh villagers as a separate "non-Kurdish" population after all. In terms of ABO blood groups, the Boués conceded "a certain unity . . . among these Kurdish populations from three regions having practically no relations with one another."[20]

In March 1955, Ali Mashoun travelled with André Boué to the city of Yazd and its surrounding villages to study Zoroastrians, a religious minority whom a number of anthropologists believed could represent the "gene pool" of the ancient Persians prior to the Muslim invasions of the seventh century CE. The Boués aimed to learn whether the Zoroastrians, who claimed never to intermarry with outsiders or accept converts, were genetically different from the Shi'i Muslims among whom they lived. André claimed that it was "very easy" to tell the two groups apart based on individuals' first names: "the Muslims borrow their name from their religion, the Guèbres, in contrast, adopt the old Persian names from before the Arab conquest or the names of their mythological figures. Also, each community gathers in clearly delimited quarters." The work of determining religious identity through names and neighbourhoods certainly fell on Mashoun. Although Boué vaguely refers to "historians" for his description of Zoroastrian resistance to conversion to Islam, only Mashoun could help him understand which names were "Muslim" versus "old Persian."[21] The Boués' interchangeable references to the Zoroastrians as *les Guèbres* (after the Persian word *gabr*), a pejorative term used only by Muslims for the Zoroastrian community, also implicates the limitations of their local knowledge. Unlike other foreign scientists in 1950s Iran who accessed the Zoroastrian community through middle-class, urban Zoroastrian scientists, the Boués apparently had no substantial direct interactions with their research subjects in Yazd or with educated Zoroastrians in Tehran.[22] Rather, their conversations with Zoroastrian villagers, and all their knowledge of Zoroastrian history and customs, were mediated by Mashoun, other Muslim colleagues at the Pasteur Institute, and/or outdated French sources that used the term *guèbres*.

In any case, the influence of Mashoun and his colleagues went well beyond terminology. Their beliefs about the history and culture of Iran's various ethnolinguistic and religious groups strongly informed the Boués' hypotheses about Iranian genetic variation. However, these hypotheses were not borne out. Although the Boués and their Muslim colleagues expected to find major genetic differences between Muslims and Zoroastrians in Yazd, they instead found that "religious separation, though it has completely isolated the two groups and persisted for more than millennium, has not induced any difference between the two communities." In fact, the frequency distribution of the Yazd region as a whole was quite distinct from the population of Tehran.[23] Contrary to their expectations, the Boués decided, most of the genetic variation between Iranian populations could be explained according to one major geographic barrier—not the mountains, as predicted in their Kurdish study, but the impassable "Great Salt Desert" in the centre of the country.[24] In this way, they could account for the surprising level of genetic homogeneity between the isolated and ostensibly "pure" Kurdish, Persian, and Turkic-speaking villagers and tribespeople in the north and west of Iran, which together were significantly different from the Yazdis (regardless of religion) in the southeast. In other words, the differences of language, religion, and ethnicity upon which the Boués and their Iranian colleagues had based their hypotheses could not explain the country-wide patterns of genetic difference in blood group frequencies.

What, then, did all this identity work actually achieve? It fulfilled a social function within the scientific collaboration between the Boués and the Iranian Pasteurians. Although their expertise in the lab was devalued relative to that of French or American researchers, in the fieldwork setting, the Iranian scientists effectively controlled foreigners' access to local information and shaped their perceptions of Iranian history and society. The performance of persuading their impoverished and backward fellow citizens to donate blood also enacted identitarian difference, demonstrating to their foreign colleagues that they were cultured, rational Europeans rather than relics of primitive humanity. Through this process, the Boués and Gajdusek perceived Pournaki, Bahmanyar, and Mashoun to be more reliable arbiters of ethnic and religious identity categories than the actual individuals who provided their blood. Yet, the influence of the Iranian scientists' expectations was rendered visible only because it created hypotheses that were disproven by the blood group data. The surprising results made it necessary for the Boués to clearly explain the criteria for their identity work within their publications. In cases where hypotheses are not disproven, inconsistent category criteria may remain hidden, cloaking the social and political implications of identity determination behind invisible labour(ers).

NOTES

1. For a review of relevant literature, see Elise K. Burton, *Genetic Crossroads: The Middle East and the Science of Human Heredity* (Stanford, CA: Stanford University Press, 2021), 3–9.

2. Veronika Lipphardt, "The Jewish Community of Rome: An Isolated Population? Sampling Procedures and Bio-Historical Narratives in Genetic Analysis in the 1950s," *BioSocieties* 5, no. 3 (2010): 309.

3. See, for example, Dorothy Roberts, *Fatal Invention: How Science, Politics, and Big Business Re-Create Race in the Twenty-First Century* (New York: New Press, 2011); Angela Saini, *Superior: The Return of Race Science* (London: 4th Estate, 2019).

4. Amir Afkhami, "Institut Pasteur," in *Encyclopaedia Iranica* (New York: Encyclopaedia Iranica Foundation, 2004).

5. Interview with André Boué and Joelle Boué, April 22, 2005. Interviews with Human and Medical Geneticists series, Special Collections and Archives, Cardiff University, Cardiff, UK. Hereafter: "Boué Oral History, 2005."

6. André Boué and Joelle Boué, "Étude sur la répartition des groupes sanguins en Iran," *Le Sang* 26 (1955): 707; and Boué Oral History, 2005.

7. Daniel Carleton Gajdusek, *A Year in the Middle East: Expeditions in Iran and Afghanistan with Travels in Europe and North Africa, February 4, 1954 to December 22, 1954* (Bethesda, MD: U.S. Department of Health and Human Services, Public Health Service, National Institutes of Health, 1991), 180. For a critical account of Gajdusek's career, see Warwick Anderson, *The Collectors of Lost Souls: Turning Kuru Scientists into Whitemen* (Baltimore, MD: Johns Hopkins University Press, 2008).

8. See Elise K. Burton, "Rethinking Collaboration: Medical Research and Working Relationships at the Iranian Pasteur Institute." *Isis* 112, no. 3 (2021): 461–83.

9. Gajdusek, *A Year in the Middle East*, 171, 47, 165.

10. Gajdusek, *A Year in the Middle East*, 200, 163, 167–71, 180.

11. Gajdusek, *A Year in the Middle East*, 200.

12. Gajdusek, *A Year in the Middle East*, 241–42.

13. Gajdusek, *A Year in the Middle East*, 179–81.

14. Gajdusek, *A Year in the Middle East*, 196.

15. Gajdusek, *A Year in the Middle East*, 102–3.

16. Boué Oral History, 2005.

17. Gajdusek, *A Year in the Middle East*, 358.

18. Gajdusek, *A Year in the Middle East*, 358.

19. Boué and Boué, "Étude sur la répartition des groupes sanguins en Iran," 707–9. Emphasis added. Gajdusek, too, describes Sahneh as having a mixed "Kurd and Turk populace" (Gajdusek, *A Year in the Middle East*, 433).

20. Boué and Boué, "Étude sur la répartition des groupes sanguins en Iran," 711.

21. André Boué and Joelle Boué, "Étude sur la répartition des groupes sanguins en Iran II," *Annales de l'Institut Pasteur* 91, no. 6 (1956): 905–6. Gajdusek also uses *gueber* as a synonym for the Zoroastrians in his own reports (Gajdusek, *A Year in the Middle East*, 420).

22. In contrast, see the case of James E. Bowman in Burton, *Genetic Crossroads*, 167–78.

23. Boué and Boué, "Étude sur la répartition des groupes sanguins en Iran II," 906–7.

24. Boué and Boué, "Étude sur la répartition des groupes sanguins en Iran," 908.

II

POWER

Commentary: (Em)Powering Narratives of Technology

Gabriela Soto Laveaga

The chapters in this section push us to rethink key historiographic themes about power, visibility, and erasure. Among some salient topics are hypervisibility as a form of obfuscation, invalidating knowledges, collective acceptance of the knowable, epistemological erasures, invisible forms of evidence, and the power to name. The latter point—the power to name—is a connecting thread for most chapters. The power to name, be it groups of people, varieties of plants, even the language that will be preserved in archives, is a particular form of authority; it signals to others how the world should be seen. Yet the power to define by naming has a complicated relationship with technology. The chapters in this section are focused on the ripple effects of technologies used by a diverse cadre of experts to classify and order people and non-human species. Let me use two examples separated by both geography and chronology to propose larger claims about the power to both make visible and to obscure, before I turn briefly to the chapters.

In spring 2019 the world saw the image of a black hole for the first time. As we marvelled at the universe, human ingenuity, and technical innovation, the name of the black hole was announced: *Pōwehi.* At least that is how it would be known in Hawaii.[1]

As reported in the media, the Hawaiian name was suggested by Larry Kimura, professor of Hawaiian Language at the University of Hawai'i, and came from an eighteenth-century creation chant. The Hawaiian meaning of the name—"profound dark source of unending creation"—seemed aptly suitable given the event and that two telescopes used to image the black hole were located in Hawaii.[2] Moreover, embracing a native name that spoke directly to a creation chant appeared to acknowledge a multiplicity of beliefs and understandings.

Indeed, the deputy director of the James Clerk Maxwell Telescope on Mauna Kea underscored this feel-good moment when he told the Associated Press, "This isn't astronomers naming this. This is coming from a cultural expert and language expert. This is him coming to the table and giving us a gift of this name. It's a gift from Hawaiian culture and history, not the other way around."[3]

Yet the complexity of this place, even for such a famous place of telescopes, became apparent a few months later. That same summer, protestors camped atop Mauna Kea to oppose the planned construction of the Thirty Meter Telescope (TMT), billed as the largest telescope in the northern hemisphere. Accusations of mismanagement and land rights were interwoven with claims that Mauna Kea is a sacred place to native Hawaiians or Kanaka Maoli.

Reporting on the protests tended to take a decidedly different, less "feel-good," tone. For example, when describing stalled scientific observations, a newspaper reported that "astronomers are concerned that the TMT has been hijacked by issues that have little to do with science."[4] Cultural names were a thing to be celebrated, but protestors claiming land rights that directly impacted scientific research could not be tolerated it seemed if science were to advance. While other scholars have focused on the continuing protests over the construction of TMT, I wish to instead focus on a few aspects of this exchange that tie directly with the following chapters.

We know that the boundaries of science and culture are messy, so how do we, as academics, address this muddle if we continue to have stark contradictions such as those taking place on, say, Mauna Kea, and in the spaces carved out by the chapters in this section? More work is turning to examine those untidy borders. As histories of science become more inclusive of ethnicities, genders, legal status, and ways of perceiving the world, these boundaries are increasingly rubbing up against each other forcing us to take note, re-evaluate what we know, and ponder if our preferred research tools are adequate. One such tool is the power to frame (or name) research questions. There is power in naming, but there is also power in how we frame questions.

As all researchers know, if we don't ask certain questions, answers remain unknown. In the previous example, symbolic participation was no longer satisfactory. Though the black hole was imbued with deeper meanings of a culturally powerful name, for protesting locals, the act of bestowing an ethnic name did not translate into inclusion nor respect for a different understanding of how the universe works. It was no longer enough to simply change a name, what protestors sought (indeed, seek) was for scientists to understand the equal power of *other* creation narratives. Only then could sacred land also be understood on the same valence as the instrument (the telescope) that would rest on it. As it were, these two groups seemed to speak different languages, each highlighting what seemed an incomprehensible issue for the other side: save the mountain, save the telescope.

Curiously, it was the ability to engage with multiple and differing understandings and perceptions that allowed us to finally *see* a black hole or what had never been seen before. The collaborative effort needed to *know* the world around us is brilliantly captured in *The Edge of All We Know* (2020), a film directed by Peter

Galison. An ode to scientific collaboration, the film takes us from telescopes in Mexico to classrooms at MIT to dinner table conversations, all the while reinforcing that the black hole could not have been *seen* without a multiplicity of views from different geographic locations. Most important, in order to see the black hole, physicists needed to rely on, and speak to, programmers, and all needed to translate their work to bureaucrats and funding agencies. In this particular example, to know was to collaborate, to be transparent, to share knowledge. On the other hand, the power of knowledge production also remains rooted in its ability to obfuscate.

It may be worth pausing here to complicate the meaning of obscuring knowledge by using a sixteenth-century manuscript as an example. As historian Jorge Cañizares-Esguerra writes, the Gamarra-Inga manuscript was one of hundreds of manuscripts sent to the Spanish crown in the sixteenth and seventeenth century as a petition for innovation in mining. These petitions are where "the history of science and technology in colonial Spanish America ought to be found."[5] Yet these petitions for innovation were written in code, intended for "a public of one, the king," and they remained, long forgotten, for centuries in the archives. Nonetheless, their existence, Cañizares-Esguerra asserts, is evidence of the "historiographical fiction of the 'Scientific Revolution'" from its chronology to its geography. These unearthed manuscripts do indeed call into question our foundational timeline of modern science and raise intriguing questions about the impact of hidden knowledge once it is excavated. Power in this setting meant that rival empires would not know the intricacies of mining technologies developed in the Spanish world. To confuse, to obfuscate, to minimize the circulation of knowledge was the goal. Contrary to the example of coded manuscripts, the chapters in this section raise a different, though no-less stirring, set of questions.

If we place these two vignettes in conversation with each other, we see that they are both about power but also quite explicitly—though diametrically opposite—about erasure.

The intent of the first example, naming a black hole with a Hawaiian name, if even extra-officially, was to highlight cultures colonized and once concealed by imperial tendencies. Moreover, and most excitingly, the name also claimed to be part of a move to include multiple ways of seeing the universe. This apparently transparent and inclusive action was, while deep in meaning, shallow in its embrace of different ways of understanding the world, as protestors pointed out. The second example, the Gamarra-Inga manuscript is, by contrast, about intentional erasures and our inability to understand what remains hidden. These manuscripts illustrate that the Spanish crown, long thought as a second-class empire when it came to technological innovation, was instead intent on maintaining its technology a secret from probing English, Portuguese, and French eyes. This particular erasure, intended secrecy, in the archives has multiple layers of meaning, yet power still lies in the ability of only a few to understand. This type of filtered or select knowledge as a tool of power is present in several of these chapters. Yet at the core, most of these chapters circle back to the notion of expertise and who is perceived as an expert.

For instance, the Roma as a group is made and remade by experts whose scholarly production about the group earn both academic accolades and funding from the European Union, as Mihai Surdu explains (chapter 8). Indeed it is these experts whose labour keeps a disparate group of people in a constant loop of categorization from which, it seems, they can never quite escape. Collecting data on the Roma has become somewhat of a black hole of census information, where the Roma are perennially examined, quantified, and sampled, with little hope of shaking off the label given to them by experts dependent on the existence of the Roma for their own survival as authorities. The very practice of *counting Roma* is based on an arbitrary definition of a numerically imagined population category. Numbers are concrete and tangible while identities are ephemeral, hence the Roma census instils objectivity in something that is difficult to quantify. The "invisible" markers that define the Roma persist even when people who were allowed to self-identify refused to identify as Roma. Researchers then determined that a negation of their ethnicity was in fact a common Roma trait. Even in rejecting the label, the "Roma" could not shake off their so-called identity, they could not name themselves, forge an independent self. As Audra Simpson reminds us, the generative nature of refusing to participate (not being classified) is an act of resistance.[6] But this group, Surdu reminds us, has lost the power to self-identify. It is others beyond their community, specifically Roma experts, who hold the power to make invented labels, or categories, visible. Ironically, it is the marginalized and invented identity of Roma that makes these constituents marginal to society but desirable (visible) to researchers.

Leaving the Roma in Europe to move halfway across the world, we nonetheless find a similarly disquieting situation in Brazil where the genetic composition of the nation's Northeasterners, believed to be the most miscegenated group, was often deduced via dubious research methods such as the cluster of surnames in a population. As in the case of the Roma, the invisible symbols of expert authority—in this case, that the researcher herself was of mixed race—reinforced occasionally flimsy arguments about the region's racial composition. As Ana Carolina Vimieiro Gomes points out (chapter 9), Brazilian physician and geneticist Eliane Azevedo was not creating new categories but rather working within those (problematic ones) that already existed within the geneticists' community. Her meticulously detailed research notebooks illustrate how, like the Roma, seemingly solid research practices bolstered long-held beliefs. To return to the opening vignette of the sixteenth-century manuscripts, our reliance on print culture to construct truth often obscures the weight of what did *not* make it into Azevedo's notebooks. As with the aforementioned analysis of sixteenth- and seventeenth-century ciphers, this upturns our belief that "openness" constitutes knowledge: the notebooks' *visibility* obscured more than it revealed.

Sarah Blacker's essay (chapter 10) engages as legitimate the voice of, in this case, members of the Mikisew Cree First Nation and the Athabasca Chipewyn First Nation, who strategically use translation to "avoid assimilation into paradigms of Western science." Using a "three-track methodology," which places Indigenous Traditional Knowledge about toxicology data on equal status as that of Western science, this practice seeks to ensure that traditional knowledge is not *made to fit* into Western explanations. This practice ensures that other forms of explaining the world, such as "water [as] a living thing" can be made legible to those—policy-makers, politicians, and so on—who hold the power to make a change. In other words, it falls on the members of First Nations to educate and make knowable the impact of the damage wrought on their ecosystems. This reminds me of Miranda Johnson's recent article, in which she examines Tuhiwai Smith's landmark work *Decolonizing Methodologies* but queries why the book rests on dichotomizing indigenous and non-indigenous researchers. As Johnson states, "Perhaps a new type of researcher is emerging, one that is straddling objectivity and subjectivity, identity and difference, and doing so not in the hallowed halls of academe but in the theaters of justice—institutions that were imposed by a colonization that continues to work on the present."[7]

In a similar vein, Omnia El Shakry's chapter (chapter 11) makes room for unknowability, or rather the limits of expert interpretations, in Al-Qanatir, the site of Egypt's largest women's prison. In this space of policed bodies, not all women prisoners are the same. El Shakry points out that while political prisoners enjoy a hypervisibility through the memoirs they produce, illiterate women's experiences are to be catalogued, interpreted, and explained by the experts with whom they come in contact. Is it possible, she ponders, to acknowledge the limits of knowledge and produce work grounded in the unknowability of the subjects' experience? El Shakry pushes for a methodology that does not seek transparency, but rather values the absence of knowledge and the silences in prisoners' drawings that cannot be explained, much as the uncoded, sixteenth-century manuscripts whose value to the empire rested on their unknowability to the majority. In this case, seeking not to know is a powerful interpretive tool. The drawings themselves hover in the realm of the explicit and implicit, often left to the interpretation of the observer. These imprisoned women, who are perceived to have morsels of control over their lives, can and should have drawings, El Shakry argues, that remain research enigmas. It is the site of their existence—the prison—which warrants this argument yet not all women's contributions should remain unexplained. In a recent online article about the sudden disappearance of the ubiquitous female voice on TikTok as a way to think about how women's work in tech gets erased, Mar Hicks reminds us that "women's voices and bodies can be found all throughout the history of computing—from being heard in launch countdowns to being visible in photographs—but only relatively recently have historians written these women back into the narrative by explaining what they did."[8] There is power in excavating these needed stories, but there is also powerful agency in allowing inner lives to remain unknown.

Leaping from unknowability to the topic of other-than-human labour, Susannah Chapman (chapter 12) introduces new actors to the well-known story of the Green Revolution. Beyond the seemingly miraculous power of higher yields of new varieties of wheat, rice, and maize introduced by scientists and agronomists, Chapman seeks to acknowledge the work of other beings. Taking us further than farmers displaced from the narratives of scientific innovations, and even the work performed by nature and animals, Chapman incorporates deities, saints, and other spiritual forces. Faith as invisible labour is a fascinating concept for it pushes us to rethink who—or what—is the unknown expert. In this case, jinn or the immortal spirits can and do reward generous Gambian farmers with transformed rice plants. This godly offering to deserving humans displaces years of scientists' efforts at varietal breeding and instead focuses on the kinship between humans and that which cannot be explained. In some ways this invisible labour echoes Projit Mukharji's *occulted materialities* or the unforeseen forces that make Indian engineers and computer scientists pray to a god who oversees computers.[9] The convergence of what is reasoned and known has more porous borders in certain areas of the world. It is important to know when language is not being understood, and it falls on the researcher to ensure that the meaning is grasped.

Finally, in Stuart McCook's chapter (chapter 13), we find ourselves again in farmers' fields, yet this time it is coffee, which takes centre stage, more specifically coffee genes that produce dwarf plants which, in turn, yield more berries. Yet the benefit of switching to new varieties of coffee, as seen from the point of view of scientists, did not always translate to the farmers who tabulated the cost and labour needed to transform their fields. Put differently, technological knowledge of bettered seeds did not include the hidden costs of change. Unable to see the cost burden to small-plot farmers, research organizations remained perplexed as to the lack of change and so continued to push specific varieties. It is a defensible choice, McCook shows, to opt for what is believed to be the ill-informed option.

Erasure is different to invisibility. Erasure connotes an intended action to negate existence. But if we consider these chapters through the lenses of both invisibility and erasure, we arrive at different questions. Indeed, the subjects of these chapters perform a metaphysical act of both being hypervisible yet remaining marginalized in actual life. Their marginality marks them as ideal test subjects, for instance, but does not warrant full inclusion or citizenship rights. In the same fashion, authors in this section invert the value of a category such as power. They instead urge us to ponder how the dichotomizing nature of an analytic lens such as power has led us to think in binaries—powerful/powerless—that simplify complex relationships. To think beyond binaries is certainly not new—think here, for example, of the hundreds of important writings on the agency of the enslaved, interned, or incarcerated—yet what the authors in this section are saying is, however, novel. The majority of these chapters push us to rethink sites of knowledge production and its practice. Most of these works examined specific pairings of scientists and social scientists in close interaction with individuals whom they felt needed to be defined and translated to a larger public. In short, none of the scientists' interlocutors were perceived as equals. Made aware of this constructed power imbalance, we must revisit the end product (i.e., publications, reports, field notes, etc.) and ask more questions of those who led the studies. It is in examining the researchers that we can further understand the often complicated ways that power operates in specific scientific projects.

NOTES

1. Christopher Mele, "That First Black Hole Seen in an Image Is Now Called Pōwehi, at Least in Hawaii," *New York Times*, April 13, 2019, https://www.nytimes.com/2019/04/13/science/powehi-black-hole.html.

2. Associated Press, "Powehi: Black Hole Gets a Name Meaning 'the Adorned Fathomless Dark Creation,'" *The Guardian*, April 12, 2019, https://www.theguardian.com/science/2019/apr/12/powehi-black-hole-gets-a-name-meaning-the-adorned-fathomless-dark-creation.

3. Seth Borenstein, "Picture Was Clear, but Black Hole's Name a Little Fuzzy," *Whyy*, April 14, 2019, https://whyy.org/articles/picture-was-clear-but-black-holes-name-a-little-fuzzy/.

4. Daniel Clery, "Telescopes in Hawaii Reopen after Deal with Protesters," *Science Mag*, August 13, 2019, https://www.sciencemag.org/news/2019/08/telescopes-hawaii-reopen-after-deal-protesters.

5. Jorge Cañizares-Esguerra, "Bartolomé Inga's Mining Technologies: Indians, Science, Cyphered Secrecy, and Modernity in the New World," *History and Technology*, 34, no. 1 (2018): 65.

6. Audra Simpson, "On Ethnographic Refusal: Indigeneity, 'Voice' and Colonial Citizenship," *Junctures* 9, December 2007.

7. Miranda Johnson, "Toward a Genealogy of the Researcher as Subject in Post/Decolonial Pacific Histories," *History and Theory* 59, no. 3 (2020): 421–29.

8. Mar Hicks, "The Voices of Women in Tech Are Still Being Erased," https://www.technologyreview.com/2021/08/03/1030460/women-erased-tech-history-computing/, accessed August 4, 2021.

9. Projit Bihari Mukharji, "Occulted Materialities," *History and Technology* 34, no. 1 (2018): 31–40, DOI: 10.1080/07341512.2018.1516851.

8

Categorizing Roma in Censuses, Surveys, and Expert Estimates

Mihai Surdu[1]

PRODUCTION OF ROMA-RELATED NUMBERS

Roma are believed to be the largest minority in Europe with a current population of about 10–12 million people who share a distinct social profile. Academic, journalistic, and political accounts describe the Roma as a population that migrated from India some one thousand years ago and characterize the population as having high spatial mobility and close social ties around traditional cultural practices that are often seen as deviant and opposed to those of mainstream society. But how are the numerical figures of Roma determined in this well-established narrative? By questioning Roma data production in what follows I am not claiming that Roma people do not exist; rather, I aim to demonstrate how the Roma group is made through the visible and invisible work of various experts. As I argue further, through repeated academic, administrative, and police-led investigations carried out in different political regimes, people have learned how to identify either themselves or others as "Gypsies" or Roma. Counting and categorizing Roma are two strongly interrelated aspects of expertise that have so far not served those being studied—quite the contrary. Expertise on "Roma issues" has developed over time to sustain the reification of the Roma category and has led to further measures of repression and paternalistic management of people seen as deviating from the norm.

Counting Roma is a practice that includes or excludes individuals on the basis of arbitrary criteria that undermine their numerical assemblages.[2] As a highly diverse group, Roma do not share a common language, religion, territory, lifestyle, or physical appearance. Yet, since the eighteenth century, experts have continuously counted Roma populations through four types of quantitative methods: expert estimates, police-led "Gypsy"/Roma censuses, regular general censuses, and policy surveys. In this chapter, I examine the period after 1990 most closely, but I look back at the history of quantifying Roma (and other) populations, to reflect on a range of practices invisible to those consuming those numbers: persuasion in censuses and surveys, sampling methods, and the abstraction and analysis of census results.

FIELDWORK IN A 1990s POLICY SURVEY

As a sociology student in the early 1990s, I was a fieldworker in a countrywide research project on the social situation of Roma in Romania. Together with fellow students, I received a half-day training on sample strategy and interview techniques for approximately thirty questionnaires that each of us had to carry out. My motivation was twofold: to gain an initiation into sociological fieldwork and to supplement my student income. To administer the rather long questionnaire (it took about forty-five minutes), I had to decide whether the interviewee was a Roma person or not. Our task was made easier by the fact that towns in which we had to search for Roma subjects had been earmarked by the first census after the fall of the communist regime as zones with a higher Roma population. The task of selecting interviewees was very much in the hands of individual fieldworkers like me, who had to decide whether or not to record their interlocutor as Roma. With a polite greeting and a short formula explaining the aim of the survey and informing subjects of our policies on confidentiality and anonymity, I asked potential subjects to accept participation in this Roma-related survey. As I have since learned, this is what polling agencies and research institutes in their Roma-related surveys consider to be "implicit consent," which means that if someone agrees to be interviewed in a survey pertaining to Roma identity, they implicitly accept the designation of Roma ethnicity.

As the interview progressed, I asked about the language spoken in the family, to double-check the subject's ethnicity. As a supplementary check, fieldworkers had to assess whether the clothes of the subjects were coloured in Romani-like fashion or their houses vividly painted in Romani style. Regardless, in the final section of the questionnaire, devoted to sociodemographic variables, we had to straightforwardly ask: "What is your ethnicity?" For this question I was instructed to wait for the spontaneous answer of the subject, and if it did not readily come, to help by showing or reading the list with predetermined categories: "Romanian," "Hungarian," "Roma/Gypsy," "German," "Other." Subjects were expected to choose a single designation. Except for a few cases when the answer was directly "Romanian" or "Gypsy," most of my subjects did not sort themselves into a category as envisaged by the lead researchers. Hence, when I asked, "What is your ethnicity?," I then often waited for some long and embarrassing moments. In several cases I was told, "Tick whatever you like, what you think is best to be noted." I felt it was incorrect to note my perception, rather than that of the subject. Nevertheless, I had to circle a category. So, insisting on having the subject's answer, I reframed the question as such: "To which ethnic group do you belong?" Some answered, "It's your choice. Put the right thing, write what you think is good." As I tried to clarify with people in front of me what "good" might mean for them, I was told to circle the answer that would best serve the survey, or even that might be good for me, having made a long and tiring trip to visit them. This dialogue sometimes lasted for a few minutes without further helping me to circle a single ethnic category as the questionnaire requested. In order to finalize the survey and my task of filling out the thirty questionnaires, *I had to decide myself who was a Roma and who was not.*

Since the 1990s, hundreds of Roma-related surveys and studies (on a variety of topics, such as education, health, employment, discrimination, migration, and social welfare) have been carried out in Romania and all over Europe, and their findings broadly disseminated. Certain assumptions related to a Roma population profile, as for example, marginality and poverty, have usually been set in surveys well before data gathering begins. By using various forms of external identification, social scientists have adapted their epistemic object to fit a policy target. As some researchers acknowledge, samples in social studies of Roma after 1990 occur exclusively in homogenous poor neighbourhoods contributing in this way to the reification and stigmatization of the group.[3]

EXPERT ESTIMATES AND POLICE-LED CENSUSES

Early estimates of "Gypsy" numbers in Europe were first assembled at the end of the eighteenth century.[4] The sources and methods these pioneering scholars used to calculate the number of Roma in Europe are not documented. Similarly, late twentieth-century scholars have compiled "a rough estimate of the total number of 'Gypsies' in Europe"[5] with no sources to rely on. Yet their compilations have been assimilated and widely disseminated by the international organization the Council of Europe. For those countries in Europe considered to have the highest number of Roma, surprisingly, Council of Europe estimates of Roma numbers remained the same between 1994 and 2007; in other words, the numbers of "Gypsies and Travellers" published by the Council of Europe were frozen for more than a decade. This is unexpected: within an interval of thirteen years, one would expect the numbers, if reflecting a real population, would either increase or decrease.

In addition to scholars, police have compiled Roma numbers for centuries, by carrying out "Gypsy"-only censuses based on invisible, though systematic, categorization and surveillance work. In eighteenth-century France, for example, the police approached the work of defining and controlling "dangerous classes" in a manner resembling that of bookkeeping: police duties included establishing legal employment, recruitment, hiring and firing protocols, checking certifications, surveying changes in patronage, and the division of labour, as well as the causes for refusing and leaving work. Police classed those not employed by a master or without valid unemployment certificates as vagabonds, "masterless" or *gens sans aveu*.[6] From the second half of the nineteenth century and much later, police would identify "Gypsies" by checking their papers, conducting interviews, compiling lists of suspects, and documenting family histories. Like expert surveys, police-led censuses were based on external identification and pre-determined definitions and categories while ignoring self-ascription.

Diverse types of expertise were brought to bear on the making of these numbers: the 1893 census in the Kingdom of Hungary, for example, was a matter of cooperation between the police (the Ministry of Interior), professional ethnographers, and statisticians. As for the definition and registration of "Gypsies," the 1893 census instructions mentioned:

> The observation of Gypsy descent and origin normally does not run into especially great difficulties. The public opinion, the folk-consciousness [Volkbewusstsein], keep a reliable, current record of those with Gypsy heritage; the anthropological character is a sure enough identification, surer than language, which is the only criteria of Gypsydom which appears in the general census.[7]

According to census instructions, all that was needed for someone to be counted as a "Gypsy" was to be seen as such by "public opinion" or deduced as such by visual inspection. While the calculation of percentages and the making of moral statistics was the work of statisticians, police were responsible for on-the-ground collection work.

Defining a "Gypsy" population and counting it was also a police venture in the 1895 census of "nomads, Bohemians and vagabonds" in France and in the 1905 census of *Zigeuners* in Bavaria.[8] Police-led "Gypsy" censuses have continued ever since. Under communist regimes in Romania and Bulgaria, for example, official censuses were considered unreliable, and, again, the task of counting Roma was assigned to the police.[9] Although Roma were not part of public discourse or academic inquiry in these countries during communism, they continued to be subject to police surveillance.

LABOURS OF PERSUASION AND PROPAGANDA IN INTERWAR CENSUSES

New statistics about "Gypsies" became available in the interwar period, when many countries introduced regular censuses—that is, national surveys that asked citizens a set of standardized questions, which included ethnicity. In these, the task of identification, at least in theory, was left directly to subjects. The Romanian census registered a "Gypsy" ethnicity and created a way for subjects to choose his or her own identity. For census numbers to be collected and assembled, an enormous practical feat of persuasion had to be accomplished—work that has not received much attention from scholars analysing censuses as a form of identity production. Human sciences are "solicitous sciences"—in that they often rely on researchers' techniques of persuasion.[10] In the case of censuses and surveys, fieldworkers, census takers, and statistical agents must consistently undertake persuasion work to create subjects of inquiry. Censuses not only involve the work of definition, assemblage, and data processing, but above all, they require people's cooperation. As one of the experts working on the first modern Romanian census in 1930 stated, this cooperation had to be ensured "by means of active and massive propaganda[,] timely executed so as to enlighten the people about the aim and utility of the census."[11] To create willing subjects, experts recommended that the state use printed media, leaflets, posters, cinematography, radio, as well as school and church networks. As one of the documents about the organization of the 1930 census attests, statisticians also calculated propaganda tools in numbers: Romanian state authorities printed and distributed 240,000 colour wall calendars, 250,000 leaflets for census takers, 20,000 posters for exhibition in trains and buses, and about 5,000,000 postage stamps; in addition, they broadcasted 8,000 meters of propaganda movies, introduced millions of stickers in cigarette packs, and published thousands of articles in popular media.[12]

Beyond mass-media propaganda, census organizers considered face-to-face meetings to be important tools for convincing people of the utility and benefits of the census. They saw teachers and priests as significant state agents for census-taking, given the trust they enjoyed and their ease in approaching individuals in professional and private settings.[13] One census expert explained the role of teachers and priests:

> Propaganda through schools and churches is required to accomplish great services. In schools, by awakening on the one hand the curiosity and interest of the pupils, the teacher will guide them to disseminate at home, in their families, the words of light and faith about the necessity and benefits of the census. The same great work could be easily and effectively done by priests from the church pulpit as well as in their private discussions.[14]

Coupled with this propaganda, seen as popular enlightenment, experts advised that it would help to remind people that census participation was compulsory by law and that a lack of cooperation would be punished. The census takers were seen as state agents on duty, rather than as paid fieldworkers, and the whole census process was compared to that of a well-prepared military campaign, in which the census takers were soldiers and the central statistical office designated as the commandment of the army.[15] So, Romania's counting of Roma, through its introduction of national censuses, relied on the persuasion, teaching, and disciplining of subjects.

Even beyond census-taking, proper other survey practices likely function to "teach" subjects their ethnicity—shaping responses to censuses. Just before the 2011 Romanian census, people were encouraged to declare themselves as Roma, both by NGOs advocating for Roma rights and by the chief of the Romanian Institute of Statistics. For both kinds of organizations, there were financial stakes behind this mobilization: the more Roma the country had, the more European funds would be received for their social inclusion. It is hard to estimate how much these public appeals contributed to the making of official ethnic statistics, but they were undoubtedly part of the work of persuasion and assemblage that contributed to the making of Roma numbers.

ABSTRACTING POPULATIONS

Census organizers generally do not make available to non-experts the processes of coding, abstracting, and analysing census data. Post-coding—that is, the post-hoc assembly of freely chosen answers given by subjects under a single umbrella category—has had an essential function in producing quantitative group data from free-response questions on the Romanian census. Without post-coding, the numbers simply do not stick together. The procedure of post-coding is usually not explained to the wider public, as it is considered too technical and difficult to be understood by non-experts, for whom such processes are accordingly irrelevant. Moreover, explaining this complex procedure does not lend more credibility to the data; on the contrary, it threatens to shed light on the constructed character of ethnic numbers. Side-stepping the description of statistical work that goes into censuses (and surveys) makes ethnic categories appear to be natural kinds rather than statistical working tools.

To exemplify the largely invisible work of coding of ethnic census categories, I turn now to the last Romanian census, in 2011. In this survey, ethnicity was an open and optional question, an item that subjects were free to decide whether to answer or not. Accordingly, a large number of people preferred not to assign themselves to any particular ethnicity. Social scientists and experts unanimously interpreted this preference, which seems a reasonable choice, as the "greatest anomaly" in the history of modern census-taking in Romania since 1930. The number of non-answers, which reflected a refusal to identify with a specific ethnic category, was so big that it defied all expectations. The expert conclusion was, strangely, that people of an undeclared ethnicity were in fact Roma unwilling to disclose their true ethnicity.

Thus, in the 2011 census in Romania, 621,573 people were found to be of Roma ethnicity despite the fact that not all of these people declared themselves as such. As could be found in the census manuals, the category "Roma" was a highly composite one, uniting no less than nineteen different sub-categories under the umbrella term "Roma."[16] Most of these sub-categories were names of occupations considered by some anthropologists to be specific to "Gypsies"/Roma: brick maker (*cărămidar*), musician (*lăutar*), bear trainer (*ursar*), flower seller (*boldean*), tinsmith (*spoitor*), or tinker (*căldărar*). By contrast with the nineteen sub-categories linked to Roma, the Hungarian minority had three sub-categories and the German minority just seven. As disaggregated data for sub-categories were not provided, it is, in fact, not known how many people chose the label "Roma," which was one of the nineteen sub-categories. If people had wanted to record themselves by the name "Roma," they would have done so and not given other answers to the open question related to ethnicity. As Bruno Latour states: "*Numbers* are one of the many ways to sum up, to summarise, to totalise as the name 'total' indicates—to bring together elements which are, nevertheless, not there."[17]

Scholars, international organizations, governmental representatives, and policemen, past and present, contributed to the categorization and counting of Roma. These different actors and institutions assembled a Roma population by shaping the group according to institutional perspectives in line with their positions and interests, rather than those of the people being categorized. Whereas, historically, the production of statistical knowledge about "Gypsy"/Roma people has relied largely on expert or external presuppositions, more recently, the construction of ethnic numbers has relied on survey subjects learning what answer to give to ethnicity-related questions—and in some cases, such numbers have been disrupted by subjects' refusal.

My own uncomfortable experiences in the 1990s lead me to wonder whether less essentialist notions of Roma ethnicity might be realized through the use of non-exclusive ethnic categories—by giving subjects the possibility, in other words, to choose more than one ethnic category (a procedure used, e.g., in the 2011 Hungarian census). Following individuals' subjectivities in how they choose to define themselves, instead of imposing categories on them, may be closer to the census' aims. This would rupture totalizing labels and tune the census to more granular categories. Nevertheless, so far, most of the counting methods advanced in censuses and surveys have imagined ethnicity as an essential data point and have ignored the fact that ethnic categories are not exclusive but intertwined.

Roma population estimates are epistemic objects *and* policy targets. The knowledge derived from expert estimates—whether based on police or census data—is used to create objects of knowledge and objects of political action. The numbers created by such surveys, once launched in public discourse, seem to take on a life of their own that is independent of where they are produced, of their rationale, and of methods of calculation. Roma population estimates are imported from one field to another by virtue of their purported objectivity as numbers. High numbers of Roma produced by experts serve political discourses (in the past combating the "Gypsy" nomadism and a deviant lifestyle, currently alleviating Roma poverty, unemployment, and social exclusion), which promote the perception of emergency, moral panic, and risk. There is a lot at stake for practices that remain largely hidden.

NOTES

1. In producing this chapter, I benefited from the feedback and support of several people and institutions. My deepest thanks go to Veronika Lipphardt who constantly encouraged me, engaged with my work, and opened my interest for the fields of science studies and history of science. I am grateful for the feedback and editorial work of Alexandra Widmer. While working on this paper, I enjoyed being a postdoctoral research fellow of Max Planck Institute for the History of Science, Berlin, and a senior fellow of the Institute of Advanced Study of Central European University, Budapest: my warmest thanks to staff and colleagues.

2. Mihai Surdu, *Those Who Count: Expert Practices of Roma Classification* (Budapest; New York: Central European University, 2016).

3. Cosima Rughiniş, "Quantitative Tales of Ethnic Differentiation: Measuring and Using Roma/Gypsy Ethnicity in Statistical Analyses." *Ethnic and Racial Studies*, 34, no. 4 (2009): 594–619. O. Prieto-Flores, L. Puigvert, and I. Santa Kruz, "Overcoming the Odds:

Constricted Ethnicity in Middle-Class Roma," *Identities: Global Studies in Culture and Power* 19, no. 2 (2012): 191–209; Surdu, *Those Who Count.*

4. H. M. G. Grellmann and M. Raper, *Dissertation on the Gipsies: Being an Historical Enquiry, Concerning the Manner of Life, Family Economy, Customs and Conditions of these People in Europe, and Their Origin* (London: Printed for the editor, by G. Bigg and to be had of P. Elmsley, and T. Cadell, in the Strand, and J. Sewell in Cornhill: 1787); M. Kogalnitchan, *Esquisse sur l'histoire, les moeurs et la langue des Cigains, connus en France sous le nom de Bohémiens* (Berlin: Libraire de B. Behr, 1837); J. A. Vaillant, *Les Romeś. Histoire vraie des vrais Bohémiens* (Paris: E. Dentu, Libraire-Editeur, 1857); C. P. Serboianu, *Les Tsiganes: histo ire-ethnographie-linguistique-grammaire-dictionnaire* (Paris: Payot, 1930).

5. J.-P. Liègeois *Gypsies: An Illustrated History* (London: Al Saqi Books, (1983) [1986]), 47.

6. S. Kaplan, "Réflexions sur la police du monde du travail, 1700–1815," *Revue Historique* 261, no. 1 (1979): 17–77.

7. E. Johnson, "The Gypsy Census in the Kingdom of Hungary, 1893," *Journal of the Gypsy Lore Society* 8, no. 2 (1998): 83–117, 103.

8. M. Kaluszynski, "Republican identity: Bertillonage as Government Technique," in *Documenting Individual Identity: The Development of State Practices in the Modern World*, eds. J. Caplan and J. Torpey (Princeton, NJ: Princeton University Press, 2001), 123–38; E. Filhol, "La loi de 1912 sur la circulation des 'nomades' (Tsiganes) en France," *Revue européenne des migrations internationales* 23, no. 2 (2007): 2–20; Asséo, H. "L'invention des 'Nomades' en Europe au XXe siècle et la nationalisation impossible des Tsiganes," in G. Noiriel, ed. *L'identification. Gènese d'un travail d' État* (Paris: Berlin, 2007), 161–80; I. About, "Underclass Gypsies: An Historical Approach on Categorization and Exclusion in France in the Nineteenth and Twentieth Centuries," in *The Gypsy "Menace": Populism and the New Anti-Gypsy Politics*, ed. M. Stewart (London: C. Hurst, 2012), 95–117; L. Lucassen, " 'Harmful Tramps': Police Professionalization and Gypsies in Germany, 1700–1945," in *Gypsies and Other Itinerant Groups: A Socio-Historical Approach*, eds. L. Lucassen, W. Willems, and A.-M. Cottaar (New York: St. Martin's Press, 1998), 74–93.

9. Romanian census documented in the National Archives of Romania: C.C al P.C.R. *Informare referitoare la unele probleme pe care le ridică minoritatea de țigani din România, 1978, Document 25/336.* Police census in Bulgaria, mentioned in: J.-P. Liègeois, *Roma, Tsiganes, Voyageurs* (Strasbourg: Conseil de l'Europe, 1994); J.-P. Liègeois, *Roma in Europe* (Strasbourg: Council of Europe, 2007); World Bank, *Roma and the Transition in Central and Eastern Europe: Trends and Challenges* (Washington, DC: World Bank, 2000); United Nations Development Programme (UNDP), *Avoiding the Dependency Trap* (Bratislava, Slovakia: UNDP, 2002).

10. Sarah E. Igo, "Subjects of Persuasion: Survey Research as a Solicitous Science; or, the Public Relations of the Pools," in *Social Knowledge in the Making*, eds. C. Camic, N. Gross, and M. Lamont (Chicago, IL: The University of Chicago Press, 2011), 285–307.

11. Quotes from Romanian language publications are author's translations. L. Colescu, "Recensâmăntul populațiunii," *Buletinul Institutului Economic Românesc* 11–12 (1930): 807.

12. Direcțiunea Recensământului General al Populațiunei (DRGP), *Populațiunea actuală a României* (București: Publicațiile DRGP, 1931).

13. See, for example: L. Colescu, "Recensâmăntul populațiunii," *Buletinul Institutului Economic Românesc* 11–12 (1930): 807; M. Sanielevici, "Technica recensământului, observații, reflecții și sugestii," *Arhiva pentru Știința și Reforma Socială* 4 (1931): 576–99.

14. L. Colescu, "Recensâmăntul populațiunii," *Buletinul Institutului Economic Românesc* 11–12 (1930): 819.

15. L. Colescu, "Recensâmăntul populațiunii," 795–857.

16. In *Nomenclatorul etniilor şi limbilor materne* (Bucureşti: Institutul Naţional de Statistică, 2011) are given nineteen sub-categories. Even more sub-categories for Roma (thirty-two) are presented in *Manualul Personalului de Recensământ* (Bucureşti: Institutul Naţional de Statistică, 2011), 98.

17. Bruno Latour, *Science in Action: How to Follow Scientists and Engineers through Society* (Cambridge, MA: Harvard University Press, 1987), 234, italics in original.

9

Situated Knowledge and the Genetics of the Brazilian Northeastern Population, 1960–1980

Ana Carolina Vimieiro Gomes[1]

Research notebooks are important sources for studying the material culture of science. They have the potential to yield information on the work and creative processes undergirding the production of scientific knowledge.[2] They are evidence of scientific bookkeeping, of experiments on paper, and of the workings-out that are effaced in published reports. Sometimes, they serve as a diary of the research, as a mnemonic device for the researcher or research group.

I have been pondering one of the research notebooks created in the 1980s by the Brazilian human population geneticist Eliane Elisa de Souza e Azevedo (1936–). Eliane Azevedo kindly shared one such notebook with me during a personal interview in 2017, though no reproductions are now available for publication. Its yellowed pages show how she manually sorted by racial class and counted the occurrence of the surnames found in the Bahian population she studied. At the top of the table, she used column names to represent the abbreviation of the racial groups "white," "light mestizo," "medium mulatto," "dark mulatto," and "black." She marked the most frequent surnames with a number indicating their occurrence in the population. Handwritten tally marks and red dots show how she counted and checked each surname, with circled numbers signalling the total of surnames in the column, or of a specific surname. The word *Santos*, for example, is heavily annotated as the most common devotional (name of saints, religious ceremonies, and festivities) surname observed in so-called black and mulatto groups.

Azevedo's notebook is a "paper tool," that is, a written resource to create order in the data collected and (for me) a visual representation of her epistemic labour in progress.[3] It shows some of the practical and conceptual work effaced in her publications. We see lists, categories, markings, and calculations written in the geneticist's handwriting, the workings and product of a series of abstractions that transformed data from individuals into data representative of groups. These and other notes provide the data that Azevedo then used to produce tables and graphs with the results of her studies. Azevedo and her research team analysed and published these data in scientific articles for journals such as *Human Biology*, *Current Anthropology*, and the *Annals of Human Genetics*.

Since the first decade of the twentieth-century, genetic research depended on doing work on paper.[4] Azevedo's notes show her inscriptions in a survey of patients' surnames that she carried out in a hospital in the city of Salvador, in the Northeastern Brazilian state of Bahia. Azevedo's study was related to the topic of racial ancestry and genes. What interest me in this analysis are the now-invisible processes of abstraction that were key to Azevedo's practice of racial classification and its epistemological deployment.

My chapter considers the operations by which Azevedo and contemporary population geneticists, from the mid-1960s to the 1980s, defined Brazilian Northeasterners as a single population—operations that she left unacknowledged in published work but which are evident from her notebooks and narratives. I offer evidence of how the social place and experience of Eliane Azevedo gave her special epistemological status on her research team. She was a woman who self-identified as *mestiça*, or mixed race, who came from a middle-class background, was a trained physician, and was herself a Brazilian Northeasterner.[5] I argue that she and her colleagues believed that her social identity gave her tacit (and tactical) knowledge about how to classify the region's "racial groups." Her identity among her geneticist colleagues qualified her as a credible interpreter of the racial identities of their research subjects (see also Burton, chapter 7). She and her co-workers valued this type of intimate knowledge as a means of guaranteeing a supposedly greater objectivity in the data on racial classification.

This is not to say that her social standing gave Azevedo a special capacity or a privileged epistemological standpoint to make assertions regarding racial issues involving the Brazilian population. Nevertheless, she was perceived, by her scientific colleagues (and presumably by her scientific audiences) to have such a privileged standpoint.[6] Moreover, her personal experience gave her intellectual tools to suggest a study design and imagine an explanatory model on the formation of the Northeastern people with the purpose of giving meaning to some of their population genetic data.

THE GENETICS OF THE NORTHEAST POPULATION

The use of the Brazilian Northeastern population as the subject of biomedical research was not new in the 1960s. Starting in the early twentieth century, the region was commonly represented in the United States as a place of poverty and underdevelopment, home to a society that was behind the times. Residents were characterized as lazy, feeble, promiscuous, violent, and unproductive mestizos, and they were considered by many biologists, anthropologists, and doctors to be unique objects of medical and scientific investigation.[7] Much of that research assumed that the population of the Northeast region was homogenous, not various, and defined by biological peculiarities that contrasted with other regional populations—particularly to the South and the Southeast.[8] What motivated studies of the genetic composition of the Brazilian Northeastern population was the idea that Northeasterners were the principal miscegenated group in Brazil.[9] To prove this hypothesis, U.S. geneticist Newton Morton (1929–2018) embarked on a series of genetic studies of people in Northeastern Brazil.[10] He was interested in a rural and endogamous population with a specific genetic and ethnic composition associated with poverty. As Azevedo herself recalled:

> What [Newton Morton] needed for his theory was a population living in poverty (because if the population were poor, it would be more substantially subject to the pressures of its environment), with a very high reproductive rate (to be able to have more children to analyze), with high mortality (to see whether those who died did so in a different way), and which did not use contraceptives. The Brazilian Northeastern population possessed all of those qualities.[11]

Azevedo's research on genetics and the Brazilian Northeastern population started in the mid-1960s during her PhD program at the University of Hawai'i, supervised by Morton. Azevedo's studies were part of a wider research program on population genetics headed by Morton at that time. This larger project was one of several branches of population genetics devoted to the study of so-called primitive peoples as biological and demographic phenomena.[12] Brazil was viewed by geneticists as a hub for global theoretical discussions and field studies on the micro-evolutionary process of the human species, as well as for the categorization of the population's genetic composition. They framed the Northeastern population as primitive owing to its largely pre-industrial economy. This apparently made it a valuable object of study on human genetic adaptability and variability.[13] Morton's presence in Brazil was fully supported by U.S. public funding agencies such as the National Institutes of Health (NIH) and the U.S. Public Health Service. That initiative was therefore strongly tied to broader US Cold War health and science programs in so-called Third-World countries.[14]

AZEVEDO'S CONTRIBUTIONS TO THE GENETIC STUDY OF NORTHEASTERNERS

When Eliane Azevedo started working with Morton, she had recently graduated from a medical school in Bahia State, and she was interested in genetics. She had first met him during his visit to the region to explore ideal research conditions and find samples for his population genetics research at the University of Bahia in 1962. She was the only woman and the only physician on the team, so was responsible for the clinical evaluations of the people studied. Azevedo's work consisted of medical examinations on research subjects, which included questions about contraception to allow for inferences on fertility. She performed physical examinations to look for minor congenital malformations. She also measured blood pressure and conducted liver and spleen examinations searching for signs of endemic diseases. At the end of her assessments, she provided racial classifications of the people she had studied.

As the only Brazilian Northeasterner researcher, Azevedo held a special social standing within Morton's research team. She was understood by the group to know the Northeastern population structure through personal experience, which was why (along with her medical training) she was responsible for physical exams and racial classification. Later recalling this period, Azevedo seemed proud of her role in the research. In remembering Morton's first visit to Brazil to evaluate the possibility of using Northeastern families as subjects, she recalled her contribution to the study design—a contribution that was not explicitly acknowledged in publications.[15] Born in the city of Tanquinho, located in the semi-arid Caatinga biome of Bahia State, Azevedo offered the evidence needed to corroborate Morton's reasoning and convinced him of the potential of the Brazilian Northeastern population as a subject of populational genetics research:

> I knew the structure of the population from living in it and I said, no, this isn't a problem; let's go to Tanquinho today and I'll show you around; I know the city—it wasn't even a city at the time—and I'll show you families with a huge number of kids, you know. And I took him! There were only two families. One had eighteen children and the other had twenty-one [laughs]. Pregnancies, I mean. Pregnancies.[16]

Ultimately, at the suggestion of Oswaldo Frota Pessoa, a Brazilian geneticist from the University of São Paulo, the study was not carried out in Bahia but instead in the city of São Paulo, at the *Hospedaria de Imigrantes*, a public housing institution that served as a sort of almshouse for migrants who came to São Paulo to work in factories or on plantations farther inland.[17] Due to the logistical difficulties confronted by the geneticists in the Brazilian Northeast, the *Hospedaria* offered them a unique opportunity to gather a varied and significant number of people to be sampled.[18] The researchers recruited more than a thousand migrant families as subjects for those genetic studies.

The geneticists engaged in these studies created their own means to estimate and classify the genetic composition of the population: they used the frequency of genes of every "racial class" correlated with the use of phenotypic indexes, such as anthropological traits.[19] They thus defined "racial class" along a scale based on traditional physical anthropology assessments of human traits: "abdomen, hair colour and type, and conformation of the nose and lips" resulting in seven classes, from "most caucasoid" to "most negroid." Azevedo recalled that it was Morton who brought this racial assessment model to the country, and she helped to refine the phenotypical classification criteria.[20]

In the team's scientific publications, Morton made a point of noting (in the acknowledgements section) that "the examining physician was herself a native of Bahia and her judgements [sic], although necessarily subjective, were based on a lifetime of personal experience."[21] In Morton's eyes (and presumably those of his readers), Azevedo's identity as a local with lived experience of the social groupings in the region qualified her as someone who could make scientifically meaningful judgements. By framing Azevedo as a powerful interpreter of racial identities in this region, Morton's published statement obscured the operations and abstractions that Azevedo was engaged in, thus lending scientific credibility to the data on "racial classes" and highlighting these data as objective. For example, she was apparently best placed to choose the best body part to proceed with skin colour determination: "As rural workers they were then too much sunburned. . . . For women, it was on the belly. And for the men we looked for that body part more preserved from environmental influences, understand?, from the sunburning."[22]

Azevedo herself recognized the subjective nature of this classification—she later explained it was "obviously full of mistakes, full of deviations."[23] But at the same time, she recalled her ability to determine the statistically significant racial composition of the sample due to her firsthand knowledge of Northeasterners. According to her, the statistical distribution of that classification did not exhibit any bias. It was on a normal curve, according to the statistical mean and standard deviation. What's more, Azevedo considered this classification method to be precise in statistical terms:

> Those quantitative methods to measure melanin, et cetera, aren't worth anything, because it's the person as a whole that matters. You can look and really see, in the person as a whole, how to categorize that person. Of course, there's a very big margin of error, but there's no other way. But at the end, because it's a big sample, this issue corrects itself. In a small sample or with an isolated individual, this isn't worth anything. Now, in a large sample . . . what a large sample has is substantial power of correction.[24]

Azevedo wrote her PhD dissertation using the data she collected in Brazil and later analysed in the United States. Her dissertation *Erythrocyte Isozymes, Other Polymorphisms, and the Coefficient of Kinship in Northeastern Brazil* (1969) sought to study the application of some tests of genetic variation in the analysis of population structure. Its results on inbreeding, kinship, and "isonymy" (the incidence of shared surnames) among Northeasterners were the starting point for her later research on surnames and genes.[25] Following her return to Brazil in 1969, Azevedo became a professor at the Federal University of Bahia, where she continued her work on population genetics at the Laboratory of Medical Genetics. Throughout the 1970s, her main research project was financed by the Organization of American States (OAS) program for genetic "improvement" in Latin America.[26] It lasted seven years, and it concerned the anthropogenetic classification and the study of the racial ancestry of the Bahian population.[27]

SURNAMES, GENES, AND RACIAL ANCESTRY

The study of the association between genes and surnames was a procedure used by geneticists studying population structure to determine whether biological inheritance corresponded to surname inheritance. "Isonymy" was a method subsequently developed by the American geneticist James Crow in his studies to investigate the levels of inbreeding among marriages between people of the same surname.[28] In her PhD dissertation, Azevedo applied Crow's method to the genetic data collected from the Northeastern sample population in São Paulo, and "it all went wrong." Contrary to what she had imagined, Azevedo found that surnames in the Northeastern population in general, and in Bahia in particular, did not indicate a common ancestor, nor were they (contrary to what she had assumed) reflective of kinship.

It was therefore necessary to explain her results, which meant trying to fit the reality of the Northeastern population with Crow's findings. She explained that the discrepancy in the data was due to the history of the Northeastern population:

> So then I started to think. I said, "Look, Brazil, the Brazilian population . . . it is principally the Northeast; it's a confluence of three distinct racial groups. Europeans from the Iberian Peninsula, our Indigenous peoples, who were already here, and the Black people who were brought over with slavery. And we have neither African surnames nor indigenous surnames, aside from very rare exceptions." Then another thing came to me. I started to see the light at the end of the tunnel.[29]

Azevedo reasoned that the apparent discrepancy in her results could be accounted for by the Afro-Brazilian tradition of the adoption of family names. Based on that insight, through the end of the 1980s, Azevedo produced various publications discussing the use of surnames as markers of the populations in which they occurred—she even participated in an event organized by the American Anthropological Association in 1980 devoted to the topic.[30] She obtained data on surnames, race, and genes from countless people from schools, maternity wards, other hospital departments, and blood banks in Rio de Janeiro and Bahia, as well as from blood banks in Portugal.

Her main claim regarding isonymy was that family names in Northeasterners were not correlated to genes and ancestry.[31] She interpreted the frequency of people's surnames according to family naming systems used over the course of Brazilian history, especially among the Black population. Before the abolition of slavery, it was common for enslaved people to adopt the master's family name or to use devotional names. Azevedo described her reasoning to justify the study design:

> There was a moment that they [freed slaves] needed a surname. So I imagined: If I were black, slave, here with my religiosity, they only calling me Eliane and, suddenly, I needed a surname in this doggy world of prejudices; I would chose according to my faith a name of a saint, because it would give me protection within what I believed. So I imagined: Look, the Afro descendants should have a preference for religious surnames.[32]

By analysing eighteenth and nineteenth-century documents on the emancipation of slaves, Azevedo confirmed two of the most common procedures for surname adoption: some freed slaves remained without a surname, and among those who acquired a surname, the preferred method was adopting a devotional name, such as

Santana, Jesus, Nascimento, Conceição, Teresinha, or Santos. She then found an increase in devotional surnames in populations with greater Black admixture, and even concluded that people with a white phenotype, and carrying a devotional surname such as Santos, were more likely to have Black admixture (just as was evident in her notebook).[33] Azevedo and her collaborators considered surnames to be racial markers that enabled the mapping of genetic origins and the cultural traditions of certain population groups from the Northeast of Brazil, thus using social historical narratives to base their genetic data.[34] Azevedo even concluded that the surname was a more powerful indicator of racial ancestries than phenotype classifications.[35]

"SITUATED KNOWLEDGE"

The results of Azevedo's surname studies in Bahians clearly reinforced the notion that the Northeastern population was mixed, and it reflected the persistent racialist tenor of population genetic investigations throughout the 1980s. It is important here to express concern about the power of racialism and its ability to reinvent itself in classification practices in various fields within the biomedical and human sciences—the story I have presented here is yet another example of how such practices operated.[36] The present case study gives Eliane Azevedo a centre-stage place in the history of human population genetics in Brazil: her embodied knowledge and "situated knowledge" as a Northeastern woman were unseen forces giving a supposedly greater objectivity and validity to her research data on racial ancestry and genetics.[37] Azevedo's inscriptions and narratives reveal the abstractions (invisible to her readership) used to define the Brazilian Northeastern population in biocultural terms—practices that had significant epistemological effects on her production of knowledge in the field of population genetics at the time. Relevant to all of her practices was Azevedo's own social place to her research, which went hand in hand with the tacit knowledge that she embodied. These factors guaranteed her a special epistemological (albeit not marginal or critical) status in Newton Morton's group that allowed her to address medical issues and perform racial classifications.

NOTES

1. Research for this chapter was supported by the Conselho Nacional de Desenvolvimento Científico e Tecnológico (CNPq) and the Consortium for the History of Science Technology and Medicine (CHSTM). My thanks to Eliane Azevedo for the interview, and to Ricardo Ventura Santos and Vanderlei Sebastião de Souza, who provided me access and permission to cite their interview with Azevedo. I am also grateful to Rosanna Dent and workshop participants whose insights helped me to develop the present case study.

2. Atia Sattar, "The Aesthetics of Laboratory Inscription: Claude Bernard's *Cahier Rouge*," *Isis* 104 (2013): 63–85. https://doi.org/10.1086/669883; Frederic L. Holmes, "Scientific Writing and Scientific Discovery," *Isis* 78, no. 2 (1987): 220–35.

3. On the historical relevance of "paper tools" as an analytic category for studying material culture in science, see: Ursula Klein, *Experiments, Models, Paper Tools: Cultures of Organic Chemistry in the Nineteenth Century* (Stanford. CA: Stanford University Press, 2003), 3.

4. For more on this point with respect to heredity analyses of individuals and populations: Jenny Bangham, *Blood Relations: Transfusion and the Making of Human Genetics* (Chicago, IL: University of Chicago Press, 2020).

5. Eliane Azevedo. Interview with Ana Carolina Vimieiro Gomes, trans., Salvador, December 12, 2017.

6. Alison Wylie, "Why Standpoint Matters," in *Science and Other Cultures: Issues in Philosophies of Science and Technology*, eds. Robert Figueroa and Sandra Harding (New York: Routledge, 2003), 26–48.

7. Durval M. Albuquerque Jr., *A invenção do Nordeste e outras artes* (São Paulo: Cortez, 2001); Stanley S. Blake, *The Vigorous Core of Our Nationality: Race and Regional Identity in Northeastern Brazil* (Pittsburg, PA: University of Pittsburg Press, 2011).

8. Albuquerque Jr., *A invenção do Nordeste*; Barbara Weinstein, *The Color of Modernity: São Paulo and the Making of Race and Nation in Brazil* (Durham, NC; London: Duke University Press Books, 2015).

9. Henrique Krieger, Newton Morton, M. P. Mi, Eliane Azêvedo, Ademar Freire-Maia, and N. Yasuda, "Racial Admixture in North-Eastern Brazil," *Annals of Human Genetics* 29 (1965): 123.

10. Eliane Azevedo, "Nordeste: genética e população." *Estado de São Paulo*, January 22, 1978.

11. Azevedo, Interview, 2017.

12. Ricardo Ventura Santos, Susan Lindee, and Vanderlei S. de Souza, "Varieties of the Primitive: Human Biological Diversity Studies in Cold War Brazil (1962–1970)," *American Anthropologist* 116, no. 4 (2014): 725.

13. Santos, Lindee, and Souza, "Varieties of the Primitive," 727–28.

14. Vanderlei Sebastião de Souza and Ricardo Ventura Santos, "The Emergence of Human Population Genetics and Narratives about the Formation of the Brazilian Nation (1950–1960)," *Studies in History and Philosophy of Biological and Biomedical Sciences* 47 (2014): 97–107. https://doi.org/10.1016/j.shpsc.2014.05.010.

15. See, for example, Morton's work debated in Cold Spring Harbor Symposia in 1964. Newton E. Morton, "Genetic Studies of Northeastern Brazil," *Cold Spring Harbor Symposia on Quantitative Biology* 29 (1964): 69–80.

16. Eliane Azevedo. *Interview to Ricardo Ventura Santos e Vanderlei Sebastião de Souza*; Salvador, July 31, 2012.

17. Odair da Cruz Paiva and Soraya Moura, *Hospedaria de Imigrantes de São Paulo* (São Paulo: Paz e Terra, 2008), 14.

18. The *Hospedaria de Imigrantes* was before used as a place of studies on population genetics in Northeastern Brazil. See: Saldanha, P. H. "Race Mixture among Northeastern Brazilian Populations," *American Anthropologist* 64 (1962): 751–59.

19. Morton. *Genetic Studies of Northeastern Brazil*, 69–80.

20. Azevedo, *Interview*, 2012.

21. Morton, *Genetic Studies of Northeastern Brazil*, 73; Krieger et al. *Racial Admixture*, 115.

22. Azevedo, *Interview*, 2017.

23. Azevedo, *Interview*, 2017.

24. Azevedo, *Interview*, 2017.

25. See: Eliane S. Azevedo, "The Anthropological and Cultural Meaning of Family Names in Bahia, Brazil," *Current Anthropology* 21, no. 3 (1980): 360–63. https://doi.org/10.1086/202462; José Tavares-Neto and Eliane S. Azevêdo, "Family Names and ABO Blood Group Frequencies in a Mixed Population of Bahia, Brazil," *Human Biology* 50, no. 3 (1978): 361–67.

26. According to the report on the Twelfth Meeting of the Advisory Committee on Medical Research from the Pan American Health Organization in 1973, genetics was among the three areas (with food and nutrition and biochemistry) relevant to health in the OAS program for Latin America. See: Pan American Health Organization, "Health Research in Latin America—Recent Developments." *Twelfth Meeting of the Advisory Committee on Medical Research*, 1973.

27. Eliane Azevedo, *Características antropogenéticas da população da Bahia, Brazil* (Research report presented to Organization of American States, 1979).

28. See, for example: Crow, James F. and Mange Arthur P., "Measurement of Inbreeding from the Frequency of Marriages between Persons of the Same Surname," *Biodemography and Social Biology* 12, no. 4 (1965): 199–203.

29. Azevedo, *Interview*, 2017.

30. Gabriel Lasker. *Surnames and Genetic Structure* (Cambridge: Cambridge University Press, 1985), 6–11.

31. José Tavares Netto and Eliana Azevedo, "Racial Origins and Historical Aspects of Family Names in Bahia, Brazil," *Human Biology* 49, no. 3 (1977), 287–88.

32. Azevedo, *Interview*, 2017.

33. José Tavares Netto and Eliane Azevedo, "Family Names and ABO Group Frequencies in a Mixed Population of Bahia, Brazil," *Human Biology* 50, no. 3 (1978): 361–67; Eliana Azevedo, "The Anthropological and Cultural Meaning of Family Names in Bahia, Brazil," *Current Anthropology* 21, no. 3 (1980): 360.

34. On "bio-historical narratives," see Veronika Lipphardt, "The Jewish Community of Rome: An Isolated Population? Sampling Procedures in Bio-Historical Narratives in Genetic Analyses in the 1950s," *Biosocieties* 5 (2010): 306–29.

35. Netto and Azevedo, *Family Names and ABO group*, 361.

36. For an analysis of the persistent use of race and contentious debates about the category in genetics, see: Jenny Reardon, *Race to the Finish: Identity and Governance in an Age of Genomics* (Princeton, NJ: Princeton University Press, 2004).

37. Donna, Haraway, "Situated Knowledges: The Science Question in Feminism and the Privilege of Partial Perspective," *Feminist Studies* 14, no. 3 (1988): 575–99.

10

The Invisible Labour of Translating Indigenous Traditional Knowledge in Canada

Sarah Blacker

Settler colonialism is not only an economic and political regime but also an epistemological one. This chapter examines a methodological response to the politics of settler colonial science in Canada—a science that excludes and invalidates Indigenous Traditional Knowledge. These epistemological erasures create the need for additional scientific and translational labour in Indigenous communities—labour that, I argue, is rendered invisible through the persistent repudiation of Indigenous knowledges by the scientific infrastructure, including funding agencies, universities, and federal science. This chapter is part of a larger project investigating the role of evidence practices in settler colonial science in maintaining hierarchies of knowledge. Here, I situate an innovative evidence practice—known as the "three-track methodology"—that has been designed to make knowledge claims in a climate in which significant resources are mobilized towards delegitimizing Indigenous knowledges. In developing and enacting the three-track methodology in 2013, members of the Mikisew Cree First Nation and the Athabasca Chipewyan First Nation choose to translate their own knowledge on the scale and consequences of contamination in their community.[1] In considering this labour, this chapter does not paint a picture of Indigenous victimhood, but instead emphasizes the resilience and resurgence of Indigenous communities in developing strategic forms of translation that enable the communities to retain autonomy and avoid assimilation into the paradigms of Western science.

It is difficult to overstate the extent of the violence towards Indigenous Peoples, Knowledges, and lifeways that has been enacted through settler colonialism—or to enumerate the measures needed to undo these harms. As the environmental justice scholar Kyle Powys Whyte shows, the effects of the destruction perpetrated by settler colonialism must be understood on multiple levels simultaneously, as settler colonialism is "literally seeking to erase Indigenous economies, cultures, and political organizations for the sake of establishing their own."[2] Considering for a moment only one aspect of settler colonial violence—that of ecological destruction—and temporarily bracketing others, it is crucial to understand that this harm is perpetrated both materially and epistemically.[3] Ecological destruction threatens the survival of Indigenous communities not only through the contamination of water, air, soil, and food but also by threatening (or cutting off entirely) Indigenous communities' relations to the land, which is vital to Traditional Knowledge (TK). This chapter centres the agency and resilience of First Nations communities in taking action against threats posed to TK and Indigenous lifeways. It concerns the decision by the First Nations to embark upon the slow and labour-intensive process of partially translating their knowledge into the terms of settler colonial science.

ETHNOGRAPHY OF DOCUMENTS

Rather than using the more familiar method of participant observation, I have carried out document-based ethnography, supplemented by interviews. In taking this approach, I have drawn inspiration from many other scholars who are thinking about documents ethnographically, including Angela Garcia, Annelise Riles, Helen Verran, Matthew Hull, Chris Kelty, and Hannah Landecker. Document-based ethnography is well suited for conducting ethnographic research in situations of restricted access, particularly when access to field sites and other research is hampered by state policies and censorship. Governmentality is increasingly enacted through the proliferation of documents and reports as "governmental technologies," but as the making of these documents is outsourced to consultancy firms, researchers lose access to the sites where documents and reports are made.[4] When faced with these barriers, ethnographers can learn from the documents that they do have access to, by studying the conditions of their making and their circulation, and the politics of the knowledge claims they are making.

Applying these insights, I focus here on the role of documentary practices in understanding invisibility and labour in relation to Indigenous communities' agency in transforming settler colonial environmental science in Canada. Following Kelty and Landecker's suggestion to position "the literature as an informant," I have sought to illuminate data practices that are "not necessarily visible through the lens of single actors, institutions or key papers."[5] The central practice analysed here is the adoption of the three-track methodology. The following analysis focuses on the "Water Is a Living Thing" report, the named author of which is the non-Indigenous environmental scientist Stéphane McLachlan. The stated authorship does not necessarily indicate a form of erasure or exclusion of First Nations community members. Rather, under the particular conditions in Harper-era Canada, the practice of leaving some study participants unnamed and unidentified may be a strategic and planned measure taken to provide protection from retaliation. Indeed, invisibility can play an important role in providing marginalized communities with a form of protection.

Invisibility should not be understood solely as synonymous with erasure and exclusion but as something that may also be strategically sought out and planned for in the interests of justice and agency.

POLITICAL AND GEOGRAPHICAL CONTEXT

The Peace-Athabasca delta in Alberta, Canada—now contaminated by the Athabasca oil industry—has long provided food and drinking water for two Indigenous communities that live along the shore of the delta: the Mikisew Cree First Nation and the Athabasca Chipewyan First Nation. Since 1967, this industrial site has been extracting bitumen from a geographical area larger than 79,000 square kilometres, and it is currently the largest-scale industrial project in the world (by land cover). One of the most important by-products of the bitumen extraction process in northern Alberta's Oil Sands are "tailings ponds," which contain 1.2 trillion litres of water contaminated with substances such as bitumen, naphthenic acids, cyanide, and heavy metals.[6] It is estimated that millions of litres of waste water leak from the tailings ponds into the groundwater each day.[7] However, the measurements of contaminants and data produced on their relation to elevated rates of disease remain contested; scientific uncertainty persists regarding the quantity of non-sensory, yet toxicologically detectable, contaminants that seep into the First Nations' drinking water.

The land adjacent to Lake Athabasca in northern Alberta, now known as Fort Chipewyan, was settled in 1788 by the fur trading company North West Company. Fort Chipewyan is home to between 900 and 1000 members of the Athabasca Chipewyan First Nation and the Mikisew Cree First Nation. The nature of the contamination of groundwater, plants, animals, and human health in the region remains contested. Prior to 2011, Environment Canada funded studies designed to measure contamination in and around the Fort Chipewyan First Nations communities, but these studies did not incorporate Indigenous TK into their data collection, nor did these scientists gain adequate knowledge of the land's history to contextualize their findings.[8]

In 2013, a "three-track methodology" was collaboratively developed by the Indigenous communities together with McLachlan, a trusted ally of and advocate for Indigenous communities affected by environmental contamination in Canada. The methodology was designed to enable a fundamentally non-hierarchical form of knowledge production in which Indigenous TK is positioned as the epistemological equal of Western science.[9] The three-track methodology presents toxicology data in three distinct forms. The first "track" collects conventional Western scientific toxicology measurements, the second documents the presence of toxins through Indigenous TK, and the third synthesizes the first two forms of evidence, thereby "grounding" the toxicology data in TK.

Its designers insisted on a non-hierarchical form of knowledge production to guard against the tendency for collaborations to "scientize" TK by "testing and validating relevant knowledge using scientific criteria."[10] The three-track methodology is both a material object and an epistemic tool specifically designed to work against such "scientization" and assimilation. Juxtaposing TK (in narrative form) with numerical toxicological data, the three-track methodology enacts a compromise by presenting TK in relation alongside a type of data that can be recognized as credible evidence by the federal government and by industry actors alike.

In addition to the scientific reasons behind the decision to have TK direct the three-track methodology, the First Nations communities were acutely aware of the "politics of evidence" under the Canadian federal government led by the Conservative politician Stephen Harper between 2006 and 2015.[11] One of the Harper government's priorities was to support the growth of industry, and it made extraordinary efforts to restrict the circulation of scientific data that would present evidence of harm that the oil industry caused to humans or to the environment. Under that government, scientists who conducted research as part of one of Canada's federally funded research institutions had their communication with journalists and other members of the public monitored and censored by government media relations officers.[12] Journalists wishing to interview Canadian scientists were required to submit interview questions for government approval.[13] Some of the most highly censored research topics under this government included climate change, the environmental and health impacts of the oil industry, and the pollution of waterways. The Harper government also destroyed scientific data held at federally funded libraries and archives in Canada, a phenomenon that federal scientists only spoke about publicly following their resignations. Public engagement with government policy and political resistance to those policies were significantly hindered by this restricted access to scientific data.

During this period, one of the areas that became most contentious for Canadian scientists was the intersection of environmental science with industry. And despite the importance of TK about the land and environment in this context, it was also becoming increasingly difficult for Indigenous voices to be heard. Facing rapidly rising rates of disease, Indigenous communities needed a way to render their experience into a form of evidence that would be recognized by the federal government and provide access to much-needed resources. The decision by members of the Mikisew Cree First Nation and the Athabasca Chipewyan First Nation to invite McLachlan was made after the community was impressed by the results of a two-track collaborative project that McLachlan had worked on with an Indigenous community in Manitoba.[14]

The Athabasca Chipewyan First Nation and the Mikisew Cree First Nation invited McLachlan to propose possible methods of producing credible evidence of harm. McLachlan suggested that the three-track methodology would guard against the tendency of the Western science to overdetermine the TK, or to render the TK as a "quaint artefact" that would ultimately be subservient to the Western science.[15] Thus, the three-track methodology has a tension at its heart: it must strategically translate TK into the language of data *just enough* to enable recognition, while also holding back from total translation, thereby actively preventing the assimilation of TK into data by reserving space for the TK to speak autonomously in its own track (there are resonances here with Spencer, chapter 25).[16] In this sense, the three-track methodology can be understood as a "conceptual structure" designed to work against what Kristie Dotson calls "epistemic oppression" by opening up participation in science to "other epistemological accounts and commitments."[17]

THE "WATER IS A LIVING THING" REPORT: SYNTHESIZING QUALITATIVE AND QUANTITATIVE DATA

Begun in 2011 and completed in July 2014, the three-year study was funded by Health Canada and the Social Sciences and Humanities Research Council of Canada and was peer-reviewed by Health Canada's scientists. The study collected accounts of environmental changes as observed by First Nations community members; measured levels of Polycyclic Aromatic Hydrocarbons (PAHs), as well as arsenic, cadmium, mercury, and selenium in groundwater, plants, and animals; and documented elevated rates of cancer in Indigenous communities affected by the chemical contamination left behind by bitumen extraction processes. The study's methods and results were published in a 242-page report entitled, "'Water Is a Living Thing': Environmental and Human Health Implications of the Athabasca Oil Sands for the Mikisew Cree First Nation and the Athabasca Chipewyan First Nation in Northern Alberta." The report, designed to be accessible to non-specialists, was authored by McLachlan in consultation with members of both First Nations.

The "Water Is a Living Thing" report was, at the time, a rare collaboration between Western environmental scientists and an Indigenous community.[18] McLachlan thinks of his own role as that of assistant, rather than director. The study design, the identification of collection and sampling sites, and the data analysis were carried out by community members themselves (rather than by trained scientists), including First Nations youth who expressed interest in participating in the study. McLachlan took a "back seat," acting as translator to ensure that study results would be legible to policy-makers. In this sense, the decision to carry out the study using the three-track methodology was understood by those working on the project as a form of strategic translation. They believed that the knowledge and experiences of the First Nations communities would continue to be marginalized and delegitimized in this struggle to produce credible evidence unless the knowledge and experiences were *articulated through the Western scientific language of data.*

The "Water Is a Living Thing" report is exceptional in the field of environmental science in that it presents evidence in three distinct forms: first, an articulation of TK about environmental contamination; second, measurements of contamination levels taken using current industry standards; third, a synthesis of the first two forms of evidence.

McLachlan acted as a knowledge mediator in negotiations between First Nations actors and government policy-makers concerning the politics of knowledge production and what counts as evidence. The First Nations participants in the Community-Based Monitoring project made it clear that they could not endorse a study of regional contamination that did not engage with TK. The First Nations communities decided, together with McLachlan, that any translation of knowledge into data would need to be *led* by the communities' TK, placing the Elders' insights at the forefront. The Western scientists with whom the communities collaborated would act merely in a supporting role, rather than directing the study design or the interpretation of data collected.

The first track documented the First Nations' evidence of contamination through TK presented in narrative form, while the second track documented the toxicological measurements of contamination presented in bar graphs. At first glance, it is not evident how the findings of the first track can be brought to bear on the findings in the second (figure 10.1). In the first track, evidence is presented as based on oral knowledge, held collectively, through references to how community members have

6.3.2 BEAVER

Generally speaking, some community members had noticed a decline in the health of beaver, to some degree in the Peace Athabasca Delta but especially in closer proximity to the Oil Sands,

> Nov 13: *"And like Lawrence said, even the animals don't taste the same. He noticed that. That's all from Industry. Even McKay, my son-in-law, killed a beaver there. And then we smoked the beaver meat. And the meat wasn't rich and red like it used to be? It's just kinda pale. And then we smoked it, and then we had some of it, and it didn't even taste the same. Way different. And then they took another beaver there - I left it over there, I didn't bother."*

Typically, beaver kidneys had the highest contaminant levels followed by liver. In contrast, beaver meat (muscle) typically showed the lowest levels of contaminants.

BEAVER MEAT (CONSUMPTION LIMITS RELATED TO ARSENIC)

Typically, beaver kidneys had the highest arsenic levels followed by liver (Fig 6.6). In contrast, beaver meat (muscle) typically showed the lowest levels of contaminants.

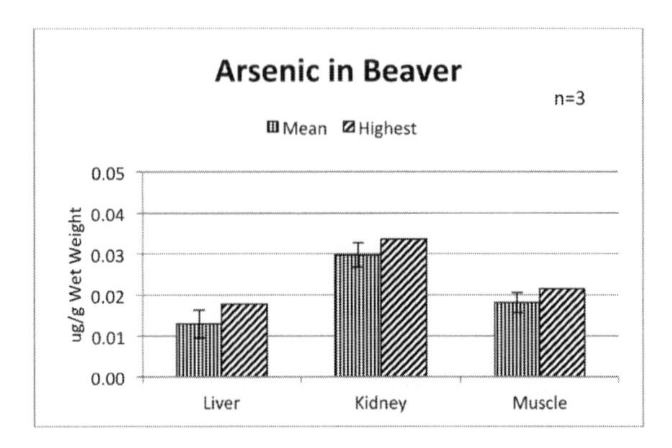

Figure 10.1. Tracks one and two documenting contamination in beavers. Published in McLachlan, S.M. 2014. 'Water is a living thing': environmental and human health implications of the Athabasca oil sands for the Mikisew Cree First Nation and Athabasca Chipewyan First Nation in Northern Alberta', p. 60. Reprinted with permission.

experienced changes over time and through descriptions of how these changes manifest visually. Documenting the changes in the bodies of beavers, for example, one programme participant observed that the animal's flesh "wasn't rich and red like it used to be"; it was "just kinda pale" and "didn't even taste the same."[19] In the report, this documentation of TK on contamination in beavers is placed alongside a bar graph presenting the levels of arsenic found in the beavers' liver, kidney, and muscle. While the report makes an effort to contextualize these arsenic measurements, the significance of the data is difficult to assess for people who are not trained as toxicologists. Conversely, the meaning of the TK is immediately clear, even to readers unfamiliar with the region.

The third track creates a space of synthesis between this narrative and numerical data, enabling dialogue between the "incommensurable" first two tracks. Here, insights provided by the TK (in narrative form) are brought to bear upon quantitative measurements, and vice versa. In the third track, quantified data gains flexibility, as it is situated as a necessarily partial form of accounting for a level of complexity that cannot be captured by numbers alone. In this sense, the third track enables exchange and transformation: the numerical data presented in the bar graphs can be altered in the third track to reflect the insights provided by the TK in the first track, just as the numbers can be made to speak to the narrative, sometimes lending additional specificity to the TK in terms of timelines, concentrations, and site specificity. As McLachlan notes, the TK contextualizes or even "corrects" the numerical data presented in the second track. The third track provides a space in which a strategic unveiling of data enables a calculated *holding back* of knowledge, with the aim of protecting TK and retaining Indigenous control over the data.

The inclusion of the third track in the report ultimately resulted in the lack of acceptance of the results of the "Water Is a Living Thing" study as reliable toxicological data under the Harper government. But however contentious, the third track remains the most important aspect of the methodology, as it stands as the site of non-assimilation. Without the presence of the third track, the study results would almost certainly only be valued for the toxicological data as presented in numerical form—the insights from TK would likely be ignored or written off as scientifically insignificant.

The third track was preserved as a space of synthesis between narrative and numerical data, enabling dialogue between the "incommensurable" first two tracks. Within the third track, quantified data gain flexibility, as they could be interpreted as a necessarily partial form of accounting for a level of complexity that cannot be captured by numbers alone. In this sense, the third track enables exchange and transformation as TK is allowed to adjust numerical measurements, and simultaneously the numbers are made to speak to the narrative, sometimes lending additional specificity in terms of timelines, concentrations, and site specificity to the TK.

As McLachlan notes, the TK contextualizes or even "corrects" the numerical data presented in the second track. The reasons why the numerical data might need to be "corrected" speak to the fundamental points of divergence between the TK and the science of toxicology: TK brings pasts and futures to bear on the assessment of the present, while toxicology considers a snapshot measurement to be the only type of accurate measurement; TK allows for the long-term observation of environmental changes and the interactions between different types of changes taking place as relevant and objective guides for the selection of a sampling site, while toxicology considers this informed selection of a sampling site to constitute bias. In light of the decision made by all participants in the "Water Is a Living Thing" study, the TK directed both the collection and interpretation of data at all times. Reflecting this, the third track provides space for the narrative knowledge to shape the numerical data. In other words, the numerical data presented in the bar graphs can be altered in the third track to reflect the insights provided by the TK in the first track.

THE CONTINUING CONTESTATION OF SETTLER SCIENCE

During the Harper era in Canada, collaborations between Western scientists and Indigenous Traditional Knowledge keepers—and particularly the three-track methodology—faced significant barriers that prevented the study results from attaining credibility and legibility in this political climate. In particular, government and industrial actors resisted the idea that "amateur" scientists lacking professional scientific training could produce toxicological data, and they suspected community-monitoring program participants to be "biased" and incapable of producing "objective" data. The First Nations' knowledge of the land and their desire to protect it were seen as barriers to attaining what conventional toxicologists understood as "objectivity." Quantified data were privileged and likely appealed to the Harper government due to their aura of objectivity, the ways in which they enable the smoothing out of frictions and complexities, "appearing pragmatic and instrumental rather than ideological."[20] McLachlan countered that the participants' scientific training had been robust, and that the First Nations participants' findings should reflect their knowledge and worldviews (as is the case for all scientists' making and interpreting data).

While the three-track methodology ultimately failed to persuade policy-makers under the Harper government, the First Nations communities were undeterred, well aware that the work of changing both cultural and political norms and scientific standards simultaneously is slow in the settler colonial context, and would likely meet with multiple forms of resistance before it gained acceptance. And indeed, over the past five years, collaborations between TK and Western science have gained scientific validity and funding opportunities in the Canadian context—prompted by the Truth and Reconciliation Commission of Canada's Calls to Action (2015) and the Trudeau government's much criticized, belated, and inadequate implementation of legislation to meet the requirements of the United Nations Declaration of the Rights of Indigenous Peoples.[21] However, the crucial question underlying these collaborations remains that of the potential for TK to retain autonomy within the collaboration. Elizabeth Povinelli's observation made more than twenty years ago remains lamentably pertinent today:

> The evaluative apparatus of national or international economic policy has been little influenced by non-Western understandings of human-environment relations. Until it is, indigenous groups will always lose the war of need. Some wider perspective will always be generated that puts their lives "in context."[22]

It was precisely because bureaucrats under the Harper government were so firmly closed to "non-Western understandings of human–environment relations" that the First Nations communities were determined to create a place at the table for their knowledge. In doing so, the First Nations communities succeeded in opposing the settler colonial state's efforts to render Indigenous peoples invisible and voiceless.

INCOMMENSURABLE KNOWLEDGES

I have considered the generative possibilities of a methodology that mobilizes conflict between incommensurable knowledges—instead of attempting to smooth over these tensions. The chapter illustrates a practical, innovative way of producing evidence of harm in a political context in which extensive resources are being directed towards censoring this evidence. My aim has been to shed light on the regimes of imperceptibility operating in settler colonial Canada; the study I have described is a practical response to a context in which certain knowledges are recognized as forms of evidence and others are delegitimized.[23]

The three-track methodology presented a form of resistance in an era of relentless quantification and datafication, demonstrating that making-things-into-data can reflect the exercise of agency rather than merely an instance of succumbing to external pressure. Furthermore, the three-track methodology troubles the persistent binary dividing participation-as-assimilation from romanticized notions of TK as premodern and void of agency (see also Bruchac, chapter 4). As the translation of TK into the language of data is fraught and unending, incommensurability remains as an artefact of incomplete translation. By keeping this incommensurability visible, the three-track methodology models how marginalized communities can produce scientific knowledge on their own terms.

NOTES

1. This chapter draws on material published in the following article: Sarah Blacker, "Strategic Translation: Pollution, Data, and Indigenous Traditional Knowledge." *Journal of the Royal Anthropological Institute* 27, no. S1 (2021): 142–58, https://doi.org/10.1111/1467-9655.13485.

2. Kyle Whyte, "Settler Colonialism, Ecology, and Environmental Injustice," *Environment and Society* 9 (2018): 135.

3. Whyte, "Settler Colonialism," 135–36.

4. Nikolas Rose and Peter Miller, "Political Power beyond the State: Problematics of Government," *The British Journal of Sociology* 61, no. s1 (2010): 271–303.

5. Christopher Kelty and Hannah Landecker, "Ten Thousand Journal Articles Later: Ethnography of 'The Literature' in Science," *Empiria. Revista de Metodología de Ciencias Sociales* no. 18 (October 2, 2009): 177.

6. A. A. Holden, R. B. Donahue, and A. C. Ulrich. "Geochemical Interactions between Process-Affected Water from Oil Sands Tailings Ponds and North Alberta Surficial Sediments," *Journal of Contaminant Hydrology* 119 (2011): 55–68.

7. Jenny Uechi, "Canada's Campaign to Block NAFTA's Oil Sands Tailings Pond Probe Slammed by Critics," *National Observer*, January 12, 2015; Vincent McDermott, "Leaking Suncor Pond Given Government Warning over 2011 Incident," *Fort McMurray Today*, April 1, 2013.

8. Kevin Timoney, "A Study of Water and Sediment Quality as Related to Public Health Issues, Fort Chipewyan, Alberta." A report prepared for the Nunee Health Board, 2007; David J. Tenenbaum. "Oil Sands Development: A Health Risk Worth Taking?" *Environmental Health Perspectives* 117, no. 4 (2009): A150–56.

9. Blacker, interview with Stéphane McLachlan, May 7, 2015.

10. Arun Agrawal, "Indigenous Knowledge and the Politics of Classification," *International Social Science Journal* 54, no. 173 (2002): 287–97; Stephen C. Ellis, "Meaningful Consideration? A Review of Traditional Knowledge in Environmental Decision Making," *Arctic* 58, no. 1 (2005): 72.

11. Jody Berland, "Editorial: The Politics of Evidence," *Canada Watch*, 2015, 1–5.

12. Pallab Ghosh, "Has Canada's Government Been Muzzling Its Scientists?," April 2, 2013, BBC.com, Science & Environment, https://www.bbc.com/news/science-environment-22005706.

13. Stephen Buranyi, "The Fight to Unmuzzle Canada's Scientists," *Vice*, August 27, 2015; Myers, Natasha. "Amplifying the Gaps between Climate Science and Forest Policy: The Write2Know Project and Participatory Dissent," *Canada Watch*, 2015, 18–21; Myers, Natasha, and Max Liboiron, "Write2Know," 2015. http://write2know.ca/.

14. Blacker, interview with Stéphane McLachlan, May 7, 2015.

15. Blacker, interview with Stéphane McLachlan, May 7, 2015.

16. The form of the three-track methodology was designed to meet Indigenous communities' desire and need for data sovereignty—to possess and to control the uses of their data on their own terms. Despite the work carried out by the National Aboriginal Health Organization and the First Nations Information Governance Committee in developing principles of ownership, control, access, and possession (OCAP) for Indigenous self-determination in research and data management, the struggle for Indigenous communities to own and control their own data is ongoing, as many communities lack the structural and material resources that would enable data ownership. A broader discussion of OCAP is beyond the scope of this chapter. For more on OCAP and Indigenous Data Sovereignty, please see Deborah McGregor, "Traditional Knowledge, Sustainable Forest Management, and Ethical Research Involving Aboriginal Peoples: An Aboriginal Scholar's Perspective." *Aboriginal Policy Research Consortium International (APRCi)* 10 (2013): 227–44. and Brian Schnarch, "Ownership, Control, Access, and Possession (OCAP) or Self-Determination Applied to Research:

A Critical Analysis of Contemporary First Nations Research and Some Options for First Nations Communities." *International Journal of Indigenous Health* 1 (2004): 80–95.

17. Lundy Braun and Hannah Kopinski, "Causal Understandings: Controversy, Social Context, and Mesothelioma Research," *BioSocieties* 13, no. 3 (September 2018): 559.

18. Such collaborations have become more common in Canada since 2015, with federal granting agencies developing initiatives that are designed to support collaborations between Indigenous communities, Indigenous scientists, and non-Indigenous scientists.

19. McLachlan, "Water Is a Living Thing," 60.

20. Sally Engle Merry, *The Seductions of Quantification: Measuring Human Rights, Gender Violence, and Sex Trafficking* (Chicago, IL: The University of Chicago Press, 2016), 4. For more information on quantification and the ways in which the circulation of data can enact the erasure of its origins, please see Mihai Surdu's chapter in this volume.

21. Also in response to the UN Special Rapporteur on the rights of Indigenous Peoples' damning report on Canada (2014).

22. Elizabeth A. Povinelli, "Do Rocks Listen? The Cultural Politics of Apprehending Australian Aboriginal Labor," *American Anthropologist* 97, no. 3 (1995): 515.

23. Michelle Murphy, *Sick Building Syndrome and the Problem of Uncertainty: Environmental Politics, Technoscience, and Women Workers* (Durham, NC: Duke University Press, 2006).

11

Invisible Bodies

Psychoanalysis, Subjugated Knowledges, and Intimate Ethics in Postwar Egypt

Omnia El Shakry[1]

Between 1957 and 1960 the National Centre for Sociological and Criminological Research commissioned a study on *Prostitution in Cairo*.[2] Overseen by the sociologist Hasan al-Saʿati, it was meant to shed light on prostitution as a complex societal phenomenon and to mitigate its effects in the wake of its 1951 ban. The project was composed of two parts: a wide-scale social survey and a clinical study. Conducted during the heyday of the regime of Gamal Abdel Nasser (r. 1954–1970), the study was typical of social scientific research at the time, placing an army of experts—sociologists, medical doctors, psychiatrists, clinical psychologists, social workers, nurses, and researchers—at its disposal. Although the results were published collectively in a 300-page report, some researchers went on to use the findings in their academic works, while the vast majority of researchers remained largely unknown. My focus here is not on the apparatus of knowledge production that renders certain researchers invisible, as important as that might be. Rather, I concern myself with the simultaneous invisibility and hypervisibility of the subjects of research, the "prostitutes" who constituted the main object of social scientific inquiry, and in particular with their role in the formulation of clinical psychoanalytic knowledge.[3]

The clinical study that is the focus of this chapter aimed to know prostitutes in their Cairo setting, describing their social and psychological characteristics and the factors responsible for their experience.[4] The enormous discursive weight of the survey, with its legion of specialists, bears the mark of the postcolonial state and its attendant figure of the expert who renders "the prostitute" transparent—calculable and comprehensively legible within a taxonomy of social forms—in the name of the postcolonial good. The subaltern female remains doubly exploited, persisting as a twofold object of knowledge. In the first instance, she is shuttled from empire to nation, taken up as a deviant social category inherited from colonial governmentality and transformed into a postcolonial object of reform; and in the second instance, shuttled yet again from the postcolonial sociological survey to later academic scholarship in European metropoles.[5] The survey thus lingers *in decolonized terrain* as an exemplar of postcolonial reason, complicating any notion of decolonization as a simple reversal of colonial exploitation.[6]

However, rather than view the female subjects of the *Prostitution* study as the unnamed and passive victims of a decolonizing social science, I imagine their labour as instead central to how we might imagine and rethink knowledge production. More specifically, I argue that the *Prostitution* study enables us to re-contextualize knowledge production in the human sciences: (i) by highlighting prisons, from which questionnaire and clinical study participants were drawn, as sites of knowledge production and praxis that refract and reinscribe carceral logics; (ii) by rethinking the psychoanalytic case study as a dense nexus of invisibility and hypervisibility, which elides the patient as a knowledge producer; and (iii) by reflecting upon the psychoanalyst and the historian's relationship to "subjugated knowledges" and "excluded socialities" as a form of intimate ethics.[7]

PRISONS, PRAXIS, AND KNOWLEDGE PRODUCTION

All participants in the clinical component of the *Prostitution* study were prisoners at al-Qanatir prison and had been variously charged for prostitution, incitement to debauchery, vagabondage, drunkenness, and illicit sexual intercourse.[8] Al-Qanatir, the site of Egypt's largest women's prison, is located to the north of Cairo and has housed many of Egypt's most well-known activists: Zaynab al-Ghazali, founder of the Muslim Women's Association, imprisoned and tortured under Nasser; Shahenda Makled, a leftist revolutionary figure active in land reform struggles; Safinaz Kazim, a journalist and writer who has spanned the political spectrum; and internationally renowned Egyptian feminist Nawal al-Saadawi.[9] Carceral experiences, particularly in the postcolonial period, have been central to modern Egypt and "much of modern Egyptian intellectual history was born in Nasser's prisons."[10]

While a celebrated tradition of prison memoirs exists, including those penned by women political prisoners, decidedly less attention has been paid to other types of prisoners.[11] The tendency has been to view prison as a site of education for political prisoners—a venue for self-growth; knowledge consumption, production, and exchange—while a site of degradation for non-political prisoners. And yet, surveying the larger literature written by women political prisoners, Marilyn Booth shows how political writers reflect on political and non-political prisoners alike as indexical of larger societal structures, both political-economic and social, that serve to subordinate women. "Prison is praxis," she argues, "it is also self-education. By its nature, the prison system exemplifies starkly the boundaries against which one must struggle, the structures which appear immutable."[12] Thus, while differences in class, education, and opportunity divided prisoners, the distinction between political and non-political prisoners should not be overstated. Political and non-political prisoners alike refracted carceral regimes as knowledge creators.

In what follows I explore the creative labour of the prisoners of the *Prostitution* study as indispensable to the production of psychoanalytic knowledge. In doing so, I query the role of populations that psychoanalysis seldom addresses—illiterate individuals and incarcerated subjects. How might we transform our view of knowledge production in order to consider the labour of such individuals conscripted into the intellectual orbit of psychoanalysis? Following Marilyn Booth and Dylan Rodríguez, I suggest we view the prisoners' labour as *radical prison praxis*—a form of knowledge production that reinscribes carceral practices.[13] I attempt to do so without fetishizing the prison as a site of resistance or centring radical subjectivity. Rather, I lay bare the carceral forms that inhere in the process of psychoanalytic interpretation, while demonstrating how such interpretations are repeatedly subverted and destabilized in the analytic encounter. I begin by first outlining the clinical work of psychoanalytic interpretation embedded in the *Prostitution* study. I then reimagine such interpretive work by highlighting its carceral context and by realigning it with a view that emphasizes the inherent unknowability, rather than transparency, of the human subject.

THE VIOLENCE OF INTERPRETATION

The clinical component of the *Prostitution* study included a social interview or case history; physical examination; psychiatric examination; and a psychological examination comprised of two intelligence tests and three personality tests: an association of ideas test, a Rorschach, and a drawing test.[14] Sami Mahmud Ali (b. 1925), the Arabic translator of Sigmund Freud's *Three Essays on the Theory of Sexuality*, and later the author of a large body of psychoanalytic writings, was responsible for the exposition of the individual case studies and the preparation of the clinical report for final presentation.[15] The drawing test was conceptualized, administered, and interpreted by Sami-Ali himself and entailed the drawing of four items, each with a symbolic weight: a plant, an animal, a human, and a building. The test shifted registers from the large-scale prism of the sociological to the subjective and intersubjective register of the psychoanalytic.[16]

The test was based in large part on Françoise Dolto's psychoanalytic work with children, in which drawing provided access to the unconscious image of the body, referring not to the body's proprioceptive relation to the world, but rather to the psychic history of the desiring subject.[17] Following this method, Sami-Ali allowed each subject the freedom to draw and to spontaneously explain each drawing, which was then interpreted according to psychoanalytic principles.[18] Years later, once established in metropolitan France as a professor of psychology at l'Université de Paris VII, Sami-Ali would continually return to these case studies in his theoretical work, illustrating that his clinical encounters with incarcerated women formed the very condition of possibility of his theoretical writings.[19]

What can we learn from the subjects of the drawing test, who after all were not Sami-Ali's own patients or analysands? Significantly, what can we learn while at the same time acknowledging that within Sami-Ali's oeuvre the extraction of surplus value from incarcerated third-world prostitutes resulted in the production of metropolitan theory?

Waguida, a pseudonym, is subject number four. She provides an associative case history, recounting her carefree childhood in the countryside, games with her younger sister, the early death of her mother, and the subsequent devotion of her father. She relates a dream she had shortly before her arrest: she is sitting on her father's lap and he penetrates her sexually, and then offers her a cake that he makes her promise to keep. Her comrade tells her the dream portends a pregnancy or a coming misfortune (which she interprets as her subsequent arrest). Her drawings are likewise dramatic. In her first drawing, she stages a scene: a fellow prisoner is disguised as a ghost, playing the tambourine. But the manner in which the figure is rendered is noteworthy: the thick contoured lines, the green colouring, all evoke woody vines (liana) that uniformly enclose an empty body surrounded by a collar (figure 11.1a). The dream, Sami-Ali notes, helps decode the drawing as the effacement of a sexuality intimately linked to incestuous desire.[20] He interprets her drawing in which not just her sexual organs but also her body proper is replaced by a plant, as a narcissistic substitution that renders her repulsive, a being whom no one can desire.[21] This is ultimately, he argues, what she strives to signify by the hysterical conversion of her own image into a vegetal body.

Figure 11.1. Originally Figure 7 in Sami-Ali, *L'espace imaginaire* (Paris: Gallimard, 1974), p. 104

For Sami-Ali, the subject's vine woman, alongside her second drawing of a kite (bird) which hovers over the prison, represents the negation of a guilty sexuality (figure 11.1b). The vine woman, he observes, expresses a guilty sexuality and the kite a fantasy of escaping from the flesh; the unconscious fantasies of childhood were repeated in the drawing sessions. It was, Sami-Ali later explained, in the transference between analyst and analysand, in which the patient unknowingly repeated with the analyst an element repressed from her childhood, that the forgotten past intruded into the present of the individual. And, thus, it was often practice and demeanour (e.g., crossing out a drawing, garrulous speech), rather than memories, that became the locus of a repressed past.[22]

Sami-Ali postulates that such modalities of drawing (repressing, voiding) constitute latent projections of one's own body. *Projection* refers to a process "whereby qualities, feelings, wishes or even 'objects,' which the subject refuses to recognise or rejects in himself, are expelled from the self and located in another person or thing."[23] Within the drawings, projection "merges with the very fabric of vision" drawn from lived bodily experience.[24] Here, the very form of corporeal life is conjoined to the representation of the world—right and left, top and bottom, front and rear, inside and outside—as in the reproduction of bodily comportment within the drawings.[25] At the same time, "the fulfilment of desire is not confined to creating an image."[26] Desire may entail erasures or suppressions, in whole or in part, as in Waguida's drawing.

Projection, according to Sami-Ali, may thus take place through the work of the negative, and absence becomes a generative space where something can emerge.[27] The gesture of drawing now has the dual function of manifesting the visible and of suggesting what is beyond the visible.[28] Rather than a projection that summons the visible to arise where it does not exist, the projection summons emptiness or a void through the familiar tactics of condensation and displacement.[29] To take several other examples of drawings from the *Prostitution* study, a sketch of a man is turned upside down and becomes a pot of flowers; the absence of the human form is a negation, which enables the visible to exist as such (figure 11.2a); a human form is emptied out of its insides (figure 11.1a); a child is discussed as an angel that cannot be represented, a spectral absence, a transmogrified absent presence (figure 11.3).[30] Perhaps unsurprisingly, Sami-Ali reads virtually all of these examples through the primal structuring of incest fantasies: for example, the prostitute's angel-child incarnates the desire for paternal incest, which is spiritualized through a *haj* painting and a reference to her father's pilgrimages in order to abolish sexual and aggressive Oedipal impulses (figure 11.3c).[31]

PSYCHOANALYSIS AND CO-CREATION

I turn now to the complex and elusive case study of Waguida discussed previously, in order to pursue a few lines of inquiry. The first is to re-read the manifest content

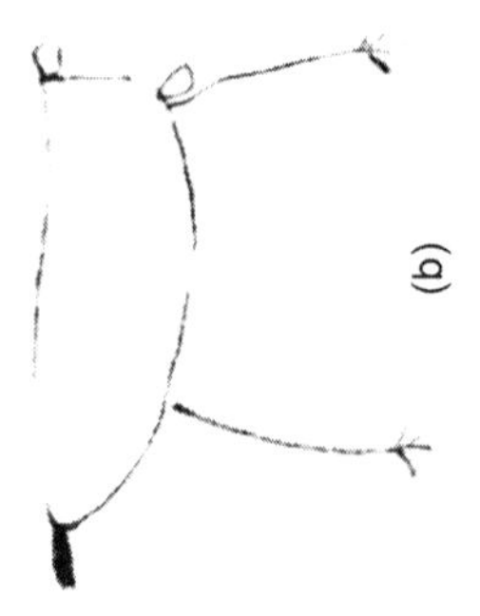

Figure 11.2. Originally Figure 4 in Sami-Ali, *L'espace imaginaire* (Paris: Gallimard, 1974), p. 97 © Éditions Gallimard. All copyrights reserved. Unless authorized, any use of *L'espace imaginaire* other than for individual and private consultation is prohibited. www.gallimard.fr.

of the (dream) drawing in order to highlight its larger carceral context. The second is to attend to the latent logic or form of the drawing test, in order to emphasize the unknowability (against a presumed transparency) of the human subject. Making space for unknowability helps us understand the limits of interpretation, while also recognizing psychoanalytic knowledge production as a co-creation between subject and analyst. Finally, I conclude by contemplating the ethical quandaries posed for the historian engaging these case studies.

Viewing the drawing through the lens of its manifest content dramatizes both a carceral logic—signified by the collar and even the vegetal encasement of the body—and a fantasy of flight—the raptors with which the prisoner identifies herself (figure 11.1). Such a dramatization might be viewed as the creation of a fantasy of flight from structures of domination and oppression.[32] In this sense, the drawing might be read as critically reinscribing the logic of incarceration within which its production was embedded and conscripted.[33] Sami-Ali's analysis of the drawing, however, strove to decode a more universal operation of the unconscious. By his

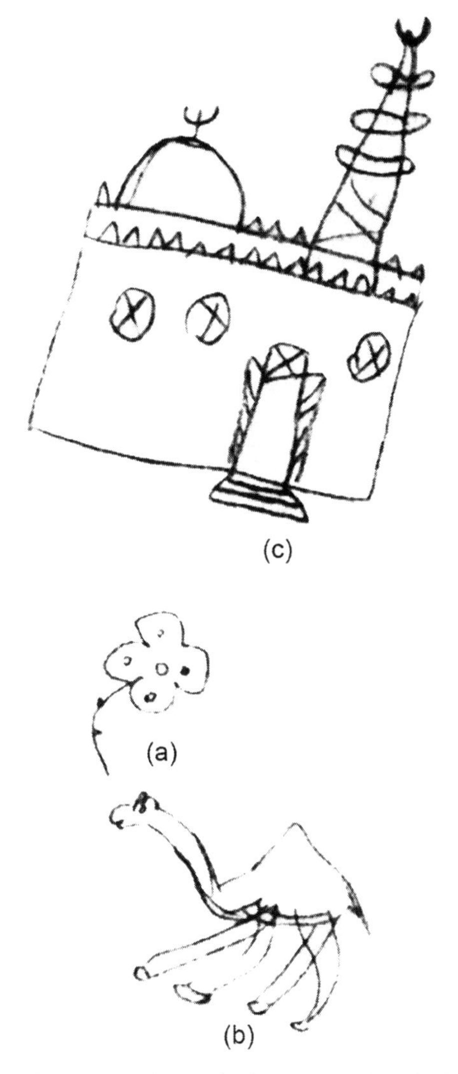

Figure 11.3. Originally Figure 8 in Sami-Ali, *L'espace imaginaire* (Paris: Gallimard, 1974), p. 106

reading, every hidden meaning and every invisible object or body pointed back to infantile incestuous desire. Sami-Ali fills the empty space of the drawings with his own interpretation, thereby interning the subject within the framework of the Oedipus complex.

If we turn now to the latent logic or form of the drawing test, we might view Waguida's drawing—in which an invisible or absent body occupies the centre of the drawing—not as indexical of a clichéd and repressed Oedipal sexuality (as Sami-Ali saw it), but instead as the visual representation of the unknowability at the core of the human subject. Such a void, lacuna, gap, or emptiness constitutes a nucleus that might be thought of as akin to the navel of the dream—"its point of contact with the unknown."[34] In place of reading such gaps and redactions as the symbolic representations of unconscious incestuous desires, then, I suggest a conceptual and methodological shift.

What if we were to depart from the stereotypical content of Sami-Ali's interpretations and our own impulses to render these subjects transparent? What if we focused in its place on the form, process, and artefacts of the drawing test? This might allow us to set aside the creation of static interpretations by the analyst, interpretations that by their very nature render subjects invisible as knowledge producers and simultaneously hypervisible as seemingly passive objects of discussion. To do so, we might focus on the empty spaces of the drawings themselves; the void of self-disclosure is no longer exclusively the sign of an infantile Oedipus, nor of a secret that hides the absolute truth of the desiring subject, but rather of something *other*, something that opens up onto a constitutive ontological gap through which the human is unknowable.

How, then, might we view the knowledge produced from these encounters as co-creations? We would do well to take our cue here from the British paediatrician and psychoanalyst Donald Winnicott, who in referring to analytic interpretations and the problem of creativity was forthright in his assertion that analysts learn the substance of their theories directly from their patients.[35] What if, in other words, we no longer understood the subjects of the *Prostitution* study as merely objects of knowledge, but rather as co-creators of psychoanalytic theory within the clinical encounter itself? Specifically, it is the patient's spontaneity and the "fluid unfolding of the deliberateness of contact" that challenges the analyst's alleged mastery of interpretation.[36]

I argue that it is only in attending to the void of non-representation and the spatiotemporal cut initiated by the hesitations, silent gestures, and phantasmatic reveries that take place while drawing that we may re-centre the subjects of the *Prostitution* study as co-creators of knowledge. Arguably, it is such absences and silent gestures that led Sami-Ali to his insights into the process of projection. Traditional psychoanalytic theories of projection view it as a process in which the subject expels feelings or wishes from the self onto a person or thing. By contrast, Sami-Ali's theory of projection holds that it is the fulcrum through which an inner and outer world come together; inextricably linked to perception, it is itself a creative mode of apprehending reality.[37] Such a view of projection, he asserts, circumvents the false dichotomies between inside and outside, perception and projection, reality and unreality.[38] Projection, in other words, is a creative rather than merely a defensive process.

From whence does such a theorization arise? It arises, in large part, from the female subjects' projection of empty spaces as creative acts. In Sami-Ali's rendering, projection does not only refer to projecting something onto an object or person (as in the conventional view); projection may also take place through the work of the negative. Projection is just as much about what is not there as what is there. In the previous drawings, projection summons emptiness or a void, a no-thing: the invisible body in Waguida's drawing, or the absent child in figure 11.3. Absence is viewed as a generative space where something may emerge. Sami-Ali, however, interprets such empty spaces by reducing them to the satisfaction of an unconscious desire.[39] But what if rather than view such absences as the hysterical expressions of incestuous desire, we see them quite simply as the right not to communicate.

I am interested here in how the *act* of drawing suggests new ways of thinking about the operations of subjectivity as *negation*, *gap*, *pause*, or *void.* In many of these examples, it is the body, through its gestures, which provides interpretive clues: hesitation ("I can't draw," an unsure hand) or changing the orientation of the paper by turning the page horizontally or vertically to complete a new drawing; drawing in an act of reverie—all demonstrate the subject as "*an isolate, permanently non-communicating, permanently unknown, in fact unfound.*"[40]

The silent gestures, corporeal hesitations, and movements of the subjects in question, may of course be viewed as individual acts of refusal. However, rather than a liberal feminist signal that would reclaim these gestures as the agentive acts of refusal of an undivided subject, I parse these gestures and hesitations alongside Winnicott's claim that the subject has the "right not to communicate . . . a protest . . . to the frightening fantasy of being infinitely exploited. In another language this would be the fantasy of being eaten or swallowed up."[41] The work of the negative—whether in the void of non-representation present in virtually all of the drawing artefacts or in the withholding of "the personal core of the self that is a true isolate"—thereby retains the kernel of the unknowability of the human subject.[42]

Stated more directly, Sami-Ali represents and speaks for "the prostitutes" while claiming to render visible and transparent the operations not only of consciousness but also of unconsciousness. By contrast, if we linger in the silent gestures, empty spaces, and aporias of interpretation, we acknowledge the sexed subaltern subject as irretrievably heterogeneous.[43] To do so is not to fill absence with the plenitude of an agent of desire, nor to access the unconscious of a purportedly undivided subject within the fullness of speech. Rather, it is to dwell in the archival and aesthetic artefacts that reveal the condition of (im)possibility that structured psychoanalysis in the postcolony and that fracture the space of encounter—between the metropolitan postcolonial intellectual and the sexed subaltern subject and between the analyst and the analysand.

INTIMATE ETHICS

I want to conclude by returning to the question of knowledge production in the human sciences through a reflection upon the ethical exigencies of these intimate case studies. There are three senses in which I refer to the intimacy of these clinical encounters. The first is the realm of intimacy and state power. Scholars have explored the intimate nature of state power—intruding as it does into the domain of "conjugality, family, domesticity, sexuality—in colonial and postcolonial times."[44] Nowhere is this more readily exemplified than in the *Prostitution* study. At the same time, Sertaç Sehlikoglu and Asli Zengin point to the need "to conceptualize intimacy beyond the limits of sexuality," defining intimacy, instead, as the way in which "people, bodies, and objects meet and touch—and the zones of contact they create."[45] The drawing test may be viewed as one such intimate zone of contact.

The second sense in which I use the term is to refer to the shared intimacy of the subject–analyst, even in a carceral testing situation. Most pertinently, Asli Zengin has explored the "intersection of intimate contact and mandated encounters with medicolegal institutions and the bodies of sex/gender transgressive people."[46] Paying novel attention to the sensorium, and touch and tactility in particular, as a domain of violent intimate state power, Zengin investigates "the use of corporeality and sensorium as a technology of sex and gender."[47] Such affective and visceral domains of state power demonstrate the scope and reach of biopolitics into the intimate realms of embodiment. I have tried to explore what happens when sexuality, intimacy, and state power come together in unanticipated ways. How do we make sense of intersubjectivities that both supersede and exceed the reach of state power and violence while remaining nestled with its orbit?

Finally, there is the intimacy of the historian, who in this instance functions as an interloper, perhaps even a voyeur. How shall we assess the ethical role of the historian involved in the unearthing of such "subjugated knowledges"? Here, I use the term "intimate ethics" to refer to that ethical relationship in which historian and historical subject are psychically implicated. To be clear, my use of the term is derived largely from a psychoanalytic perspective. Psychoanalysis, as I have demonstrated elsewhere, oscillated between ethical ideals centred on the opacity of the human subject (her resistance to intelligibility and understanding) and the belief in the transparency of humans and the possibility of their instrumentalization.[48] Such divergent views marked the difference between the prospect of a psychoanalysis that would be "*at the service* of those who suffer, and not an instrument of power or mastery over them."[49]

Such an instrumentalization of the other as an object of knowledge might take the form of a comprehensive attempt to re-narrate the individual lives of female subjects. Here, I neither engage in such a reconstructive history nor do I reinterpret the empirical data collected in order to provide a recuperative reading in which I restore the figure or person of "the prostitute" to her rightful social place: one in which she rebels against patriarchal authority, revolts against social norms and mores, and nobly inhabits a fragile space of social and economic vulnerability. To do so risks mimicking the terms of engagement of the sociological study itself, namely, aspiring to the totality, transparency, and knowability of the human subject, a mode of knowledge production that seeks to lay bare the forms of life, to enumerate, and to expose. Instead, I have done my utmost to avoid the iterative violence of enumerative, diagnostic, and etiological logics as they attach themselves to particular individuals.

By retaining the fundamental unknowability of human subjects, as against their transparency, I attempt to avoid simply recovering agential capacity or "voice." This, however, has not meant that I lack personal psychological investment in the material at hand. Psychoanalysis, both as theory and practice, presents the possibility of enjoyment in the use of the other as an instrument or object, while at the same time offering a means of undermining that sovereign pleasure, precisely by critically analysing one's own psychic implication in it. Indeed, my readings of the subject–psychoanalyst relationship are informed by my own experiences as a patient—as someone who has consistently been plagued by dreams in which I experience psychoanalytic therapy as a violent (albeit necessary) cutting up and representing of psychic reality. My own role as patient, therefore, necessarily informs my reading of the case studies as avenues in which subject and psychoanalyst alike co-create psychoanalytic knowledge.

At the same time, we must acknowledge the asymmetries of power between academic and health care professionals, on the one hand, and mental health service users, on the other. Indeed, some scholars argue "co-production in mental health is likely impossible in privileged sites of knowledge production."[50] Clearly, the clinical study participants under consideration here are not service users in any traditional sense, having been conscripted into the research whether or not their participation was consciously volitional. Nevertheless, I have contended that there may be another angle from which we may approach the knowledge produced from the drawings left to us by the subjects of the *Prostitution* study, namely, that of co-creation. Such a co-creation of psychoanalytic knowledge would not be viewed in terms of naive collaborative or dialogical practitioner–client models, which would be wholly inappropriate in this carceral context. Rather, I have argued that we focus on "alternative ways of producing knowledge" while attending to "environments that are not usually seen as sites of knowledge generation."[51] While the carceral setting highlights the violence of the case study as a methodology, which contributes to longer histories of patient erasures, I have instead explored the form, process, and artefacts of the drawing test as providing an avenue through which we might understand the co-creation of psychoanalytic knowledge.

NOTES

1. **Acknowledgements**: The reflections here have benefited from feedback from the organizers (Edna Bonhomme, Shehab Ismail, and Lamia Mognieh) and participants at the Max Planck Institute for the History of Science (MPIWG) and Europe in the Middle East—The Middle East in Europe (EUME) workshop on "Power in Medicine: Interrogating the Place of Medical Knowledge in the Modern Middle East," April 11–12, 2019 in Berlin, Germany. Edna Bonhomme, in particular, encouraged me to think about prisons as sites of knowledge production. I am especially grateful to Jenny Bangham and Xan Chacko for insightful feedback and editorial suggestions on this chapter. This chapter draws, in part, on material derived from Omnia El Shakry, "Psychoanalysis and the Imaginary: Translating Freud in Postcolonial Egypt," *Psychoanalysis and History* 20, no. 3 (2018): 313–35. Reproduced with the permission of The Licensor through PLSclear. The Version of Record is available at: https://doi .org/10.3366/pah.2018.0271.Images are reproduced from: Sami-Ali, *L'espace imaginaire*

(Paris: Gallimard, 1974), 97, 104, 106.

2. Hasan al-Saʿati et al., *al-Bighaʾ fi al-Qahira: Mash Ijtimaʿi wa-Dirasa Iklinikiyya* (Cairo: Manshurat al-Markaz al-Qawmi li-l-Buhuth al-Ijtimaʿiyya wa-l-Jinaʿiyya, 1961).

3. The "prostitute" is an overdetermined figure, to be sure. Henceforth, the term is used in keeping with the sources of the historical time period under consideration. Durba Mitra traces the history of the concept as a gloss for deviant female sexuality and provides a fruitful reflection upon terminology in *Indian Sex Life: Sexuality and the Colonial Origins of Modern Social Thought* (Princeton, NJ: Princeton University Press, 2020), 1–9.

4. Al-Saʿati et al., *al-Bighaʾ fi al-Qahira*, 87.

5. On the struggle between colonial administrators and Egyptian nationalist reformers over prostitution, see Hanan Hammad, "Regulating Sexuality: The Colonial–National Struggle over Prostitution after the British Invasion of Egypt," in *The Long 1890s in Egypt: Colonial Quiescence, Subterranean Resistance*, eds. Marilyn Booth and Anthony Gorman (Edinburgh: Edinburgh University Press, 2014), 195–221. On the centrality of sex work to the history of modern Egypt, see Francesca Biancani, *Sex Work in Colonial Egypt: Women, Modernity and the Global Economy* (London: IB Tauris, 2018); and on the wider context of prostitution in the Middle East, see Liat Kozma, *Global Women, Colonial Ports: Prostitution in the Interwar Middle East* (Albany: State University of New York Press, 2017).

6. Gayatri Chakravorty Spivak, "Woman in Difference: Mahasweta Devi's 'Douloti the Bountiful,'" *Cultural Critique* 14 (1989–1990): 105–6, emphasis in original.

7. The reference to "subjugated knowledges" and "excluded socialities" is from Elizabeth Povinelli, "The Woman on the Other Side of the Wall: Archiving the Otherwise in Postcolonial Digital Archives," *differences* 22, no. 1 (2011): 151.

8. Al-Saʿati et al., *al-Bighaʾ fi al-Qahira*, 94.

9. Marilyn Booth, "Women's Prison Memoirs in Egypt and Elsewhere: Prison, Gender, Praxis," *MERIP Middle East Report* 149 (Nov.–Dec. 1987): 35–41; Zainab al-Ghazali, *Return of the Pharaoh: Memoir in Nasir's Prison*, trans. Mokrane Guezzou (Leicester, UK: Islamic Foundation, 1994); Safinaz Kazim, *ʿAn al-Sijn wa-l-Hurriya* (Cairo: Al-Zahraʾ li-l Aʿlam al-ʿArabi, 1986); Nawal al-Saadawi, *Memoirs from the Women's Prison*, trans. Marilyn Booth (London: The Women's Press, 1986).

10. Robyn Creswell, introduction to *That Smell and Notes from Prison*, by Sonallah Ibrahim, trans. and ed. Robyn Creswell (New York: New Directions, 2013), 9; see also Stephen Cox, "Mihna and Mission in the Muslim Brotherhood" (PhD diss., University of California, Davis, 2019).

11. For important exceptions that centre women prisoners and gendered conceptions of social deviance, see Nefertiti Takla, "Barbaric Women: Race and the Colonization of Gender in Interwar Egypt," *International Journal of Middle East Studies* (2021): 1–19; Hannah Elsisi, "'They Threw Her in with the Prostitutes!': Negotiating Respectability between the Space of the Prison and the Place of Woman in Egypt (1943–59)," *Genre & Histoire* 25 (2020).

12. Booth, "Women's Prison Memoirs in Egypt," 36. See also Barbara Harlow, "Sectarian Versus Secular: The Case of Egypt," in *Barred: Women, Writing, and Political Detention* (Hanover, NH: Wesleyan University Press, 1992), 118–32.

13. Dylan Rodríguez, "Against the Discipline of 'Prison Writing': Toward a Theoretical Conception of Contemporary Radical Prison Praxis," *Genre* XXXV (Fall/Winter 2002): 411, emphasis in original.

14. Al-Sa'ati et al., *al-Bigha' fi al-Qahira*, 88–91.

15. In his Arabic writings his name appears as Sami Mahmud Ali, and in his subsequent French writings as Sami-Ali. I retain the latter to avoid confusion.

16. Al-Sa'ati et al., *al-Bigha' fi al-Qahira*, 91.

17. Françoise Dolto (b. 1908) was a French physician, psychoanalyst, and member of Jacques Lacan's circle, see *Theory and Practice in Child Psychoanalysis: An Introduction to the Work of Françoise Dolto*, eds. Guy Hall et al. (London: Karnac Books, 2009).

18. Sami-Ali, "Rusum al-Baghaya," *al-Majalla al-Jina'iyya al-Qawmiyya* 1 (1958): 104–8.

19. Sami-Ali, *De la projection: une étude psychanalytique* (Paris: Payot, 1970); Sami-Ali, *L'espace imaginaire* (Paris: Gallimard, 1974).

20. Sami-Ali, *De la projection*, 78–80; *L'espace imaginaire*, 103–5.

21. Sami-Ali, *L'espace imaginaire*, 103.

22. Sami-Ali, *De la projection*, 78–80; *L'espace imaginaire*, 103–5; cf. Sami-Ali, "Kalimat al-Mutarjam," in Sijmund Fruyd, *Thalath Maqalat fi Nazariyyat al-Jinsiyya*, trans. Sami Mahmud Ali (Cairo: Dar al-Ma'arif, 1963), 18–19.

23. Jean Laplanche and Jean-Bertrand Pontalis, *The Language of Psycho-analysis*, trans. Donald Nicholson-Smith (New York: W. W. Norton & Co. [1967] 1973), 349.

24. Sami-Ali, *L'espace imaginaire*, 121.

25. Sami-Ali, *De la projection*, 217; *L'espace imaginaire*, 87.

26. Sami-Ali, *L'espace imaginaire*, 120–21.

27. I use the phrase in the sense of André Green, *The Work of the Negative* (London: Free Association Books, 1999).

28. Sami-Ali, *L'espace imaginaire*, 120.

29. Sami-Ali, *L'espace imaginaire*, 96.

30. Sami-Ali, *L'espace imaginaire*, 97–107.

31. Sami-Ali, *L'espace imaginaire*, 107.

32. Rodríguez, "Against the Discipline of 'Prison Writing,'" 427.

33. Rodríguez, "Against the Discipline of 'Prison Writing,'" 411.

34. Sigmund Freud, *The Interpretation of Dreams* (First Part), in *The Standard Edition of the Complete Psychological Works of Sigmund Freud*, trans. and eds. James Strachey et al. (London: Hogarth Press, 1981), 4:111n1.

35. Donald W. Winnicott, "Communicating and Not Communicating Leading to a Study of Certain Opposites," in *The Maturational Processes and the Facilitating Environment* (London: Karnac, [1963] 1990), 182.

36. Margherita Spagnuolo Lobb, "Therapeutic Meeting as Improvisational Co-Creation," in *Creative License: The Art of Gestalt Therapy*, eds. Margherita Spagnuolo Lobb and Nancy Amendt-Lyon (Vienna: Springer, 2003), 38.

37. Sami-Ali, *De la projection*, x.

38. Sami-Ali, *De la projection*, xii–xiv; 43–68.

39. Sami-Ali, *De la projection*, 196.

40. "Communicating and Not Communicating," 187, emphasis in original.

41. Winnicott, "Communicating and Not Communicating," 179.

42. Winnicott, "Communicating and Not Communicating," 182.

43. Gayatri Chakravorty Spivak, "Can the Subaltern Speak?" in *Marxism and the Interpretation of Culture*, eds. Lawrence Grossberg and Cary Nelson (Urbana: University of Illinois Press, 1988), 271–313.

44. Sertaç Sehlikoglu and Asli Zengin, "Introduction: Why Revisit Intimacy?" *The Cambridge Journal of Anthropology* 33, no. 2 (2015): 21.

45. Sehlikoglu and Zengin, "Introduction," 20; Asli Zengin and Sehlikoglu Sertaç, "Everyday Intimacies of the Middle East," *Journal of Middle East Women's Studies* 12, no. 2 (2016): 139; Geraldine Pratt and Victoria Rosner, "Introduction: The Global and the Intimate," *Women's Studies Quarterly* 34, nos. 1–2 (2006): 17.

46. Asli Zengin, "Violent Intimacies: Tactile State Power, Sex/Gender Transgression, and the Politics of Touch in Contemporary Turkey," *Journal of Middle East Women's Studies* 12, no. 2 (2016): 225.

47. Zengin, "Violent Intimacies," 227.

48. Omnia El Shakry, *The Arabic Freud: Psychoanalysis and Islam in Modern Egypt* (Princeton, NJ: Princeton University Press, 2017), 14–15.

49. Dominique Scarfone, *The Unpast: The Actual Unconscious*. Translated by Dorothée Bonnigal-Katz (New York: Unconscious in Translation, 2015), 92, emphasis in original.

50. Diana Rose and Jayasree Kalathil, "Power, Privilege and Knowledge: The Untenable Promise of Co-Production in Mental 'Health,'" *Frontiers in Sociology* 4, no. 57 (2019): 1.

51. Rose and Kalathil, "Power, Privilege and Knowledge," 5–6.

12

The (In)visible Labour of Varietal Innovation

Susannah Chapman

How do new crop varieties—those relatively discrete groupings of plants often classified below the taxonomic level of "subspecies"—come out in the world?[1] This chapter provides an account of how new plant variation emerges and eventually becomes a new variety. While a reader of English might best recognize this as the work of "plant breeding," with all its implications of mastery and control, what I relay here would be poorly captured by that term.[2] Instead, I chronicle a wide range of beings-and-doings carried out by humans, God, spirits, and plants, as described by rice farmers in a predominantly Mandinka-speaking district of The Gambia between 2010 and 2012. In transmitting these accounts, I reflect on the different ways in which labour is rendered visible and invisible in stories about *who works* and *how work proceeds* in the "breeding" of new crop varieties. I thus weave farmers' accounts with three others: the legacies of colonial agricultural science, research on farmers-as-breeders, and work attentive to more-than-human agency. Different processes, modes, and degrees of invisibility run through each of these accounts (my own included)—erasure, elision, and translation, to name a few.[3] Attending to these processes, I argue, is not only important for understanding the workings of power in the production of scientific knowledge, it also provides a vital space in which to grapple with the ethical and economic worlds brought into being by the making of people and plants.

A POLITICS OF VISIBILITY IN FARMER INNOVATION

Questions of who breeds plants and how plant breeding happens matter because they structure contemporary agricultural development initiatives. The breeding of new crop varieties emerged as a key focus of agricultural "modernization" projects during the twentieth century, with the Green Revolution of the 1950s and 1960s promising new high-yielding varieties of wheat, rice, and maize bred by scientists at International Agricultural Research Centres. Since the 1990s, "New" Green Revolution and agricultural development initiatives across sub-Saharan Africa have sought to encourage the innovation of new crop varieties via the elaboration of public–private partnerships, biological and economic standardization, and intellectual property law.[4]

In The Gambia, recent efforts to reshape seed production and varietal development culminated in a National Seed Policy, which established a system for the production and registration of certified seed with the aim of building a domestic seed industry comprised of farmer-entrepreneurs, farmer seed cooperatives, and seed companies.[5] Then, in 2015 The Gambia became one of the first countries within Anglophone sub-Saharan Africa to sign the Arusha Protocol, an agreement of the African Regional Intellectual Property Organization that set forth a standardized framework of plant breeders' rights for all member states. Intended to incentivize and support the innovation of new plant varieties, the Protocol outlines a system of private, exclusive rights for the creators of new, uniform, and stable crop varieties.[6] While the framework established by the Protocol has not yet been incorporated into The Gambia's domestic law, questions of who breeds plants and how such work proceeds nonetheless shape the parameters of emerging forms of private property. How such questions get answered can both influence the distribution of credit within scientific communities and subtend property claims within increasingly valorized "knowledge economies."[7]

Yet answers to questions about who breeds plants, and how, turn upon historic processes of erasure. For much of the twentieth century, agricultural development initiatives presented crop varietal development as something that occurred largely beyond farmers' fields, incentivized and introduced from science, markets, and laws found elsewhere. In many parts of the world, this rendering drew upon a much longer history of colonial expansion and its racializing assemblages.[8] With the establishment of colonial Agricultural Departments throughout West Africa in the early twentieth century, for example, European claims about unskilled or non-innovative African farmers provided justifications for colonial interventions into local agricultural practices and economies.[9] At the same time, European assertions about peoples' inclination for creative labour relayed entrenched ideas about race and cultural hierarchy.[10] African subjects were positioned as lazy, non-innovative, and less-than-human even as colonial administrations and plantation economies profited from the knowledge and plant innovations produced by African peoples.[11]

Ideas about racialized difference continue to suffuse projects for "capacity building" across sub-Saharan Africa, in what Jemima Pierre calls the "racial vernaculars of development."[12] In this context, narratives about varietal innovation and its relationship to notions of social progress seep into the language and planning of agricultural modernization projects (McCook, chapter 13). Even the classification of seed at agricultural research institutes evokes associations between human difference and the extent to which a crop variety is perceived to have been laboured upon: As artist-theorist Elaine Gan has pointed out, distinctions between "modern," "elite," "improved" "native," and "landrace" varieties "represent degrees and hierarchies of human intervention."[13] In this context, claims about who does certain types of innovative agricultural work index longstanding hierarchical boundaries that separate the domesticated from the wild, the primitive from the modern, and the developed from the "underdeveloped."[14]

Where agricultural development initiatives have looked to international agricultural science and capitalist markets as sources of "improved" seed, research in critical agrarian studies and farmer seed systems has underscored the varietal experimentation carried out by farmers across sub-Saharan Africa.[15] This body of work foregrounds the science and economy of already existing farmer practice within a policy environment that emphasizes the need for scientific and economic intervention. At a time when intellectual property law is increasingly embraced as a tool for agricultural development, research on farmer seed systems can be understood as an exercise in making the varietal innovation of farmers—long effaced by colonialist agricultural policy and claims of human difference—visible to agricultural science, economic development projects, and intellectual property law.[16] It is within this representational terrain, wherein farmer practice has moved from the denigrated and effaced to the (more) celebrated and articulated, that I am interested in the forms of labour that remain or are rendered invisible, even as the work of African farmers is increasingly recognized as innovative in contests over intellectual property.

FROM GENES TO JINN

The Gambia's district of Jarra West contains vast stretches of lowland tidal rice swamps. From the nineteenth into the early twentieth century, this region experienced four major interlinked transformations: widespread conversion to Islam, the spread of British colonial rule, the rise of commercial groundnut production, and the expansion of rice cultivation.[17] The shifts in agricultural production also fell along gendered lines, with men expanding groundnut cash cropping and women extending rice production, primarily for foodstuffs, into the lowland swamps. Those gendered labour relations largely persist today, and women's rice cultivation, unlike men's groundnut fields, depends upon immense varietal diversity. Rice fields in Jarra West's northernmost town, Jenoi, are closely clustered, often planted with more than one variety and, in some instances, sown with admixtures of two or more varieties. In 2010, farmers in Jenoi were cultivating at least fifty-four varieties of rice, as compared to three of peanut.[18] Of these rice varieties, about a third were introductions—traceable to colonial seed distributions, seed relief programs of the 1970s and 1980s, or twenty-first century agricultural development projects. Many of the remaining varieties were of local and regional provenance, named after people who had developed them or the places from which they hailed.[19]

There are many backstories to the development of rice varieties in West Africa—The Gambia included. Some of the most extensive research on farmer experimentation has featured rice farmers who cultivate the rice agroecosystems stretching from Sierra Leone to Senegambia.[20] In The Gambia, research on contemporary practices of seed management has shown, in terms translatable to the science of plant breeding, that rice farmers select and develop unique "off-types" (plants that possess different characteristics from their surrounding population) into new rice varieties.[21] In building a case that farmers are experimenters and innovators of new crop diversity, this body of research has highlighted agronomic strategies used by farmers that encourage gene flow and hybridization within rice—practices that generate the variability upon which varietal selection can proceed. These strategies include the close planting of varieties with similar flowering times, the acceptance of some admixture in seed stocks, and the intentional cultivation of admixed varieties.[22]

While all of these planting strategies are used by farmers in Jenoi, farmers explained that unique off-types appear under specific circumstances, conditioned by the actions of both people and plants. Most explained it thus: Allah (God) sends jinn, or spirits that sometimes take human form, to sprinkle off-types into farmers' fields. Off-types thus possess a gift-like quality. They may be sent to farmers who display hard work, generosity, and kindness, or they may be thrown into rice fields that are exceptionally beautiful. Although this beauty is partly the result of human hands, it is also the product of rice plants themselves. While rice "sometimes just changes," many women explained that rice senses the beauty of its *Oryza* neighbours and works to become lovelier, especially in cases where a woman cultivates admixed varieties. The presence of off-types thus signifies spiritual recognition of a human life lived well and the hybrid beauty of rice fields. In many parts of The Gambia, to have such an amicable relationship with jinn can indicate spiritual (and hence social) power and can bolster claims to land.[23] In this sense, to speak of receiving gifts of off-types from God and jinn is a powerful social and spiritual statement about moral personhood, ethical practice, and agricultural skill that activates broader forms of social, spiritual, and economic value.

From gifted off-types, a farmer may choose to select and develop a new variety. Farmers used the Mandinka word *tomboŋ* (lit., "to pick out") to describe the physical act of selecting one type of seed out of another. But these same farmers discussed the labour of turning an off-type into a new variety as a process of making varieties organized (in Mandinka, *topatoo*) and well-mannered (*kúluu*). Both *topatoo* and *kúluu* relay an element of discipline and care, much as one would tend a stubborn or naughty child. And much like child-rearing, making a wily rice off-type well-mannered is hard work that it is often, though not always, done by women.[24] Over many seasons, a farmer tests what the nascent variety needs and likes: where it grows, when it ripens, and how it behaves. She selects, tends, observes, culls, saves, and replants.

Farmers—even those who had never done varietal selection—were adamant that *topatoo* and *kúluu* require extended bouts of work from one person. Only when the variety is trained and the seedstock is sufficiently large is it given to others. It is in the distribution and widespread planting of this seed that a variety fully develops (*yiriwaa*). But people also spoke of how, in the act of sharing (whether as a gift, exchange, or sale), the trainer of a new variety might gain a floral namesake—which was a source of fame. Faburama maanoo, Aja Jonkong maanoo, Sherifu maanoo—these are all rice varieties in Jenoi, some lost and some extant, that carry the name of the persons who selected it. In this way, off-types, which are invited by the doings of farmers and plants, created by Allah, scattered by jinn, trained by a few, and shared by many, emerge in the world as fully fledged new crop varieties.[25] As plants become varieties, people may become famous. Plants and people emerge together.

To talk to farmers about varietal innovation, then, means talking about many different kinds of physical, spiritual, and affective labour that, in different ways, is invisible (figure 12.1). Some of the labour is "invisible" because it is carried out in the margins of fields or the recesses of personal seed stores, such as the mundane human work required to make a plant variety well-mannered. Still other parts of this labour are "invisible" because, for some, they are hard to see. They require a certain kind of skilled vision or are discernible only by proxy. Quasi-surreptitious contests among beautiful rice or gifts of seed that arrive when no one is looking are best discerned in the arresting beauty of a lush field or in the finding of an entirely unknown, novel off-type during harvest.

Yet beyond Jarra West, some of this labour is invisible because it appears to exist outside of models of mutation, gene flow, or hybridization offered by plant science. Even as studies of farmer plant breeding find that the outcomes achieved by Gambian farmers are "not much different from formal crop development," they also have

Figure 12.1. Some of the invisible labour of which I write. Beautiful and closely-clustered rice fields ripen in shades of black and gold, with some plots still shimmering green. Can you see their beauty? In the foreground sits a pair of shoes that were left on the swamp's edge by a woman, now invisible on the distant horizon, who has gone to check on her rice plot. Photograph by author.

held that farmers, citing jinn and not genes, "may not have a clear understanding of what exactly causes change in rice."[26] In these analyses, the work of God, jinn, and rice is rendered invisible because it seems to exceed what anthropologist Marisol de la Cadena refers to as "the limits" of the real.[27] Under such an accounting, the actions of plants and spirits lurk in the realm of impossibility or, at best, only come into view under the rubric of cultural belief: Farmers may believe that God, jinn, rice, and their own social actions are enmeshed in the emergence of novel off-types, but "in fact" those very farmers foster the conditions that produce off-types through their management of gene flow, hybridization, and varietal admixture. In such a translation, the enmeshed, relational "causes" that are God, jinn, plants, and sociality are enrolled into the explanations of the world as understood by formal plant breeding. In making some labour more visible, this translation elides the affective ecologies that shape seed emergence (see Chacko, chapter 16) and the meaning that is produced in the process.[28]

To be clear, I am not arguing against studies focused on farmer-as-plant-breeder. Rather, I want to make a case for paying close attention to all of the forms of work that enable varietal development: admixing, interplanting, hybridizing, selecting, hard work, generosity, gifting, and beauty contests. Of course, such a project wields its own elisions. Bringing the contributions of God, jinn, and plants into the fray of varietal development gives and redistributes credit for a particular kind of creative work. Does this not have the potential to dampen the sheen of human creativity within a context where that creativity was long denied? Answering this question has forced me to consider all the other ways of being that become more visible in bringing these more-than-humans into the account. It brings me to questions of signification, meaning, and value.

FROM AGENTIVE TO AFFECTIVE LABOUR

In his article "Inventing Seed," Thom van Dooren reflects on how intellectual property law turns upon distinctions between invention and discovery. He shows how, in a context where innovation is prioritized as a tool for agricultural modernization, research that details the innovative agricultural practices of farmers is a way to make visible labour and practices that were long erased by colonial science and development policy. Where the drive to spur innovation hinges on the creation of new property regimes, to show how farmer varietal selection is commensurable with formal plant breeding is a "potent political claim" that can enable a more equitable distribution of rights and benefits under the law.[29] Making farmer labour visible to agricultural science, in other words, has provided leverage for making their innovations more visible to intellectual property law.

At the same time, van Dooren argues that "efforts to simply extend 'cultural' influence (creative ability) to indigenous peoples and peasant farmers continues to position 'nature' . . . as a passive background in which nonhumans possess no meaningful agency."[30] Rather, for van Dooren, acknowledging human and more-than-human contributions to varietal development would enable a more thorough critique of the law, because it would necessitate "an argument against the philosophical (and legal) coherence of IP frameworks" that regulate property in plant varieties.[31] It would reveal, in other words, the artifice of plant invention, making it harder to foreground the work of the plant breeder "at the expense of nonscientist (including nonhuman) labour."[32] Such a disruption would be one step towards building a case against intellectual property in plants. At the very least, recognizing the "agency" of more humans and of nonhumans could be a potent argument against the most exclusive iterations of the law.[33]

But what is this agency? How is it to be understood? For farmers, is it the work of *tomboŋ*, *topatoo*, and *kúluu*, roughly translatable into selection practices used in some plant breeding? Or is it also the relational work of generosity and kindness that invites gifts of off-types? And for plants, is it the mutability of genes and the recombinative power of hybridization? Or is it the friendly competition of rice beauty contests and affective gifts from God and jinn? While van Dooren does not specify, farmers' accounts from Jarra West are a reminder that "agency" can be reckoned in multiple ways: it can be read through an endless number of lenses, and how it gets read can be just as consequential as whether it is read. Indeed, when plants were first incorporated into intellectual property law in the United States—in the form of plant patents—one of the major deliberations facing lawmakers was over how plants could qualify as inventions when nature so obviously "played a key role in the creation of new plant varieties."[34] As legal historians Alain Pottage and Brad Sherman have shown, the law addressed this question not by cutting nature's work out but by redefining what counted as an invention. Plant breeders became inventors in the context of their ability to recognize and reproduce nature's own creations. And thus breeders and plants (and arguably plant labour) found their first place within the growing remit of intellectual property law.

Allowing more-than-human labour into the fray does not necessarily seem to stave off the encroachment of private property. This is because the mere recognition of agency—without attention to relationality, sociality, and place—flattens the contributions of more-than-humans. It is possible to say that rice is agentive because it hybridizes and changes, and at the same time, it is possible to stubbornly refuse to see such changes as gifts, invited by hard work and beauty. As work by Indigenous and feminist scholars has shown, more-than-human agency read this way—firmly within the limits of the "real" according to plant science—erases local epistemes and embodied governance.[35] In such a register, previously effaced work and contributions come into focus, but not the meanings, obligations, and relationalities that emerge in the process of turning gifted off-types into well-behaved new varieties. Thus when farmers describe the contributions of God, jinn, and rice, they are not only offering explanations of practices otherwise translatable as hybridization, random genetic mutation, or gene flow. Rather, they are talking about ways of being and doing that produce plants as well as relations, obligations, and people.[36] In other words, farmers are describing both biological emergence and their economy of innovation. Their accounts of new biological matter thus *matter* socially, economically, and ethically.

Perhaps it is telling that, in Jarra West, the relative invisibility of difficult tasks such as *topatoo* and *kúluu* stand in stark contrast to the potential fame conferred by a varietal namesake. Farmers who had disciplined a new variety often spoke of the reward conferred by naming. Naming is a testament of hard work, generosity, and gifts from God and jinn. It also makes certain types of varietal labour more visible—it amplifies some tasks over others—and in doing so, it works as a conduit through which praise and prayerful wishes from other farmers flow back to the human namesake. This conduit can operate even between two people who have never met. If you plant Sherifu maanoo and you love it and you give thanks for it, the blessing travels back to the namesake: God keeps track of that accounting. Varietal naming in Jarra West, like the concept of invention, is thus a way of foregrounding the work of some over others. But unlike inventions in intellectual property, where attribution still goes to the human, farmers who select varieties consistently point to many other human and more-than-human contributors. Acknowledging the contributions of God and jinn shifts the meaning of what it means to have an off-type (it is a gift for hard work and generosity!) and what it means to select and discipline it (it is blessed hard work!). It brings to the fore the work and deeds that happened long before the plant came to be.

Paying attention to the more-than-human contributions of varietal development provides space to think about all the different relations that constitute "plant breeding." How these relations are made visible within accounts of farmer experimentation is consequential for the types of political and economic relations that are imagined as being vital for scientific innovation. At a time when intellectual property law is increasingly embraced as a tool of innovation, not paying attention to these contributions—and their signification—continues to render invisible the value-practices that shape both biological experimentation and the economic worlds that sustain it.

NOTES

1. Portions of this chapter draw on material presented in a previously published article, "To Make One's Name Famous: Varietal Innovation and Intellectual Property in The Gambia," *American Ethnologist* 45, no. 4 (November 2018). This work is indebted to the insights and generosity of the people with whom I work in Jenoi, especially Kajally Samura and Fanta Dibba.

2. Marianne Elisabeth Lien, Heather Anne Swanson, and Gro B. Ween, "Introduction: Naming the Beast—Exploring the Otherwise," in *Domestication Gone Wild: Politics and Practices of Multispecies Relations*, eds. Heather Anne Swanson, Marianne Elisabeth Lien, and Gro B. Ween (Durham, NC: Duke University Press, 2018), 9–10.

3. Gabriela Soto Laveaga raised questions of erasure during the workshop for the papers in this section. My attention to erasure is indebted to her insightful comments.

4. Raj Patel, "The Long Green Revolution," *The Journal of Peasant Studies* 4 (2013): 37–42.

5. David J. Reece, Daniel N. Dalohoun, Essa Drammeh, Paul Van Mele, and Saidu Bah, "The Gambia: Capturing the Media," in *African Seed Enterprises: Sowing the Seeds of Food Security*, eds. Paul Van Mele, Jeffery W. Bentley, and Robert G. Guéi (Wallingford, UK: CAB International, 2011), 109–32; Amadou Jallow, "Farafenni Hosts Second National Seed Fair," *The Point*, June 19, 2019.

6. ARIPO (African Regional Intellectual Property Organization), *Arusha Protocol for the Protection of New Varieties of Plants* (Arusha Tanzania, 2015).

7. Marilyn Strathern, "Cutting the Network," *Journal of the Royal Anthropological Institute* 2 (1996): 523–25; Thom van Dooren, "Inventing Seed: The Nature(s) of Intellectual Property in Plants," *Environment and Planning D* 26 (2008): 676–97.

8. My use of racializing assemblages here comes from Alexander Weheliye, *Habeas Viscus: Racializing Assemblages, Biopolitics, and Black Feminist Theories of the Human* (Durham, NC: Duke University Press, 2014), 4.

9. Sara Berry, "The Concept of Innovation and the History of Cocoa Farming in Western Nigeria," *The Journal of African History* 15 (1974): 84; Joanna Davidson, *Sacred Rice: An Ethnography of Identity, Environment, and Development in Rural West Africa* (Oxford: Oxford University Press, 2015).

10. Rebecca Herzig and Banu Subramaniam, "Labour in the Age of 'Bio-Everything,'" *Radical History Review* 127 (January 2019): 104–5.

11. Sara Ives, "'More-than-Human' and 'Less-than-Human': Race, Botany, and the Challenge of Multispecies Ethnography," *Catalyst: Feminism, Theory, Technoscience* 5 (2019): 2; Judith Carney, *Black Rice: The African Origins of Rice Cultivation in the Americas* (Cambridge, MA: Harvard University Press, 2002).

12. Jemima Pierre, "The Racial Vernaculars of Development: A View from West Africa," *American Anthropologist* 122 (2019): 87.

13. Elaine Gan, "Sorting Seeds into Racialized Futures and Pasts," *Catalyst: Feminism, Theory, Technoscience* 5, no. 2 (December 2019): 2.

14. Lien et al., "Naming the Beast," 10–13.

15. Ian Scoones and John Thompson, "The Politics of Seed in Africa's Green Revolution: Alternative Narratives and Competing Pathways," *IDS Bulletin* 42 (2011): 7–9.

16. Rosemary Coombe, "Works in Progress: Traditional Knowledge, Biological Diversity, and Intellectual Property in a Neoliberal Era," in *Globalization Under Construction: Governmentality, Law, and Identity*, eds. Richard W. Perry and Bill Maurer (Minneapolis: University of Minnesota Press, 2003), 19–20; van Dooren, "Inventing Seed," 676–97.

17. Bala Saho, "Appropriation of Islam in a Gambian Village: Life and Times of Shaykh Mass Kah, 1827–1936," *African Studies Quarterly* 12 (2011): 2–3.

18. In that year I interviewed sixty-three rice farmers (fifty-seven women) about crop repertoires.

19. Susannah Chapman, "To Make One's Name Famous: Varietal Innovation and Intellectual Property in The Gambia," *American Ethnologist* 45 (2018): 485–86.

20. See, for example, Paul Richards, *Coping with Hunger: Hazard and Experiment in an African Rice Farming System* (London: Unwin Hyman, 1986); Walter de Boef, Kojo Amanor, Kate Wellard, and Anthony Bebbington, eds., *Cultivating Knowledge: Genetic Diversity, Farmer Experimentation, and Crop Research* (London: Intermediate Technology Press, 1993).

21. Edwin Nuijten and Paul Richards, "Pollen Flows within and between Rice and Millet Fields in Relation to Farmer Variety Development in The Gambia," *Plant Genetic Resources* 9 (2011).

22. Edwin Nuijten, "Gender and Management of Crop Diversity in The Gambia," *Journal of Political Ecology* 17 (2010): 42–58.

23. Assan Sarr, *Islam, Power, and Dependency in the Gambia River Basin: The Politics of Land Control, 1790–1940* (Rochester, NY: University of Rochester Press, 2016), 97–98.

24. Although this work was sometimes done by men, women generally managed rice seed stocks.

25. Chapman, "To Make One's Name Famous," 487–90.

26. Edwin Nuijten, "Farmer Management of Gene Flow: The Impact of Gender and Breeding System on Genetic Diversity and Crop Improvement in The Gambia" (PhD diss., Wageningen Universiteit, Netherlands, 2005), 218–225.

27. Marisol de la Cadena, *Earth Beings: Ecologies of Practice across Andean Worlds* (Durham, NC: Duke University Press, 2015), 14–15.

28. Claims about "actual causes" are striking given that the source of biological novelty often elude the understandings and predictions of plant science. See, Helen Anne Curry, *Evolution Made to Order: Plant Breeding and Technological Innovation in Twentieth-Century America* (Chicago, IL: Chicago University Press, 2016).

29. Thom van Dooren, "Inventing Seed: The Nature(s) of Intellectual Property in Plants," *Environment and Planning D* 26, no. 4 (August 2008): 688.

30. Ibid., 688.

31. Ibid., 689.

32. Ibid., 690.

33. Ibid., 690.

34. Alain Pottage and Brad Sherman, "Organisms and Manufactures: On the History of Plant Inventions," *Melbourne University Law Review* 31 (2007): 539.

35. Vanessa Watts, "Indigenous Place-Thought and Agency amongst Humans and Non-Humans (First Woman and Sky Woman Go on a European World Tour!)," *Decolonization: Indigeneity, Education, and Society* 2, no. 1 (2013): 28–30; Zoe Todd, "An Indigenous Feminist's Take on the Ontological Turn: 'Ontology' Is Just Another Word for Colonialism." *Journal of Historical Sociology* 29, no. 1 (March 2016): 9.

36. Chapman, "To Make One's Name Famous," 489–91.

13

Coffee Breeders, Farmers, and the Labours of Agricultural Modernization

Stuart McCook

Coffee farming in Latin America was transformed by a gene, a simple mutation that was discovered in Brazil in the 1930s. The mutation produced "dwarf" coffee plants, which were smaller than traditional varieties but bore the same amount of fruit. Coffee breeders in Latin America used these dwarf plants to develop new, high-yielding coffee varieties. These varieties were designed to help farmers increase the productivity of their farms, by increasing yields per acre. They were the keystone of a new, intensive production system (a "technified" system, in the parlance of coffee experts), which included dense planting and the heavy use of fertilizers and other off-farm inputs. The breeders and other coffee technocrats saw technification (and increased productivity) as a self-evident good, which would unquestionably benefit farmers. But the history of technification in Latin America suggests that farmers did not always see the benefits in the same way. By exploring this gap between innovation (by scientists) and use (by farmers), we can build a nuanced picture of the invisible labour involved in technification.

The gap between coffee breeders and farmers is somewhat surprising, given how coffee breeding in Latin America was organized. In most of Latin America, coffee breeding was a public good. Until the 1990s, coffee breeding was conducted primarily by Latin American scientists. It was conducted in public or semi-public coffee institutes, operated by governments or producer associations. It was not directly influenced by foreign corporations or foundations. And yet critical aspects of coffee farming remained invisible, or only partly visible, to the breeders. The fundamental challenge was ideological. Like many plant breeders in the mid-twentieth century, coffee breeders took a technocratic approach to addressing larger social, political, and economic problems. They assumed that technified production was self-evidently worthwhile and that farmers would embrace it. They expected farmers to transform the way they produced coffee.

This blinded breeders to the tremendous amount of labour (both visible and invisible) necessary to transform a conventional farm into a technified farm and to sustain the technified farm once it was in operation. Constructing and maintaining an agricultural ecosystem built around dwarf coffees required the labour of many kinds of people, both on and off the farm. On the farm, this included the labour of uprooting the old coffee trees, clearing shade trees, planting the dwarf coffees, pruning them, applying the fertilizers and other inputs, and harvesting the larger crop. Off the farm, it included the nurseries that propagated and distributed the new coffee varieties; the agronomists and extension agents who tested plants and showed farmers how to use them; and the bureaucrats in government agencies and banks who provided the funding to support technification. These challenges, among other things, made many coffee farmers reluctant to technify their farms.

There was a second, more subtle kind of invisible labour at play, a long-term feedback loop between farmers and breeders. Farmers were not passive consumers of dwarf coffees specifically, or technification more generally. Their planting decisions were shaped by shifting political, social, environmental, and economic concerns—many of which were not initially visible to breeders. Based on those concerns, they made choices about whether to technify, when to technify, and how to technify their farms. By the 1990s, this had turned the coffeelands of Latin America into a patchwork of traditional, semi-technified, and fully technified farms. In light of these practices, breeders gradually modified their breeding programs to meet the changing needs of the farmers.

INNOVATION WITHOUT USE: BREEDERS AND DWARF COFFEES, 1930–1970

Scientific coffee breeding, based on the principles of Mendelian heredity, began in earnest in the 1930s. The world's cultivated Arabica coffee (*Coffea arabica*) rested—and continues to rest—on an unusually narrow genetic base. This species, unlike most species of *Coffea*, self-pollinates, which limits its genetic variability. Latin America's diverse coffee industries were all built on just two closely related varieties of Arabica: Typica and Bourbón.[1] In the nineteenth century, farmers in Brazil's vast coffee fields did, however, start finding new Arabica varieties. Most of these variations were cosmetic, displaying subtle differences in the size, shape, or colour of the leaves. None of these "off-types" offered any significant agronomic or commercial advantages over Typica or Bourbón, and so remained curiosities.

The first program for scientific coffee breeding in Latin America was established at the Instituto Agronômico de Campinas (IAC), in the heart of Brazil's coffeelands. The IAC had conducted pioneering coffee research since the 1880s. In 1930, the IAC organized a Genetics Section dedicated to using the principles of Mendelian genetics to develop improved coffees. It would become, and remains, the world's most important centre for coffee breeding. In the 1930s, scientists from the IAC began searching Brazil's coffee farms for interesting off-types that could be used in breeding programs. In 1937, an agronomist visiting a farm in the state of Espirito Santo noticed an unusual mutation of the Yellow Bourbón variety. The plant, later baptized "Caturra," was shorter than the traditional varieties, and its fruit grew more densely on the branches. Breeders later determined that its physical characteristics were defined by a dominant gene, which they later named the Ct gene. Caturra was commercially interesting because of its "low stature and the tendency it presented to produce large harvests."[2] The breeders had found a potentially important off-type.

Dwarf coffees did not yield significantly more coffee *per tree* than Typica or Bourbón. But they could be planted more densely than the traditional varieties, and so allow farmers to produce more coffee per unit area. In a rare discussion of labour, the breeders also noted that Caturra would be easier to harvest than the traditional "tall" varieties, which were so tall that pickers required ladders to reach the upper branches.[3] In field trials, however, the breeders found that Caturra was not hardy enough to withstand the droughts that frequently struck Brazil's coffeelands.[4] So they used Caturra as a parent plant in breeding programs aimed at producing a high-yielding dwarf coffee that was adapted to Brazil's environmental conditions. They crossed Caturra with the Mundo Novo variety (itself a cross between Bourbon and Typica), and then selected and tested the progeny of the initial cross over four generations. The Brazilian Coffee Institute (IBC) released the new dwarf variety, named Cautaí, for distribution in the early 1970s.[5]

These dwarf cultivars captured the imagination of coffee experts across Latin America. The IAC, which had developed Caturra and Cautaí, formed part of an informal global network of coffee research institutes that regularly exchanged seeds and seedlings of wild and cultivated coffee varieties. Some reports suggest that the Caturra variety had reached Guatemala in the early 1940s, just three years after it had first been collected in Brazil.[6] In the early 1950s, the IBC sent Caturra seedlings to the two other major coffee research centres in Latin America: Colombia's coffee research institute (CENICAFE) and the Inter-American Institute of Agricultural Cooperation (IICA) in Costa Rica. National coffee institutes in Central America, including Costa Rica's ICAFE, also received the new dwarf coffees.[7] By the 1960s, these dwarf coffees were available at coffee institutes across Latin America, but availability alone did not immediately trigger a broader revolution in coffee farming.

This lag could not be attributed to farmers' opposition to agricultural innovation. In Northern Latin America (i.e., Colombia, Central America, and Mexico), coffee farmers experimented with a wide range of technical innovations in the decades after World War II: other "tall" coffee varieties, chemical fertilizers, and shade management. Some farmers did experiment with the new dwarf cultivars. In Costa Rica, traditional varieties such as Typica were planted at a density of about 1,570 per hectare; the dwarf coffees could be cultivated at densities of more than 7,000 plants per hectare.[8] In Colombia, early experiments with dwarf coffees also looked promising: the 1969 *Colombian Coffee Growers' Manual* praised Caturra as being "well adapted to the various climates and soils of Colombia's coffee zones, and has achieved high productivity in the various departments."[9] For some coffee farmers in northern Latin

America—especially those with access to capital and technology—the new model of intensive production based on dwarf coffees was viable.

For most other coffee farmers, though, the volatile economic and geopolitical conditions of the 1950s and 1960s discouraged them from technifying their farms—or made it impossible for them to do so. Global production did boom briefly in the early 1950s, when prices spiked because global demand for coffee (mostly from post-war Europe) had exceeded global supply. But the boom lasted just a few years, and by the late 1950s coffee prices began to fall as global production once again exceeded demand. In the context of the cold war, stagnant demand and chronically low prices became a political problem. Coffee exports provided many of NATO's allies in Latin America with badly needed hard currency. To stabilize global coffee prices, the United States, its NATO allies, and the major coffee-producing countries brokered an International Coffee Agreement. The agreement sought to stabilize global coffee prices by balancing global demand with global supply, through a system of national export quotas.[10] Given these export quotas and satisfactory albeit stagnant global coffee prices, most coffee farmers had little incentive—and often few resources—to renovate their farms.

ALIGNING FARMERS TO DWARF COFFEES, 1970–1990

Two crises in the 1970s helped bring dwarf coffees into wider production across much of Latin America. In the 1970s, Brazilian agronomists detected the coffee leaf rust—a devastating crop disease—in the state of Bahia. They were unable to contain the disease, and over the next fifteen years it spread across Latin America's coffeelands. The epidemic presented more than just an ecological challenge; it was also a deep economic and political problem. Political leaders in Latin America feared that the coffee leaf rust could, by disrupting coffee production, trigger larger political unrest among impoverished farmers and labourers.[11] A second ecological shock—the Black Frost of 1975—killed billions of coffee plants in the Brazilian state of Paraná, which precipitated a global shortage of coffee for several years thereafter. In response, the International Coffee Organization suspended coffee export quotas, allowing all producing countries to export as much coffee as they wanted.[12]

These changed ecological and economic conditions opened a space for technification. Governments and coffee institutes encouraged farmers to technify their farms, as a way of managing the rust, and also taking advantage of the new export opportunities. Brazil's bureaucratic authoritarian government responded quickly to the rust. Breeders at the Instituto Agronômico de Campinas had been testing rust-resistant Arabicas since the 1960s, but they were not ready for widespread distribution. Once it became apparent that the rust could not be contained, the IBC recommended that farmers replace Typica and Bourbon with the dwarf Catuaí—which had just been released for general use a few years before—and a taller variety called Mundo Novo. Both of these improved varieties were highly susceptible to the coffee leaf rust, and so farmers would also have to use chemical sprays to manage the rust. But in principle the higher productivity of these technified farms—along with higher global coffee prices—would offset the cost of fungicides, and the fertilizers and labour necessary to keep the farm productive.

The rust triggered two critical changes that made technification viable. First, the rust epidemic gave farmers an urgent motive to change their farming practices. Second, and equally important, the coffee technocrats at the IBC deployed an army of agronomists and other experts to assess the infected farms and to encourage farmers to renovate and reinvigorate their farms. The IBC helped underwrite the cost of uprooting of unproductive coffee varieties, and the cost of new seedlings, fertilizers, and fungicides. IBC experts also provided Brazilian farmers with detailed advice on where and how to cultivate Catuaí and Mundo Novo.[13] These technification initiatives transformed coffee cultivation in Brazil; by the mid-1990s, about 90 percent of Brazil's farmers cultivated one of those two varieties.[14]

As the rust swept through Northern Latin America, coffee institutes sought to replicate the Brazilian model. As in Brazil, the epidemic helped create a space for dwarf coffees. Colombian coffee expert Jaime Castillo Zapata saw this crisis in moral terms; he described Caturra as a "catalyst for good [i.e. technified] agricultural practices."[15] In 1978, the Central American and Mexican coffee institutes organized a regional centre, PROMECAFE, to coordinate research on the modernization of coffee cultivation. National coffee institutes—especially Colombia's FEDECAFE, Costa Rica's ICAFE, and Mexico's INMECAFE—emulated the Brazilian model as far as they could. They propagated and distributed dwarf coffee seedlings and provided farmers with access to the agricultural chemicals and the technical advice necessary to renovate their farms. By 1980, about a third of Colombian farmers had technified their farms using Caturra. Those technified farms yielded more than twice as much coffee per acre than traditional farms and accounted for more than half of Colombia's coffee production.[16] By the late 1980s, Costa Rica's technification programs had made the country among the most productive in the world, at about 1,700 kg of coffee per hectare, more than four times greater than it had been at mid-century.[17]

By the mid-1990s, about 90 percent of Brazilian, 66 percent of Colombian, and 40 percent of Costa Rican farmers had fully technified their farms. But elsewhere the numbers were much lower. In Mexico, just 9 percent of coffee farmers had fully technified their farms—mostly in the state of Veracruz, where the Mexican Coffee Institute had been particularly active.[18] The Colombian scientist Jaime Castillo Zapata criticized farmers who had chosen not to technify their farms, complaining, "It was difficult to overcome the conservatism, to give it a name, without political implications, to backward mind-sets in technical matters. There are many enemies of the new agricultural practices."[19] Some prominent coffee experts still failed to grasp that the problem was not "conservatism" as such.

Most coffee farmers in smaller countries—with Costa Rica being a significant exception—did not have access to the technical, administrative, and financial infrastructures necessary to sustain intensive production. For example, the Guatemalan government "did train agronomists to assist with state diversification and technification," writes the historian D.C. Johnson, but "the pay was so low that attrition made the programs ineffective."[20] Many farmers—especially small farmers—were functionally illiterate and could not read the technical manuals and other documentation necessary to implement technification programs. In Guatemala and elsewhere, farmer outreach programs failed to reach the smallest, most vulnerable farmers. In places like Guatemala, El Salvador, and Nicaragua, large-scale technification programs were also impeded by chronic civil wars in the 1970s and 1980s.[21] Full technification was simply out of reach for the countless, typically poor, coffee farmers who had limited access to information, technical support, credit, seedlings, or inputs.

The decision to technify was not an all-or-nothing choice. Many farmers practiced what the geographer Robert Rice has called "semi-technified" farming, selectively adopting some facets of intensified production, but not others.[22] They tailored their technological choices to the ecological, economic, and social circumstances of their farms, in countless combinations and permutations. Some farmers continued to cultivate Typica or Bourbon but managed their shade more strategically and used chemical fertilizer and sprays as necessary. In places where the coffee leaf rust was less severe—especially on farms at higher elevations—farmers cultivated dwarf coffees but did not apply fungicides. Others wanted to take a more agroecological approach, focusing on longer-term economic and ecological sustainability, and on the *quality* of the coffee they cultivated.[23] By the late 1980s, the coffeelands of Latin America were a patchwork of fully technified, semi-technified, and traditional coffee production. The labour of farmers—and in some cases their alternative visions of modern coffee production—were made visible and concrete in the evolving shape and structure of the continent's coffee landscapes.

VISIBLE LABOUR: CONVERGENT CRISES AND NEW FORMS OF INTENSIFICATION SINCE 1990

A new set of shocks struck the global coffee industry starting in the late 1980s. These threw the plight of coffee farmers—especially smallholders—into sharp relief. After the Cold War ended, coffee's geopolitical importance declined. The International Coffee Agreement's quota system, which had sought to balance global supply and demand, collapsed after the United States withdrew in 1989. In the absence of export restrictions, several key producing countries—including Brazil and Colombia—dramatically increased production. As global production once again approached and sometimes exceeded demand, coffee prices entered a new cycle of booms and busts. During periods of low prices, many farmers struggled to cover their costs of production. Many responded to the crisis by de-intensifying production, to reduce the costs of inputs and labour. They applied fertilizers and fungicides less frequently, or eliminated them altogether. They delayed replacing aging, less productive coffee plants. Farmers were also grappling with new environmental problems: soil erosion, diseases and pests (including the coffee leaf rust and the coffee berry borer), and, increasingly, climate change. A growing chorus of voices—planters, development agencies, NGOs—began to criticize the dominant model of intensification as being ecologically, economically, and socially unsustainable.[24]

Breeders adapted their programs to these new circumstances, but they continued to promote technified production. Increased productivity could help offset the challenges of volatile coffee prices. They combined productivity with other goals focused on enhancing the economic, environmental, and social sustainability of the farms—one of which was rust-resistant coffees. In the 1990s and early 2000s, breeders released new rust-resistant dwarf coffees, which promised to reduce the costs of chemical control. These coffees were known as Catimors, based on crosses between Caturra and the rust-resistant Timor Hybrids. Colombia's CENICAFE released the first of these, known as the Colombia variety, in 1980. Over the following decades the variety was adopted primarily by farmers at lower altitudes where the rust was severe. In the 1990s, coffee institutes in Central America released localized Catimor selections, such as Oro Azteca (Mexico), IHCAFE 90 (Honduras), and Costa Rica 95. Some Catimors developed a reputation for poor cup quality—a significant disadvantage for farmers who produced for specialty markets, where quality was paramount.[25]

Breeders also placed increasing emphasis on cup quality, in response to the growing importance of Starbucks, and the high-value specialty coffee market. In the early 2000s, breeders in Brazil, Colombia, and Central America all released new generations of Catimors that combined high productivity, resistance to the coffee leaf rust, and competitive cup quality. Colombia's CENICAFE released an improved Catimor, the Castillo variety, in 2002. In a 2014 cupping competition, Castillo scored just as well as Caturra, which cuppers used as a benchmark. "Castillo can produce a good cup of coffee," wrote the coffee expert who had organized the trial, "and occasionally an extraordinary one. Against a backdrop of stubborn resistance in some segments of the specialty market to Castillo and other hybrid varieties, this is no small thing."[26] Still, Colombian farmers were relatively slow to adopt the Castillo variety until a massive outbreak of the coffee leaf rust struck Colombia in 2007. The National Federation of Coffee Growers, in co-operation with the national government, offered coffee farmers financial and technical support if they renovate their farms with rust-resistant dwarf coffees, especially Castillo. By 2017, more than 80 percent of Colombia's coffee farms had been renovated with rust-resistant coffees.[27]

But dwarf coffees do not dominate everywhere. A significant and persistent minority of Latin American farmers continued to cultivate traditional tall varieties such as Bourbon and Typica, as an agroforestry crop on mixed, ecologically diverse farms. By the 1990s, breeders were realizing that these traditional farmers had neither the resources nor the inclination to cultivate dwarf coffees. They have started developing new "tall" varieties tailored to agroforestry production, rather than expecting farmers to adapt the ecosystem to the variety. Colombia's CENICAFE released the first of these—the Tabi variety—in 2002. Tabi was tall rust-resistant coffee, a cross between the Timor Hybrid and Typica and Bourbon coffees.[28] In Central America, breeders at PROMECAFE (Central America's regional coffee breeding organization), in cooperation with the French agricultural research and development agency (CIRAD), used modern breeding techniques to develop varieties (F_1 hybrids) suited to shaded agroforestry systems. One of these, the *Centroamericano* variety, is a high-yielding, rust-resistant coffee that can produce good-quality coffee in an agroforestry system. It released in 2010, and has been slowly gaining favour among some farmers and in the specialty coffee markets. Still, these new varieties are not magic bullets; seedlings of these new cultivars can cost twice as much as a traditional arabica seedling, making the cost of renovating a farm much higher.[29] Nonetheless, these new varieties represent a significant paradigm shift in coffee breeding and the relationship between breeders and farmers. The "Brazilian" model of technification, based around dwarf coffees like Caturra and Catuai, expected farmers to reconstruct their farms—and their livelihoods—around the improved variety. This newer model reverses these priorities. The breeders have designed the varieties around the needs of the farmers and the existing farm structure. This is a tacit acknowledgement of the invisible labour that the dwarf coffees required for their growth and a recognition that this labour was beyond the reach of many farmers.

Dwarf coffees have replaced Bourbon and Typica coffee as the dominant varieties across many of Latin America's coffeelands. Since the 1950s, farmers have remade many of Latin America's coffee landscapes to suit the needs of the dwarf coffees. Caturra has become the standard benchmark for cup quality and productivity. Farmers in Brazil—the world's largest coffee producer—cultivate Mundo Novo, Cautaí, and newer dwarf varieties almost exclusively. In Colombia, most farmers cultivate Caturra, Castillo, or Colombia. In the rest of Latin America, dwarf coffees are widely cultivated, although not always as part of a fully technified system. The new F_1 varieties represent a potential paradigm shift, which offers farmers the promise of increased productivity while producing coffee in an agroforestry system. This paradigm shift—or at least the emergence of this alternative paradigm—reflects the often-invisible labour of farmers who have chosen not to adopt the now-dominant paradigm of technification. And in doing so, they have changed the ideology, and the practice, of coffee breeding and coffee research.

NOTES

1. A. B. Eskes and Thierry Leroy, "Coffee Selection and Breeding," in *Coffee: Growing, Processing, Sustainable Production: A Guidebook for Growers, Processors, Traders and Researchers*, ed. Jean Nicolas Wintgens (Weinheim, Germany: Wiley-VCH, 2004), 58–59.

2. C. A. Krug, J. E. T. Mendes, and A. Carvalho, "Taxonomia de Coffea arabica L.: II—Coffea Arabica L. Var. Caturra e sua forma Xanthocarpa," *Bragantia* 9, nos. 9–12 (1949): 157–63, https://doi.org/10.1590/S0006-87051949000300001.

3. Krug, Mendes, and Carvalho, 162.

4. A. Carvalho et al., "Genética de Coffea. XXVI. Hereditariedade do porte reduzido do cultivar Caturra," *Bragantia* 43, no. 2 (1984): 444.

5. A. Carvalho and L. C. Monaco, "Transferência do fator caturra para o cultivar Mundo Novo de Coffea arabica," *Bragantia* 31, no. 1 (1972): 379–99, https://doi.org/10.1590/S0006-87051972000100031.

6. Anacafé, *Manual de caficultura*, Third Edition (Guatemala: Anacafé—Asociación Nacional del Café, 1998), 30.

7. Mario Samper K., "Trayectoria y viabilidad de las caficulturas centroamericanas," in *Desafíos de la caficultura en Centroamérica*, eds. Benoît Bertrand and Bruno Rapidel (Agroamerica, 1999), 37; Federación Nacional de Cafeteros de Colombia, *Manual del cafetero colombiano*, Fourth Edition (Bogotá, Colombia: Federación Nacional de Cafeteros de Colombia, 1979), 13.

8. Samper K., "Trayectoria y viabilidad," 30–42.

9. Federación Nacional de Cafeteros de Colombia, *Manual del cafetero colombiano*, Third Edition (Bogotá, Colombia: Federación Nacional de Cafeteros de Colombia, 1969), 19.

10. John Talbot, *Grounds for Agreement: The Political Economy of the Coffee Commodity Chain* (Lanham, MD: Rowman & Littlefield Publishers, 2004), chap. 3.

11. Jonathan Morris, *Coffee: A Global History* (London: Reaktion Books, 2019), 141–46; Stuart McCook, *Coffee Is Not Forever: A Global History of the Coffee Rust*, Series in Ecology and History (Athens: Ohio University Press, 2019), chap. 7.

12. David C. Johnson explains this succinctly in "The International Coffee Agreement and the Production of Coffee in Guatemala, 1962–1989," *Latin American Perspectives* 37, no. 2 (March 2010): 40–41, https://doi.org/10.1177/0094582X09356957; On the coffee rust in Latin America, see McCook, *Coffee Is Not Forever*, chap. 8.

13. On the initial responses to rust in Brazil, see Jonathan Earl Coulis, "Marching Rows of Coffee: The Pursuit of Modern Agriculture in Brazil, 1950–1990" (PhD diss., Emory University, 2019), chap. 3.

14. Coulis, "Marching Rows of Coffee"; Eskes and Leroy, "Coffee Selection and Breeding," 75–76.

15. Jaime Castillo Zapata, "Mejoramiento genético del café en Colombia," in *50 años de Cenicafé 1938–1988: conferencias conmemorativas*, eds. Federación Nacional de Cafeteros de Colombia and Centro Nacional de Investigaciones del Café (Chinchiná, Colombia: Centro Nacional de Investigaciones de Café, 1990), 51.

16. Castillo Zapata, "Mejoramiento genético del café."

17. Samper K., "Trayectoria y viabilidad," 50.

18. Robert A. Rice, "A Place Unbecoming: The Coffee Farm of Northern Latin America," *Geographical Review* 89, no. 4 (October 1999): 554–70.

19. Castillo Zapata, "Mejoramiento genético del café," 48.

20. Johnson, "The International Coffee Agreement and the Production of Coffee in Guatemala, 1962–1989," 41.

21. Johnson, 42–43; Samper K., "Trayectoria y viabilidad," 43–61.

22. Rice, "A Place Unbecoming."

23. Andrea Montero, Marc Badia-Miró, and Enric Tello, "Geographic Expansion and Intensification of Coffee-Growing in Costa Rica during the Green Revolution (1950–89): Drivers and Outcomes," *Historia Agraria*, April 1, 2021, 27–29, https://doi.org/10.26882/histagrar.083e04m.

24. Steven Topik, John M. Talbot, and Mario Samper, "Introduction: Globalization, Neoliberalism, and the Latin American Coffee Societies," *Latin American Perspectives* 37, no. 2 (2010): 5–20, http://www.jstor.org.proxy3.library.mcgill.ca/stable/20684712.

25. McCook, *Coffee Is Not Forever*, chap. 8.

26. Michael Sheridan, "382. The Castillo-Caturra Cage Match," *CRS Coffeelands Blog* (blog), January 2014, http://coffeelands.crs.org/2014/01/382-the-castillo-caturra-cage-match/.

27. McCook, *Coffee Is Not Forever*, 179–84.

28. Germán Moreno Ruiz, “TABI: variedad de café de porte alto con resistencia a la roya,” *Avances Técnicos CENICAFE*, no. 300 (June 2002), http://www.cenicafe.org/es/publications/avt0300.pdf.

29. Benoit Bertrand et al., “Création et diffusion des variétés de caféiers Arabica: quelles innovations variétales?,” *Cahiers Agricultures* 21, no. 2–3 (2012): 77–88, https://doi.org/10.1684/agr.2012.0547; Benoit Bertrand et al., “New Varieties for Innovative Agroforestry Coffee Systems,” in *The Agroecological Transition of Agricultural Systems in the Global South*, eds. François-Xavier Côte et al. (Versailles, France: AFD, CIRAD, Éditions Quae, 2018), 161–76, http://agritrop.cirad.fr/592978/.

III

PROCESS

Commentary: Invisible, Secret, and Social

M. Susan Lindee

I think in some ways the history of science as a field has been hampered and held back by the profound success of the enterprise we study. Medieval historians understood long ago that peasants mattered if the goal was to understand the past. Historians of industrialization knew long ago that shop floor workers were a critical part of the story. Labour historians don't just work on CEOs.

But in our case, scientists, engineers, and physicians built a system that explicitly devalued the perspectives of the technical equivalent of peasants, and historians of these systems have been, generally, persuaded (either self-aware or not) that these experts were right. Certainly most historians of science have been convinced that they should tell the kinds of stories that experts would find legible, the kinds of stories experts have generally told themselves about the processes of making knowledge.

We can recognize the practical incentives for experts to value their own special status as knowledge-makers. We need not take these incentives too seriously as a guide *for ourselves*, as scholars interested *in history* and in how systems work, how the gears turn, how knowledge is negotiated into being. And when you look closely at any scientific project you see a lot of people—often people who matter a lot, and who do not necessarily appear in conventional historical accounts. These include people who explain their medical conditions to experts, share their family Bibles to reconstruct genealogies, guide visiting experts to interesting field sites and organisms, and so on.

I remember my surprise decades ago when I realized how much serious work—how much time and effort—had been devoted by physician and PhD geneticist James Neel to recruiting and working with a large network of midwives in Hiroshima and Nagasaki. His goal had been to track the possible genetic effects of radiation in the offspring of those exposed to radiation from the two atomic bomb attacks. The midwives were the first line of data collection—the first people who would examine a newborn and assess whether that person needed to be brought into the laboratory. They were a fairly important part of the scientific project, and Neel himself worked to gain their trust and cooperation. He worked with them to explain the goals of the study and to negotiate compensation, rules, and protocols.

The midwives were part of the scientific project, though in my own work (drawing only on U.S. records), I was unable to assess how much they enthusiastically cooperated, and how much they acceded to the demands of a powerful scientist and scientific organization from the United States. But even in those records it was clear that they did a lot of knowledge work—they decided what to report, what qualities of an infant merited attention, what aspects of family history the U.S.-funded Atomic Bomb Casualty Commission (ABCC) needed to know. They even faced possible repercussions for reporting an infant, because many families did not want abnormality to come to the attention of the ABCC, and because abnormality in a newborn could be a source of shame for the family.

What they reported was quietly woven into the technical accounts of radiation risk—and eventually to international radiation protection standards. The midwives were significant knowledge participants in the effort to track genetic effects, and while they were not given author credit, or even much public gratitude, a historian who was interested in the ABCC and how its scientific projects worked would be unable to leave them out of the story, because they are there, meeting on Hijiyama Hill and filing reports and complaining about policies.[1] My point is that their labour might have been "invisible" in some sense, but it was right up front in the formal documents that were collected and preserved in the archives.

When I started looking at the development of human genetics after 1950, again I found many invisible informants, including Amish families who let human geneticist Victor McKusick draw information from their Bibles. McKusick reanimated family histories in new ways so that they became resources for understanding a form of dwarfism called Ellis van Creveld syndrome. In this case, Biblical genealogies became biological records. Similarly, parents explained what Familial Dysautonomia (FD) was like when the doctor was not there, because those with FD responded to the stress of the patient experience in ways that made their condition more difficult to understand. Parents experienced FD—they knew its late-night effects—in ways that were impossible for physicians to intuit. Like the midwives, they were inside the story of interpreting this complex situation of human vulnerability.[2]

I mention my own research experiences to make the point that our attention to invisible labour is fundamentally about accuracy. While it might have elements of a justice project, in the sense that it allocates "credit" to persons heretofore uncredited, for us as historians its importance is different. We want to understand how knowledge is made, and that means noticing a range of actors and resources. The everyday processes of making new knowledge involve many more people than have usually been recognized.

We should be interested in the process of explicitly or implicitly cutting people out of the historical picture. Whitney Laemmli's dancers, filmed in action for various reasons, were oddly assumed to be doing nothing relevant to their own personal identity, and therefore below the bar for recognition (chapter 18). Elena Aronova's peasant–scientist tried to carve out a space for his own ideas, while larger changes in the field provoked by the bomb defined his forms of knowledge as irrelevant (chapter 14). And Xan Chacko's seed bank curators conduct painstaking, perhaps non-charismatic, work that cannot compete for attention with freezers and polar bear security systems (chapter 16). What is cut out, eliminated, and de-emphasized can be revealing, the absence itself, a form of evidence.

In 1987, sociologist Arlene Kaplan Daniels coined the term "invisible work" to describe culturally devalued women's unpaid labour—including caring for children, housework, and volunteer work.[3] Later scholars added reproductive labour, emotional labour, and care work. More recently, the term has been extended to "dirty work" including sex work, the work deaf people perform to communicate with the hearing, and the labour that medical patients undertake in managing their own health care.[4] As this suggests, scholars have characterized many kinds of labour as invisible, even when very different mechanisms produce the invisibility.

But the labour we explore has another dimension. It is epistemologically invisible.

The wonderful chapters I consider in this commentary all focus gently on the processes involved in dividing up the work of science and choosing which aspects of that labour will attract *attention*, which is kind of the opposite of invisibility. Each of these chapters opens opportunities for discussion, teaching, and scholarly intersections. In seed banks, lost-language projects, paleontological labs, blood donation centres, citizen's seismology centres, and dance archives, forms of labour emerge that seem peripheral and that are instead absolutely central, if one is asking how power and knowledge work. Peripheral status is not an inherent quality of any form of labour. Rather, it is an outcome produced by systems that situate individuals and define their meanings in knowledge networks.

The challenges of casual or indifferent archiving, or of secrecy concerns, or of privacy laws in some ways validate the core insight of attention to invisibility: Some kinds of labour are structured to be invisible and are not recorded for purposive, meaningful reasons, rather than because it is in some simple sense unimportant.

The newer literature on secrecy in science, for example, in the work of Alex Wellerstein and of Brian Balmer, suggests how attention to strategies of invisibility can be informative precisely because they show how structures are enforced.[5] Wellerstein explores how and why secrets relating to the atomic bomb took particular institutional and ideological forms and what we learn as we follow secrecy regimes into the present. He shows that Manhattan Project managers did a pretty good job of keeping things secret—he calculates that only a few thousand of those working on the bomb knew what they were building, something close to 1 percent of those employed—so invisibility here was both internal and external. And this secrecy could work against broader goals: General Leslie Groves' well-known insistence on compartmentalization, so that workers across the sprawling Manhattan Project only knew what they "needed" to know, actually facilitated espionage. Groves cared more about controlling rumours and domestic reporting by journalists than he did about possible spies (Wellerstein says there were nine spies in the project), and the two issues (spies vs. domestic rumors) required different strategies. Since Groves did not expect spies but was deeply suspicious of Congress and of local journalists in New Mexico, he focused on the domestic rumors. This account motivates us to ask: what needs to be invisible, and to whom?

Brian Balmer's account of "shadow data" proposes that a shadow is more than something missing. It is a sign of something that cast it, something solid and real so to speak. "If the metaphor of the 'shadow' is to have analytical bite, then it must amount to more than just a substitution for more direct terms such as missing, absent, and neglected," he notes.[6] A shadow signifies something about whatever cast it. He also points out that people occupy different relationships to what is secret, or shadowed, or open. People inhabit different worlds with respect to data and sometimes secrets are hidden by being placed in plain sight—as in camouflage, or an open day at a military establishment.

My own work has considered how military projects in the Cold War fused "military" and "civilian" purposes, operating in the half-light of both publicity and secrecy. The under-ice warheads at the U.S. Army's Camp Century in Greenland were secret, but a lucky Boy Scout won a visit to the underwater village there, with its hot showers fueled by a nuclear reactor.[7] Nuclear missile silos, bases, plutonium production, or testing programs were physically dispersed in places expected to preserve secrecy and isolation. Meanwhile, radioactive wastes contradicted these expectations, moving out into the world, sending "messages" to the entire planet. It was a physical performance of the publicity/secrecy spectrum, as radioactive materials showed up everywhere, in children's baby teeth, fish, topsoil, photographic plates, reefs, and rain. They spread through water and air, invaded human and animal bodies, travelled through the food chain, settled in soils, and accumulated in streams and rivers. The Cold War remade the world physically as a bomb site and missile base, and the traces of these actions persist in contamination all over the world: What happened then will effectively never be secret, as the contamination promises to outlive the species that produced it.[8] Historian Robert Jacobs proposes that "radiation makes people invisible," by which he means that global *hibakusha* (people exposed to radioactivity) are consistently written out of the story of global weapons testing, forced to give up their land, exposed to radiation risk, discriminated against, and misled by powerful institutions. The labour they have performed in the arms races since 1945 is purposefully invisible, not seen, not acknowledged.[9]

Meanwhile at the height of the Cold War, communications and political science scholars routinely obscured the defense origins of their funding, so that ideas about public policy nurtured using Central Intelligence Agency (CIA) or Department of Defense funding were recycled as neutral, academic social scientific research. Political scientist Harold Lasswell was heavily funded by the CIA and played a critical role in the field. He and other political science and communications scholars often simply reworded or retitled projects for public consumption so that their military relevance disappeared.[10]

This Cold War quality of open scientific data—publicly known, but with origins or elements that were fuzzy or concealed or vaguely disappeared, or with details that were left out—is part of the reason that the influence of defense interests on knowledge production in general has been so poorly mapped and incompletely understood.

As in military systems, the invisibility we track often involves things happening in plain sight—the dancers, for example; they are subject to selective invisibility, experienced by some, but not by others. Secrecy and invisibility operate in the same spaces—the not-necessarily-acknowledged work of institutions and systems.

Balmer invokes invisible labour, as he explores the moral economy of secrecy: The fact that data can be separated into ostensibly "open" and "closed" data has ethical dimensions that intersect with notions of credit, as both data and contributors can disappear in the process. This disappearance is often split along the lines of seniority, gender, or scientists versus invisible technicians.[11]

I hope this diversion into military secrecy can make some of the tensions in other systems easier to see. Secrecy and invisibility are both implicated in questions of credit, and both produced by very sophisticated, negotiated structures of knowledge-making. It is not an accident that some things are acknowledged and others are not, that some things are publicly known and others obscured. These processes are part of how the system works, and any accurate account of modern science must come to terms in some way with these elements of privileging and occluding, recognizing and subsuming.

Implicated first in these chapters is identity. Blood donors, preparators, peasant seismologists, and seed curators are likely to see themselves as "humble servants of knowledge." I think dancers and the carriers of first languages are active producers of both dance and language, but not necessarily with knowledge production as the central value shaping their experiences. Darwin proposed that breeders engaged in selection processes, of which they were unaware—their long familiarity with a particular organism could make some of the breeders' own knowledge labour, their biological insight, *invisible to themselves*. He called this *unconscious selection*—not selection through ideas or goals. So I will call the dancers and the Oneida speakers, following Darwin, unconscious humble servants of knowledge (for whom knowledge might not be the primary goal).

The next process that might be worth considering is compensation: who is paid for this labour? My colleague Harun Küçük, in his new book, *Science without Leisure* (2019), proposes that after 1660 income and compensation shaped the development of science in the Ottoman empire in profound ways—that Ottoman scientists became darkly practical (giving their time to compensated work relevant to agriculture, medicine, and engineering). This was partly because they could not survive on the meagre salaries paid for academic research in pursuing pure knowledge, and it had an impact on their ability to thrive as knowledge-makers. The triumphs of Ottoman natural philosophy were built on earlier eras of financial and institutional security.

So we ask who is invisible, but also, who gets paid? And how much? The fact that so many scientists we valorize from the past were independently wealthy, able to pursue natural knowledge with family resources, makes this twentieth-century story quite novel—we are looking at communities of knowledge in the midst of a century that is recalibrating what a life in science can and should look like—a life in which someone without financial resources can be a scientific expert. That really was a new option after 1880 or so.

Here Aronova's case is particularly interesting, because the Soviet state encouraged folk-knowledge systems, and he was a paid employee. Bangham's blood donors were not paid, but the lab assistants at seed banks, that Chacko explores, and those in paleontological labs that Wylie looks at (chapter 19) are paid workers, *on the job.* So too were the Oneida workers employed in the language project, including Andrew Beechtree and others whose work was eventually rescued from obscurity: the language project's Oneida workers included two women and nine men, who were paid for eighteen months.

I mean to suggest that compensation is a structure or process that might be more important than the papers considered here. No one makes it central, even though compensation might end up being the most important structural change in science after 1900 if we were writing with the long perspective as historians a few centuries into the future. From this long perspective, the transition from wealthy experts who are autonomous and reliable because they are men with resources to a more diverse paid labour force composed of compensated people along a wide spectrum might be the most profound change in technical knowledge systems of the last century. We cannot yet fully interpret its consequences, but it almost certainly matters for what we know and how we know it. Some attention to how money matters here might be generative.

There is also the process that I will describe in terms of revealing, publicizing, making known, or calling attention. This issue is threaded through all of these chapters. Some of the blood donors Bangham considers have their names attached to antigens as much for convenience and clarity as for credit, but these names are "immortal." As she notes, such names were mobile, transportable, and unique but also more "charismatic" (like the freezers in the seed banks). One expert called the named bloods "more personal." Privacy questions affect the blood donors in ways that they do not affect preparators or curators, or even folk scientists who might crave recognition and publicity. Chacko suggests that work can be left out of public accounts because it is almost tedious. Labour can also disappear because of secrecy concerns—both military and industrial. I raise the question of publicity and outward-facing science even though I am not sure how much it matters in all these stories.

Now I want to turn the chapters in this section and raise some specific questions about them, considering them in alphabetical order.

Aronova looks at institutions making room for "amateurs" or for labourers who lack the credentials expected by the scientific community. But the room is shadowed by their status or lack of it. Mannar provides a remarkable case study partly because he wrote so many letters and they survive. Stalinist science was supposedly participatory and open to alternative explanations —more open than its Western counterparts. And he insisted on his right to define his own terms and his own ideas about what it meant to be a scientist. I was particularly drawn to Aronova's discussion of the signs of risk that locals could detect, in animal behaviour, for example, and how this anecdotal knowledge could be leveraged for support. He even wondered if there were almost unconscious signs—things people knew without knowing they knew it. As seismology changed, and became Big Science, his approaches lost their appeal. Nuclear testing changed what it meant to do this kind of intellectual labour. Mannar stayed on the payroll of the Geophysical Institute, but his vision for the involvement of ordinary citizens was swept away by the forces of modern seismology. We usually think of this moment as a crisis in the political meanings of scientific knowledge—with Lysenko as the charismatic instigator of a disaster.[12] But Mannar is not contaminated, and his challenges to conventional earthquake science are not signs of the dangers and constraints of politics *on his side*; but rather on the side of those who focused on the earth sciences as a guide to underground nuclear testing.

The invisibility in Bangham's account (chapter 15) has a different moral dimension: blood donors are humble servants of knowledge who are presumed to value their own anonymity. Their bodies stand for "biology" and humanity—they are manifestations of the biological self—rather than of their own individual identity. Even when their blood is indeed special, it is universal, deployed to tell stories about universal health and human history. Part of what makes Bangham's account so compelling—both here and in her remarkable new book—is the way she captures this mixture of morality, sacrifice, humility, and transcendence. Blood donors are invisible because they are literally not present: their bodies are just archives, historical records of biological relationships, rather than personal ones, and they donate their bloods to science out of a pure, moral commitment to knowledge. These relationships of course are also colonial, racial, hierarchical, shot through with human cruelty. The racialized qualities of blood and blood donation and human biological history have been central to collecting human bodily materials here and around the world, and while this case is far less egregious than many others, the same forces still operate here—and in genomics today.

Chacko's seed banks are sites of global security and salvation science, places where discretion and caution (about locations and resources) is dictated by the seeming survival importance of seeds. In her case, the most invisible labourers, seed curators, are erased in favor of cold storage: a technological system "stands in" for the human labour of caring, sorting, and ranking seeds. Cold storage is not the only element in an effective and secure seed banking system, but it looms large in most accounts, and it justifies public funding. The vagaries of actual, quotidian human labour—complex and imperfect—might call into question the security of the banks by revealing something about contingency and uncertainty. Like librarians or bookbinders, the seed curators are doing the work that literally makes the banking possible.[13] Yet calling too much attention to their work, she suggests, could call into question the enterprise to which they devote their lives. In practice, these workers invite a consideration of possible ways of imagining scientific careers, in terms of different layers of work, responsibility, and contribution.

Kaplan (chapter 17) is also engaged with the preservation of precious resources, in her case languages, and she points out that many engaged with these languages assume that written inscriptions will be more powerful than oral traditions. I was reminded of a very different approach regarding Japanese atomic bomb survivors, all over seventy-five years of age, who have been teaching younger surrogates to repeat in spoken language their own unique stories of the bombings, so that the oral traditions of telling these stories of suffering and survival can persist, as a part of oral tradition. I was also reminded of one of the ideas for nuclear semiotics: one way to keep people out of contaminated sites would be to create a priesthood that would preserve the stories of how dangerous the sites are, telling them over and over again.[14] In her study, Kaplan considers labour that was compensated, appreciated, publicly acknowledged, and then "lost" as a result of circumstance, time, and war. The project itself combined active inscription by native speakers and formal linguistic study—though university students might have been seen as primary if the notebooks she considers had in fact been lost. She raises key questions about the tension between oral tradition and inscription that touch on the nature of language itself. Any speaker knows that language is resolutely an experience of sound—almost like dance, something visceral and embodied. The written text does not quite capture it. How the Oneida experts thought about their roles in the project, as both speakers and writers, seems highly relevant to this question.

In Laemmli's case, the dancers are literally hyper-visible—doing very hard work that requires significant skill, and they are at the same time *individually invisible*. This work is the explicit subject of the film archive of folklorist Alan Lomax. He thought that he was making their contributions explicit, clear, and recognizable, because their contributions were emphatically not "personal." They were, in his imagination, the bearers of kinetic health, and they performed human flourishing—all very high-minded and luxuriant ideas. He thought he would make the unseen seen, but his notion of seeing very strangely did not include individual identity, names, or credit. So what was there to be seen? Some kind of universal, abstract good in the movement of bodies? The old "personal is political" question illuminates the stakes in Lomax's project. The moral universe of dance as a guide to a proper life, as Lomax imagined his undertaking, seems to elevate morality "above" the personal. The dancers are symbols of the human, rather than individual workers.

Meanwhile Wylie's paleontological preparators seem to be *invisible by choice*. Wylie proposes that we commonly assume that invisibility indicates oppression. In many cases it does of course, but what she points out is that freedom from surveillance and observation—being invisible—can work to the advantage of some workers. I was reminded of Timothy Pachirat's study of industrialized slaughterhouses, where he found that workers in the most gruesome sectors of the slaughterhouse had more "freedom," because management generally avoided those rooms.[15] Managers did not want to be in the rooms where the cows were shot any more than workers did, but the managers' absence was an advantage for those stuck working there: not being noticed or surveilled had everyday workplace advantages.

Wylie's paleontological preparators also value their ability to operate more freely, though this freedom is rather different from that of the slaughterhouse workers. Her account also calls to mind the significant power of people who may not seem particularly powerful: Anyone who has served in a leadership role in an academic department comes to realize how much quiet power administrative staff wield. Their *below-the-radar labour* can have profound consequences for students and faculty. While they do not formally have control, they make decisions all the time that shape how well a department or school runs. Wylie proposes that highly skilled workers, as in her case study, can *prefer* to operate below the radar, protecting their slags of rock from scientists who may be unprepared to recognize how materials should be handled. When they really conflict with the palaeontologists, the palaeontologists have the last word, and being invisible is not an unmitigated good. At the same time, it has partial and contingent advantages.

As I have suggested here, secrecy and security have shaped what it means to be a technical expert. In a heartfelt July 1954 letter to a powerful atomic energy commissioner, Yale University biophysicist Ernest Pollard described how he had learned to keep secrets. "Many of us scientists learned the meaning of secrecy and the discretion that goes with it during the war," Pollard said. "We had very little instruction from outside." When the war was over, he made a conscious decision to avoid secret research. He thought carefully through the problems of secrecy and security and made the decision to handle only material that was entirely open. "I returned one or two documents I received concerning the formation of the Brookhaven Laboratory, in which I played a small part, without opening them." But the outbreak of the Korean War, in June 1950, and his own concerns about the Soviet Union, led to a change of heart. He came to feel that "I as a scientist should pay a tax of twenty percent of my time to do work that would definitely aid the military strength of the United States."[16] In the process, as he was engaged in secret research during the Cold War, he learned a form of extreme social discipline that he called "the scientists morality." "I have learned to guard myself at all times, at home, among my family, with the fellows of my college when they spend convivial evenings, with students after class asking me questions about newspaper articles, on railroad trains and even in church. It has been a major effort on my part, unrelenting, continually with me, to guard the secrets that I may carry."[17]

Pollard's comments resonate with those of many other experts in the heart of the Cold War in the United States. Being a scientist often meant concealing one's work and ideas from friends, family, students, and colleagues. An enterprise founded on an ideology of openness and free exchange became increasingly oriented around keeping secrets.[18] Individual scientists could lose their jobs if they lost their security clearances.[19] And security clearance could be withdrawn for a wide range of infractions, including accepting dinner invitations from people who were members of the Communist Party.[20]

Scientists even lost jobs for refusing to testify when called before the House Committee on Un-American Activities by Senator Joseph McCarthy (R-Wisconsin).[21] The physicist David Bohm, who lost his assistant professorship at Princeton for this reason, went on to make illustrious scientific and philosophical contributions under difficult circumstances in Brazil and later, in the United Kingdom.[22] His work, oddly enough, became less visible to his colleagues in the world of Princeton, but also more free and creative. Bohm is today considered one of the most important theoretical physicists of the twentieth century, not for his bomb-making skills but for his contributions to quantum theory and to neuropsychology. Does being forced into invisibility sometimes create new options and opportunities?

Situating invisibility in this nexus of social place and political order might be a path to understanding why it matters. Just as secrecy creates shadows, invisibility leaves traces, in the work itself and in the ways experts think. I would suggest that it calls our attention to the truly social nature of knowledge enterprises.

NOTES

1. My book on the ABCC explores this situation in some detail: M. Susan Lindee, *Suffering Made Real: American Science and the Survivors at Hiroshima* (Chicago, IL: University of Chicago Press, 1994).

2. These issues are explored in detail in various chapters of my 2005 book: Susan Lindee, *Moments of Truth in Genetic Medicine* (Baltimore, MD: Johns Hopkins University Press, 2005).

3. A. K. Daniels, "Invisible Work," *Social Problems* 34, no. 5 (1987): 403–15; see also Erin Hatton, "Mechanisms of Invisibility," *Work, Employment & Society* 31, no. 2 (2017): 336–51.

4. Hatton, "Mechanisms of Invisibility," 336–51.

5. Alex Wellerstein, *Restricted Data* (Chicago, IL: University of Chicago Press, 2021); Brian Balmer, "A Secret Formula, a Rogue Patent and Public Knowledge about Nerve Gas: Secrecy as a Spatial-Epistemic Tool," *Social Studies of Science* 36, no. 5 (2006): 691–722.

6. Brian Balmer, "Shadow Values and the Politics of Extrapolation," *Science, Technology, & Human Values* 42, no. 2, Special Issue: Data Shadows (2017): 311–14, 311.

7. On Camp Century: Kristian H. Nielsen, Henry Nielsen, and Janet Martin-Nielsen, "City under the Ice: The Closed World of Camp Century in Cold War Culture," *Science as Culture* 23, no. 4 (2014): 443–64.

8. M. Susan Lindee, *Rational Fog: Science, Technology in Modern War* (Harvard University Press, 2020).

9. Robert Jacobs, "The Radiation That Makes People Invisible: A Global Hibakusha Perspective," *The Asia Pacific Journal: Japan Focus* 12, issue 31, no. 1 (2014): 1–9.

10. Christopher Simpson, *Science of Coercion: Communication Research and Psychological Warfare* (New York: Oxford University Press, 1996).

11. Brian Balmer, "Shadow Values," 2017.

12. On Lysenko: Loren Graham, *Lysenko's Ghost: Epigenetics and Russia* (Cambridge, MA: Harvard University Press, 2016).

13. Robin Elizabeth Desmeules, "The Bookbinding of Hortense P. Cantlie for McGill Library: Surfacing a Legacy of Invisible Labor in the Stacks," *Libraries: Culture, History, and Society* 4, no. 2 (2020): 139–61.

14. Rosemary Joyce, *The Future of Nuclear Waste: What Art and Archeology Can Tell Us about Securing the World's Most Hazardous Material* (Oxford: Oxford University Press, 2020).

15. Timothy Pachirat, *Every Twelve Seconds: Industrial Slaughter and the Politics of Sight* (New Haven, CT: Yale University Press, 2013).

16. Lindee, *Rational Fog*, 205.

17. Lindee, *Rational Fog*, 205.

18. For a thought-provoking discussion of this question, see Peter Galison, "Removing Knowledge," *Critical Inquiry* 31, no. 1 (2004): 229–43.

19. Jessica Wang, *American Science in an Age of Anxiety: Scientists, Anticommunism, and the Cold War* (Chapel Hill: University of North Carolina Press, 1999).

20. See discussion of Steinberg in Lindee, *Rational Fog*, 213.

21. Wang, *American Science*, 1999; Kelly Moore, *Disrupting Science: Social Movements, American Scientists, and the Politics of the Military, 1945–1975* (Princeton, NJ: Princeton University Press, 2008); Sarah Bridger, *Scientists at War: The Ethics of Cold War Weapons Research* (Cambridge, MA: Harvard University Press, 2015).

22. Freire, "Science and Exile: David Bohm, the Cold War and a New Interpretation of Quantum Mechanics," *Historical Studies in the Physical and Biological Sciences* 36, no. 1 (2005): 1–34.

14

Citizen Seismology, Stalinist Science, and Vladimir Mannar's Cold Wars

Elena Aronova

This chapter explores the sometimes invisible and sometimes highly visible labour of non-professional contributors and interveners in scientific enterprise often referred to as "citizen scientists."[1] The term, "citizen science," is most commonly used to denote activist practices in low-income, indigenous, or otherwise marginalized communities exemplified by such programs as the environmental justice movement, participatory mapping in the developed West, or popular epidemiology in the developing world. This chapter examines citizen science through the case of a Soviet amateur scientist Vladimir Mannar. His case is particularly compelling because he left a rich documentary record of his own voice, found in the letters he wrote to the Central Committee of the Communist Party of the Union of Soviet Socialist Republics (USSR) in the summer of 1953, in the wake of Joseph Stalin's death.[2] Mannar's vision of "citizen seismology" reflected in his letters provides a vantage point from which to consider a dialectics of visibility and invisibility of non-scientific experts within the epistemological and ideological framework of the so-called Stalinist science.

Stalinist science refers to the features of the system of Soviet science during the period of Stalin's regime (1927–1953), which showcased the visions of science as expert-driven and, at the same time, more participatory and open than its Western counterpart. Mannar's case reveals a form of labour in science that defies such clear-cut binaries as visible versus invisible, professional versus non-professional, or even paid versus unpaid. Through Mannar's conflicted experiences of navigating the competing pulls of Cold War seismology, I examine the complex and tangled mix of political and epistemic sensibilities that motivated a vision of the citizen seismology in the late 1940s and early 1950s in the context of the postwar and Cold War transformations of Soviet science. Mannar's expressive written record highlights the tensions, conflicts, and mediations between scientific experts and a citizen expert who insisted on the right to define his own terms of his contribution to science.

AN AMATEUR SEISMOLOGIST

On March 9, 1953, four days after the death of Joseph Stalin, the Soviet leading authority, the Central Committee of the Communist Party of the Soviet Union (CPSU), received a telegram from the remote town of Bayram-Ali in Turkmen Soviet Socialist Republic in Central Asia. The correspondent, Vladimir Mannar, reached out to report about his modest contribution to celebrate Stalin's life and achievements: the establishment of a seismological station of his own. In the correspondence that followed, Mannar described himself as *amateur* (*любитель* in Russian) in the literal sense of the word: doing what he was doing for his love of the subject rather than for a living. As a teenager, he was attracted to science by the public lectures given in revolutionary Petrograd by the scientific luminaries such as the Russian zoologist Pyotr Yulievich Schimdt. Inspired, Mannar proceeded to study physics on his own: he enrolled in distant learning courses, started to read popular and technical scientific literature, and reproduced various physical experiments using homemade instruments of his own design.

Mannar became interested in seismology in the wake of an earthquake that struck the city of Ashgabat, the capital of Soviet Turkmenia, on October 6, 1948. The 1948 Ashgabat earthquake was one of the most devastating in Soviet history. The main shock razed to the ground the city of Ashgabat, killing thousands. Yet outside the main city, only few deaths were attributed to the earthquake. As Mannar reported, people in the countryside left their homes in advance, alerted by their animals' strange behaviour and other signs.

Mannar, for whom the Ashgabat earthquake was a life-changing event, began to collect the anecdotal knowledge and the experiences of local villagers about the earthquake precursors—the pre-seismic cues (or "proseisms"). Simultaneously, Mannar began studying professional and popular literature on earthquakes—"I purchased all seismological and geophysical literature I could find in the bookstores in the cities of Tashkent and Chirchik"—and designed simple instruments "to detect and monitor other possible pre-seismic cues, such as the release of gases and charged particles, and the change of animal behavior."[3]

In 1951, after three years of intense activities, which included collecting information from local villagers, recording various data using homemade instruments, and organizing local school children into a club of volunteer observers, Mannar summarized his findings in the report entitled "On Animal Behavior before the Earthquakes."[4] On the advice of the head of Tashkent's seismological station, Mannar sent his report to the Geophysical Institute in Moscow. The report, according to Mannar, was "warmly received." A seismologist at the institute, G. P. Gorshkov, encouraged Mannar to continue data collection and gave advice on how to build a seismograph and record different ground motions. Mannar was disappointed that his report was not accepted for publication in the annual publication series of the Institute. Instead, however, he was invited to Moscow to attend seminars held by the Geophysical Institute for the technical supervisors of the Institute's seismic stations around the country. Mannar was also offered hands-on practical training and learned to operate standard instruments used at the seismic stations at the time: horizontal and vertical seismometers with galvanometers and recording drums. After the crash course in seismology, Mannar was appointed as a technical supervisor of one of the Institute's seismological stations, "Mary," in his hometown Bayram-Ali, responsible for overseeing, timing, and troubleshooting the equipment of the seismograph vault and for transmitting the data from the station in Turkmenia to the Institute in Moscow.

The case of Mannar, while extraordinary in several respects, typifies the involvement of amateurs and volunteers in earthquake observations and data prospecting worldwide. As the historian Deborah Coen pointed out, in the nineteenth and early twentieth centuries, amateurs, while "invisible" in the published record of science, were widely cultivated to produce data and knowledge on environmental disasters, often with the support of the state.[5] The very object of seismological study—earthquake, one of the most terrifying natural disasters—made the mobilization of the networks of amateur earthquake observers "a basic tactic of nation-building" in political regimes as different as republican Switzerland, imperial Austria, progressive California, and Communist China. The amateurs were absolutely central, albeit invisible, players in the processes of knowledge production regarding earthquakes.

Russia was no different in this respect. The professionalization of seismology in Russia in the last decade of the nineteenth century and, especially, after the Bolshevik revolution, did not shut the volunteer observers out of the discipline. Rather, their contribution was increasingly anonymized and their presence made invisible as they were redefined as the suppliers of data for professional scientists in the distant "centres of calculation," in Bruno Latour's sense of the term.[6] One prime example of these changes is the Seismological Institute, which engaged, since its inception in the 1920s and until the late 1950s, hundreds of volunteers into a network of the "regional correspondents" feeding the data into the Institute. Data received from the correspondents were transferred into a card catalogue by the Institute's scientists, who "cooked" them, aggregating data into the tables and stripping them of the references to their sources.

While data produced by nonprofessionals were increasingly marginalized in seismological centres of calculation, the demand for volunteers increased as seismology entered the new world of the Cold War largesse. One of the consequences of the transformation of seismology into a "big science" in the aftermath of World War II was the proliferation of seismological stations spread over the territory of the Soviet Union. The postwar heir to the Seismological Institute—the Geophysical Institute in Moscow, which was the result of a merging of the Seismological Institute and the Institute of Theoretical Physics in 1946—orchestrated a network of stations in seismically active regions of the Soviet Union. In these regions, which were also the most remote and least populated areas, volunteers were vital for staffing the stations that fed the data into the Geophysical Institute in Moscow. Mannar's story is a case in point.

Mannar's trajectory from an enthusiastic volunteer to a supervisor of a seismological station was not exceptional. Ambitious amateurs such as Mannar were recruited to work at the remote seismological stations as operators—officially titled the stations' "supervisors" (*nachal'nik stantcii*). Some background in physics and mathematics was welcomed, but otherwise regional seismic stations were staffed by nonprofessionals, usually a couple, who were trained on the job.

Mannar, however, found the experience disappointing. From this position, he plunged into a spirited critique of professional seismology, which was invested, as he claimed, in the sophisticated instrumental microseismic measurements and theoretical research focused on global processes that were far removed from the local and real-life problems such as earthquake prediction. The solutions he developed were rather unique and sprung from different sources. Lysenko's campaign against genetics, which reached its peak in August 1948 just two months before the Ashgabat earthquake, provided Mannar's own campaign with ideological underpinning.

CITIZEN SCIENCE IN THE SOVIET CONTEXT

The Russian educated intelligentsia embraced the vision of bringing "science to the people." In the late imperial Russia, the naturalist societies and scientific associations sponsored a broad range of activities that engaged ordinary citizens. After the revolution, the Soviet regime promoted the vision of bringing science to the masses as the core value of the new people's state; the basis of rational, *scientific socialist* society; and the means for cultivating new scientific elites for the socialist society.

In the late 1940s, a series of public campaigns that marked the beginning of the Cold War brought to the limelight Trofim Denisovich Lysenko (1898–1976)—a self-fashioned "peasant scientist," an agronomist who rose to power in the 1930s backed by Stalin himself. The campaign against genetics disrupted the established alliance between professional experts and non-professionals. The history of Lysenkoism and its tragic consequences for Soviet biology is one of the most researched episodes in the history of Soviet science. Yet the existing histories of "the Lysenko affair" have focused either on the political history of Lysenkoism or on the history of Lysenko's personal rise to fame.[7] Little attention has been paid so far to a larger movement to which Lysenko's individual trajectory was intimately linked: the massive mobilization of *kolkhozniki*—the workers of the collective farms, or *kolkhozy*—into scientific activities organized in the *kolkhoz* agro-laboratories, or "hut labs." Lysenko's campaign against genetics, as much as anything else, brought "scientists of the people" to the fore, presenting them as equal participants in Soviet socialist science.

The hut labs movement originated in Ukraine in the early 1920s as a response to severe food shortage and famine brought about by the chaos of revolution followed by the civil war, the "war communism," and the beginning of collectivization. The movement started as a newspaper campaign. In 1921, the newspaper *Bednota* ("The Poor") invited the peasants to send their questions to the newspaper and published the replies to the questions, serving as a mediator between the experts and the farmers, while at the same time urging the correspondents to learn the answers for themselves by organizing their own hut labs and pursuing experiments on their own. Starting in Ukraine, the movement spread to other regions. Many collective farms established meteorological stations and kept routine weather observations; for the most part, the hut labs were engaged in experimenting with crop rotation, fertilization, weed control, and stimulation of seeds and plant growth.

The production of scientific results was *not* the aim of the movement. Rather, the hut lab movement was exalted in the media as a new way of doing science: the peasant scientists were praised as daring inventors who would drive the old technical intelligentsia to the margins. Lysenko, who emphatically identified with peasant scientists, praised the hut lab movement not only for providing crucial evidence for his theories but, more importantly, for cultivating a new kind of scientist—a quasi-professional working in the hut labs who was on an equal footing with professionally trained specialists in their modern labs.

Lysenko's public campaign against genetics provided Mannar with a model and a resource, demonstrating that people like himself could gain a place for themselves as more than just the suppliers of mechanical labour to professional scientists in their labs. Instead, they could become quasi-professionals in the vanguard of truly socialist science. "A specialist is an acquired, not an inborn, quality," wrote Mannar alluding to Lysenko's life trajectory and his theories at once.[8] In his own aborted campaign to reform seismology, Mannar referred to the hut labs as a case in point.

Seismology should learn from agriculture, Mannar reasoned. Like agricultural science, seismology should become more open to ordinary citizens. Moreover, people's seismology would fill the niche in which professional scientists were seemingly failing—earthquake prediction. As Mannar argued, citizens in seismically active regions countrywide, just like farmers in starving Ukraine, should start learning their answers about earthquakes for themselves, mobilizing "a dense network of pre-seismic monitoring stations built by enthusiast volunteers in the seismically active areas. . . . Such stations would give an entirely new kind of data, for seismology as well as for people's pre-seismology."[9]

Mannar's advocacy of people's seismology could have led to the larger consequences beyond a mere correspondence between a dedicated amateur and the party functionaries. In the late 1940s and early 1950s, the beginning of the Cold War was marked by the staged public debates in different disciplines: in philosophy in 1947, in biology in 1948, in linguistics and physiology in 1950, and in political economy in 1951. The meetings differed in their agendas and the outcomes, but they all addressed a common theme: how to reconcile Soviet ideology and science in the name of socialist science.

A campaign modelled on that in biology was planned in physics, but it was cancelled. The atomic bomb placed physicists in a unique position that shielded them from the interventions from outsiders of all kind. The bomb's protective shadow extended beyond physics to the neighbouring fields, if only indirectly related to the nuclear weapons industry. Seismology was one of these fields in the shadow of the bomb. The link to nuclear weapons research and technologies not only protected seismology from a Lysenko-like scenario but also thoroughly transformed seismology, redefining the disciplinary conventions about research questions, methods, and data in seismology. Mannar's project was motivated by these transformations as much as by the ideological drivers.

THE POLITICS OF SEISMOLOGY AND VLADIMIR MANNAR'S COLD WARS

With the beginning of nuclear explosions, seismology became tightly linked to nuclear weapons research and technologies in both the United States and USSR. Seismology offered a method of uranium prospecting and nuclear test detection methods. Both areas became prioritized in the Soviet Union. The dual, scientific qua military agenda gave a major boost to both civilian and military branches of Soviet seismology. The field was thoroughly transformed through its link to nuclear weapons. In the 1950s, seismology swelled. New seismological stations were established on the territory of the Soviet Union. Mannar worked on one of these stations, collecting instrumental microseismological data, all the while criticizing professional seismology for neglecting macro-observations.

The transformation of seismology into a Cold War "big science" marginalized nonprofessional volunteers like Mannar in many different ways beyond the secrecy regime with its deliberate restriction of access to the authorized professionals only. Another feature of Cold War seismology, which Mannar experienced firsthand, was the redefinition of the disciplinary conventions around earthquake prediction. As a scientific problem, earthquake prediction was replaced with the statistical assessment of the likelihood of a future quake. As Deborah Coen has put it, in the second half of the twentieth century "the earthquake as scientific object became something unrecognizable to its victims."[10]

As Mannar navigated the complex terrain of Cold War seismology, he had grown increasingly critical of professional seismologists' style of work. Weaving together his advocacy of people's seismology with his bewilderment at what he saw as seismologists' lack of interest in earthquake prediction, Mannar concluded that Soviet big science seismology was misled, functioning de facto as a "capitalist science." Most consequentially, Mannar argued, this was reflected in the scientific instruments: the seismological stations were equipped with "commercial" instruments—that is, ready-made instruments manufactured at a factory floor.

The commercial principles of instrument-making, Mannar argued, not only made scientific instruments costly and thus limited the *number* of stations that were possible to equip with overpriced devices; they also determined the *kind* of research these stations enabled: these instruments were narrowly designed for specific scientific priorities, which did not include macroseismic observations or earthquake prediction. In contrast to the "capitalist" model, Mannar advocated a "socialist method of scientific instrument-making," characterized by modest finishing, small size, detailed instrument documentation and, above all, the possibility to assemble the devices locally, adjusting them, by scientists and amateurs alike, for their particular purposes.[11]

Mannar's proposals, although carefully reviewed, did not go any further. Following the standard procedure, the party functionaries forwarded his proposals to the Geophysical Institute and were satisfied with the assessment by the Institute's scientists that Mannar's critique was misplaced and scientifically unsound. Mannar responded with a bitter thirty-page-long typed manuscript entitled "Brief critical remarks on the work of the Geophysical Institute in general and its Seismological Section in particular," in which he denounced the Institute's geophysicists as ideologically corrupt in their pursuit of "capitalist, bourgeois science." Despite the heavy wording of the accusations, peppered with the quotes from Stalin and occasional references to the example set by Lysenko in agricultural sciences, Mannar's denunciations were archived without further consideration and his file closed for good. While Mannar gained professional status in seismology as a technical supervisor of a seismic station on a payroll of the Geophysical Institute, his vision of socialist seismology, based on the involvement of ordinary citizens in monitoring the environmental changes in seismically active regions using simple homemade tools and macroseismic observations, was swept away within increasingly technical and instrumental seismology.

CITIZEN SEISMOLOGY IN A SOVIET SOCIALIST KEY

The episode of Vladimir Mannar and his aborted campaign for socialist seismology opens a window into a world of Soviet "citizen scientists" who saw in Lysenko an exemplar of a "scientist for the people" and an inspiring model to emulate. There are some interesting parallels between the trajectories of the two. Lysenko was able to gain professional status in biology due to comparatively low barriers for "entrance" in Stalinist science in the 1920s and 1930s and to the interpretative flexibility of the evolutionary framework in these years, which made it possible for Lysenko to intervene in major academic debates in the theoretical biology of the time. Mannar, on the other hand, gained professional status in seismology at the moment of the expansion of the field, which created job opportunities for people just like him—educated but not professionally trained citizens who manned seismic stations in the remote and seismically active regions of the Soviet Union. Leading Soviet seismologists encouraged observational activities on the part of lay citizens, establishing channels of communication and seeking to include them in a dialogue with professional experts. The similarity does not go further, however. Lysenko, assisted by several canny advisers and endorsed by Stalin, plunged into ideological battles that led to an official ban of genetics with catastrophic consequences for Soviet biology. Yet this most pathological version of Stalinist citizen science was also quite unique. During Stalin's time, the would-be Lysenko surfaced periodically within disciplines ranging from linguistics to physics but always failed. Mannar, writing in the wake of Stalin's death, had even lesser a chance to have Soviet authorities on his side.

What then can Mannar's case teach us about science in a citizen science mode? Mannar's trajectory encapsulates the implicit tension between citizen science's two meanings, one that encourages amateur contributions to science within a framework defined by experts and the other that implies a critical stance towards experts' interests and goals. Unlike the Lysenko affair's "pathology," Mannar's story represented reasonably "normal" developments stemming from attempts to reconcile the two loyalties, to science and to citizenship, doing both at once—playing by the scientists' rules and at the same time maintaining a critical distance. As citizen science becomes more and more common today, Mannar's story may serve as a useful reminder that citizens can have larger agendas and aspirations than merely providing scientists with data and observations or serving as "sensors" for scientific projects.

The case of Mannar and his vision of people's seismology might be more similar to the nineteenth century's way of thinking of who is professional than to the "citizen science" practices familiar today. Throughout the twentieth century and especially after World War II, the divide between the amateur and professional science has dramatically increased in most Western societies. In the case of seismology, as Coen has pointed out, the entire "field of knowledge that depended on the self-reported observations of ordinary people in extraordinary situations"—that is, the earthquake observers—was being swept away after the onset of the Cold War, replaced by high-technology science of microseisms and risk assessments.[12] Yet, there are some material convergences between the nineteenth and the twenty-first century, such as the explosion of new communication technologies in both periods that shrinks the divide between amateur and professional scientists. Mannar's story of an amateur seismologist in mid-twentieth-century Russia thus suggests a corrective to the image of twentieth-century science as widening or even creating the gap between the experts and ordinary citizens.

The case of seismology is especially revealing in this regard. While nonprofessionals' contribution as data collectors and nature's observers is acknowledged in such disciplines as ornithology, zoology, botany, entomology, and other field sciences, such instrumental and highly technical big science enterprise as Cold War seismology is commonly regarded as a field with higher barriers for the nonprofessionals to enter. As Mannar's story illustrates, however, rather than elbowing nonprofessionals out, the transformation of seismology into a high-tech big science marginalized their data that fell outside the redefined disciplinary conventions.

Mannar's pre-digital socialist seismology in late-Stalinist Soviet Union is an unlikely comparison to the global and Internet-driven citizen seismology of today. Yet this case can offer insight into the inherent tensions and the ambiguities of citizen science. Mannar's expressive written record of his mediations with professional seismologists brings to the fore the usually silent and invisible participant in scientific practice—a technician, an amateur, a member of the public, an "ordinary citizen." Mannar's story highlights the dialogical and "two-way street" character of communication between scientists and citizens engaged in producing science while, at the same time, striving to create personal empowerment.

NOTES

1. For complete list of references and sources, see Elena Aronova, "Citizen Seismology, Stalinist Science, and Vladimir Mannar's Cold Wars," *Science, Technology and Human Values* 42, no. 2 (2017): 226–56.

2. See Rossiiskii gosudarstvennyi arkhiv noveishei istorii (Russian State Archive of Contemporary History, formerly Archive of the Communist Party of the USSR, or RGANI), Moscow, Russia, f. 5, op. 17, no. 417, l. 1–116 (hereafter Mannar File).

3. Vladimir Mannar, "Kratkoe soobshchenie ob otkrytii pervoj proseismicheskoj stantcii," n.d. (Mannar File, l. 46).

4. Vladimir Mannar i Anna Mannar, "Kratkii doklad o rozhdenii novoi vetvi cotcialisticheskoi Sovetskoi nauki—proseismologii," n.d. (Mannar File, l. 57).

5. See Deborah Coen, *The Earthquake Observers: Disaster Science from Lisbon to Richter* (Chicago, IL: University of Chicago Press, 2013), and idem, "Witness to Disaster: Comparative Histories of Earthquake Science and Response," *Science in Context* 25, no. 1 (2012): 1–15.

6. Bruno Latour, *Science in Action* (Cambridge, MA: Harvard University Press, 1987).

7. See, for instance, Nikolai Krementsov, *Stalinist Science* (Princeton, NJ: Princeton University Press, 1997); Nikolai Krementsov and William deJong-Lambert, eds., *The Lysenko Controversy as a Global Phenomenon*, in 2 vols. (Palgrave Studies in the History of Science and Technology, 2017); Dominique Lecourt, *Proletarian Science? The Case of Lysenko* (London: NLB, 1977); Nils Roll-Hansen, *The Lysenko Effect: The Politics of Science* (Amherst, NY: Humanity Books, 2004).

8. Mannar, "Kratkoe soobshchenie" (Mannar File, l. 46). Mannar alluded to Lysenko's promotion of the neo-Lamarckian idea of plasticity and the inheritance of acquired characters, juxtaposed to a Western-born notion of genes responsible for the transmission of "inborn" traits.

9. Vladimir Mannar i Anna Mannar, "Kratkii doklad" (Mannar File, l. 74).

10. Coen, *The Earthquake Observers*, 267.

11. Mannar and Mannar, "Kratkii doklad," l. 65.

12. Coen, *The Earthquake Observers*, 9.

15

Blood, Paper, and Invisibility in Mid-century Transfusion Science

Jenny Bangham

The World War II was just ending when the *British Medical Journal* announced the discovery of three new blood groups: "Willis," "Levey," and "Lutheran."[1] Blood groups affect whose blood can safely be transfused into which patients, and during the war, blood grouping had become integral to Britain's nationwide transfusion service. The Emergency Blood Transfusion Service depended on large-scale blood storage and a nationwide bureaucracy for managing donors—these had helped transform blood into a safe and reliable therapy, and simultaneously created a deluge of new knowledge about blood. The service depended on millions of volunteers across the country willing to give small but crucial regular donations of blood—donors took time out of their days, travelled to transfusion centres, were willingly punctured by needles, and patiently waited on beds while their blood was conveyed through rubber tubes to glass bottles. The wartime service and its postwar successor, the National Blood Transfusion Service (NBTS), were predicated on the notion that the identities of donors did not matter; they were largely interchangeable and subsumed into factory-scale systems of storage and mobilization.[2] Individual donors were invisible to the patients who benefitted from this therapy, and to the doctors delivering it.

Occasionally, though, donors were singled out for greater attention. After all, human blood was not *wholly* interchangeable. By the 1920s, it was well known that blood could be classified into four main groups (A, B, AB, and O), and that transfusing blood of the wrong group could be very dangerous for a recipient. By the 1930s, several more blood groups had been defined, and during the war, the complex subtleties of blood became even more apparent. Owing to robust bureaucratic procedures, adverse transfusion reactions could be followed up and investigated, and specific samples could direct attention to new serological complexities ("serology" is a field of research and practice concerned with immune reactions).[3] As a result, understanding of blood compatibility soon went far beyond the four original groups; the large number of people being tested and transfused brought into view a host of new variations. And this intensifying concern with the subtle differences of blood types went hand in hand with flourishing interest in people with "rare" blood—who might be called upon to provide donations for patients very sensitive to the more common blood types. In striking contrast to the public anonymity of most donors, individuals understood to have "rare blood" became the subject of many newspaper articles, radio plays, films, and even special clubs.[4] In concert, new blood groups began to be named after donors—such as "Willis," "Levey," and "Lutheran."

Discoveries of new blood groups were not only important for the safety of transfusion. They also promised new resources for scientists interested in the young but flourishing field of human genetics. In the 1940s, the blood groups were still some of the only human traits known to have clear-cut Mendelian inheritance (i.e., unlike most other traits, their inheritance followed a simple set of rules). New serological genetic discoveries built on the blood and paper records of donors and transfusion patients opened up all sorts of new research possibilities—including the mapping of human chromosomes and the study of "racial" differences (Burton, chapter 7; Vimieiro, chapter 9). Unlike in the United States, "racial" labels were not officially attached to donors in Britain, but the study of blood in the context of transfusion facilitated and helped to promote scientific investigations into blood and race.[5] For mid-century scientists interested in the study of human genetics, the visceral donations to the transfusion service created new resources and methods for studying human inheritance, identity and difference.

As a historian narrating the history of human genetics and blood research, I have struggled with the relative invisibility of the motivations, experiences, and social circumstances of many of those who created and laboured within this infrastructure. Many of the sources available for telling this history are themselves structured by the system at the centre of the story. Transfusion infrastructure depended on invisibilities: large-scale extraction, storage, and transportation were only possible because of the efficiencies yielded by standardization and routine.[6] The blood grouping technicians, clerks, and donors on whose labours the service depended were interchangeable as well as extremely numerous. The transfusion service kept careful track of its donors, but it mostly obscured their contributions as individuals owing to the sheer numbers of donors it recruited and massive volumes of blood it mobilized. Meanwhile, population genetic research into "racial," geographical, religious, "tribal," and national diversity, labelled, marked, and flattened donors into groupings that eclipsed other forms of personal identity.

As well as the invisibilities created by the transfusion infrastructure, archival sources have been shaped by recent privacy concerns about old paperwork pertaining to blood. The wartime transfusion services helped create the conditions for modern human genetics—a field that is now understood to offer powerful insights into our identities, history, and our health. My research depended on two vast archives carefully catalogued and made available by the Wellcome Collection, which sought to collect papers pertaining to the history of genetics. But because of the many new meanings that can now be made from pedigrees and blood tests, and because of increasingly careful protections around the identities and medical data of donors and patients, the archival records of this early history were recently re-scrutinized by Wellcome archivists and many closed. Thus, the new meanings and uses of blood, including those relevant to present-day health insurance policies, have impacted the archival reclassification of records and correspondence relating to people who are part of the historical record.

The experiences and social worlds of donors and patients are crucial to the history of transfusion and genetics. Among the sources available, a handful of people stand out—including a small number of donors whose names became attached to blood groups. Blood group naming practices serve as an aperture for reflecting on how and why the institutions and procedures of the life sciences make some people visible in retrospect and others less so.

My case study describes the circumstances under which the donors "Lutheran," "Willis," and "Levey" were singled out for scrutiny. At the end of the *British Medical Journal* paper first announcing these new groups, its authors noted their gratitude to the donors, but also to a single patient, a "Miss F. M.," who was suffering from an auto-immune condition that caused anaemia and who had received blood from all three donors (and several more).[7] In the three sections of this chapter, I outline the structural conditions that made "F. M." into such a valuable research subject; I describe why donors' names became attached to the new antigens made visible using her blood, and conjecture how those names functioned. I then briefly reflect on how the archives I used in my own research both protect and erase the identities of people who were part of this history.[8]

A LIVING ARCHIVE

How did the living body of this young patient at Oxford's Radcliffe Hospital become such a valuable resource for serological research? The conditions that positioned her as a research subject began taking shape soon after the outbreak of the war. The transfusion service first operated in London but soon expanded throughout the country. It quickly became a robust, distributed infrastructure of bottles, fridges, vans, and a vast paper bureaucracy. It depended on the labours of nurses, serologists, and clerks (figure 15.1), and of donors recruited and disciplined by national and regional publicity campaigns. Blood testing was crucial. In 1939 just a handful of blood groups were known: the clinically important ABO groups, and the less important MN and P groups. ABO testing was mandatory in Britain, necessitating the training of hundreds of serologists to work in transfusion centres across the country. But ABO groupings could not absolutely guarantee the safety of blood—sometimes unexpected reactions occurred despite careful testing. For that reason, doctors needed to be able to link the outcomes of all transfusions back to individual donations, enabling transfusion officers to investigate any problems. The transfusion service coupled donors and recipients across space and time using labels that could

Figure 15.1. A photograph of the North West London Depot, in the Slough Social Centre (c. 1940). At the centre is the well-known figure of Janet Vaughan (in glasses), one of the founders of the wartime transfusion service. She presides over clerks sorting registrations and calling up donors. To the left is a large map on which is marked the hospitals supplied from Slough. The depot was responsible for the blood supply of the North West quarter of London, which included Basingstoke, Buckingham and Aylesbury. Reproduced with the kind permission of The Bodleian Libraries, The University of Oxford.

be tied to and untied from bottles of blood. While the Regional Transfusion Centres carried out local investigations into puzzling blood, the Medical Research Council also established several expert laboratories devoted to blood research. One of these was the Galton Serum Unit in Cambridge (50 miles north of London), which set standards for testing reagents, and investigated intransigent serological problems.

The Cambridge Galton Serum Unit quickly became a passage point for puzzling samples that had been singled out by Regional Transfusion Centres. If depot serologists were unable to figure out why a transfusion had endangered a recipient, they sent samples to Cambridge for further investigation. Unit researchers would follow up particularly strange and fascinating specimens with further requests to the depots for blood, and in some cases they even visited distant parts of the country to sample donors in their own homes. The Regional Transfusion Officers became sentinels, scrutinizing for rare serological treasure among thousands of routine tests, drawing the most intransigent specimens to the attention of the Cambridge researchers. A regional transfusion officer in the Northern English town of Sheffield underlined his role when he described himself (to one of the Cambridge scientists) as "a lonely lighthouse keeper in a sea of problems . . . with between 2,000 and 3,000 samples per week passing through the laboratory."[9] He saw himself as caring for and continually repairing a sophisticated instrument for making new blood group variants and systems visible.

This was how, as the war went on, the routine testing of hundreds of thousands of investigations of curious samples yielded a remarkable array of novel blood groups. Also working in collaboration with labs overseas—especially with labs on the U.S. East Coast—British transfusion doctors and scientists transformed human blood into an increasingly complex fluid, both serologically and genetically. As such knowledge expanded, transfusion began to be used not just as an emergency treatment for shock, but also as a routine therapy in surgery, as well as in antenatal and neonatal care. By the end of the war, many hospitals were using human blood as a treatment for long-term conditions. Indeed, donated blood was now deemed safe and plentiful enough for patients suffering from chronic anaemia, like F. M., to benefit from repeated transfusions.

Multiply transfused patients opened up new lines of investigation for labs like the Galton Serum Unit. Sometime near to the end of the war, one of the unit's researchers, Robert Race (by then, known internationally for his work on the "Rhesus" blood groups), started collaborating with doctors at Oxford's Radcliffe Hospital to investigate the haematological crises experienced by F. M. The patient faced what would become a common problem for multiply transfused patients. Some people experiencing a transfusion will produce antibodies in response to the antigens in the donor blood. This may not be a problem for a first transfusion, and can be minimized with careful cross-match testing of blood types. But successive transfusions result in the build-up of antibodies in a patient's blood, narrowing the kinds of blood available to them in the future. This was a dangerous predicament for patients, but for researchers wishing to study blood group serology, multiply transfused people were also an exceptionally rich resource of antibody types.

F. M. herself appeared to have exquisitely sensitive antibody reactions, and as a result the multiple transfusions that she experienced transformed her into a veritable archive of antibodies. By studying her blood, the scientists effectively made her body into an immunological instrument that could recognize novel antigens hitherto undetected in the blood of her donors. Thus, the wartime transfusion infrastructure had not just created new therapeutic opportunities for treating F. M.'s anaemia, but had also turned her blood into a resource for discovery. The institutions of the NBTS and the Radcliffe Hospital encompassed and positioned F. M. as both a treatable patient and a research subject.

The British wartime and postwar infrastructure enabled the systematic scrutiny of large numbers of samples and people, most of whom remained invisible within that infrastructure. But the administrative ordering and tracking of paper by large numbers of clerical staff enabled the singling out of specific individuals for investigation. Then there might be a flurry of excitement as researchers rushed to test a sample behaving in unexpected ways. In one letter a scientist remarked to a colleague about a particularly intriguing donor: "I wonder if Madame Kozyreff realises what a prize she is and how many serological laboratories will be wanting to bleed her."[10] Amid thousands of routine tests, the spotlight of serological surveillance made some individuals exceptionally precious, both to doctors and researchers.

NAMING BLOOD GROUPS

Back in Oxford, the scrutiny of F. M.'s blood brought other individuals into view. During the course of her treatment, F. M. received pints of blood from several donors. Doctors regularly assayed the concentrations of specific antibodies in her bloodstream; by monitoring her reactions, the researchers detected the presence of several entirely new antibodies. By isolating those, and testing them against arrays of standard red cells, the researchers showed that F. M. had been exposed to several novel antigens carried in the blood of her donors. Those antigens were "novel" in the sense that they had not previously been defined by serologists (further research would determine whether they were common or rare). Using administrative records to find the origin of that transfused blood, Robert Race and his colleagues pinned those new antigens to individual NBTS donors, who all consented to further tests. Crucially the researchers persuaded donors' families to give samples too.

With these methods, the researchers used F. M.'s blood to define several new antigens, and trace those to three living donors. The first was named after a donor called Willis and was found to relate to the already well-documented Rhesus (Rh) class of antigens. The second, named after a donor called Levey, was found to be exceptionally rare and unrelated to any existing group. The third, from another of F. M.'s donors with the surname Lutheran, was the real prize, in that the antigen appeared to be both novel and relatively common, and represented a whole new blood group system. Tracing Lutheran through the families of several additional donors, laboratory workers, and students, the researchers concluded that the antigen was inherited as a Mendelian dominant allele.

Blood group antigens had not always been named after donors. Karl Landsteiner had named the earliest in 1900 using first two letters of the alphabet (A and B), and, reportedly, "O" for the German word "ohne," meaning "without" ("O" individuals lacked A or B antigens). In the 1920s, the practice of using letters continued with the "MN" and "P" groups. The move to naming blood groups after donors coincided with the intensification of serological research within the 1940s wartime infrastructure. Within this new world of planned, routine surveillance, some novel groups were named for the antigen-carrying donor—as was the case for "Willis," "Levey," and "Lutheran."[11] Others were named for the person who had made the relevant antibody—such as (to name three of many) "Duffy," "Lewis," and "Colton."[12]

The wartime and postwar practice of naming blood groups after individuals was directly related to the system of testing, scrutinizing, and singling out interesting samples from the thousands generated nationwide.[13] Adverse transfusions, like those experienced by F. M., could be monitored thanks to careful record keeping. The labour of thousands of clerical staff, who kept track of donors and their blood across the country, made it possible for researchers to pursue people with interesting and valuable blood. Unusual specimens, like those given by F. M.'s donors, would be investigated and moved between hospital or transfusion centre and research lab. The most promising of those samples would be shared between serologists in different labs, sometimes even between labs on different continents.

With this constant movement and scrutiny of specimens, one reliable system of naming—one that could clearly distinguish one sample from others in the cohort—was the surname of the donor, or a shortened version of it.[14] A personal name was mobile, transportable, and (often) unique and potentially more immediately legible than a number or combination of letters.[15] It also speaks to what researchers found charismatic. One blood group serologist later spoke about his experiences of handling donor names in laboratory settings: "It made it more personal that you were working with a real person's specimen. Quite different from, say, sample 4567–89."[16] And of course, just as with the donors Lutheran, Willis, and Levey, it was necessary that researchers often struck up ongoing relationships with donors' families, obtaining repeat samples for genetic analysis. Thus, a name likely had affective resonances to the scientists engaged in the study of blood. Besides, if specimens *had* been labelled using arbitrary numbers or letters they could only have been shared successfully if institutions had decided on a system of standards. The name of the donor provided a readily distinguishable marker that was evocative, memorable, and (usually) easy to write and say aloud.[17]

This was a very partial kind of visibility. For example, the donors whose blood yielded new groups that took their names were treated differently from donors who were able to provide "rare" blood under emergency conditions. The latter often provided sensational stories of pursuit and redemption. However, the visibility accorded to Lutheran, Levey, Willis, and others was important within the British public projection of the altruistic donor.[18] Since the outbreak of war, those organizing the modern, large-scale, highly distributed transfusion service had simultaneously projected an image of precisely the opposite: donation that was local, face-to-face, and personal.[19] In this respect, the move to an expansive infrastructure of transfusion cohered with the practice of naming groups after (relatively invisible) donors.

DONORS IN THE ARCHIVE

I came to the story of blood groups and transfusion via an interest in the history of human genetics. My historical field of vision was shaped by the acquisitions department of the Wellcome, which, in the late 1990s and early 2000s, amassed a formidable collection of papers from the blood-grouping labs of Robert Race, and his colleagues Ruth Sanger and Arthur Mourant. The archival spotlight on the lives and careers of these scientists brings out from the shadows particular individuals who contributed to the labour of studying blood. In 2010, ten years into the "post-genomic era," the Wellcome put new emphasis on the genetic dimension of those papers, when it incorporated them into a program to make its materials relating to human genetics freely available online, a digitization effort that it dubbed "Codebreakers: Makers of Modern Genetics."[20]

This venture resonated with other efforts by the Wellcome to make its buildings and collections (in both the museum and library) more widely accessible; it was also consistent with the Wellcome's promotion in the 1990s and 2000s of freely accessibly genomic data and its highly public leadership of the open access publishing movement. But as commentators of genomics have pointed out, efforts to be more "open" (and visible) in one respect often create the conditions for new kinds of closure.[21] Just as the free sharing of genomic data is managed within structures of governance that include funders, data storage infrastructure, ethical laws, and institutional review boards, so the archives relating to those endeavours are subject to data protection. The Wellcome is one of the richest and most influential biomedical research funders in the world, and is particularly sensitive to the privacy conditions pertaining to biomedical data (Keuck, chapter 21, on the archival sensibilities of diverse institutions). As the Wellcome archivists started the process of digitizing the "Codebreaker" papers, they were rigorous in their re-evaluation of the privacy conditions around the papers relating to blood groups.

This historiographic reframing combined with biomedical developments to impact archival policies and create new partial invisibilities, whereby some records were no longer accessible as empirical sources for historical research. The Wellcome now framed the Race and Mourant papers as central to the history of human genetics. Moreover, now that the papers were accessible online, they had a far wider visibility and potentially much bigger audience, so the archivists were rigorous in their assessment of "sensitive personal data," in compliance with the Wellcome's access policy.[22] During that labour-intensive revision process, the archivists changed the access conditions of a large number of the papers that I had used during the earlier phase of my research. They marked some records "restricted access," meaning I could look at but not photograph or quote them, and others as "closed," which meant I was unable to access them, in some cases, for several more decades. The latter included letters regarding scientists' pursuit of certain donors and family pedigrees. The archivists perceived that in some instances a whole series of letters might be capable of attaching a blood group to a disease, and then to a personal name, pedigree, and family.

Many countries have legally enshrined the right to privacy regarding medical conditions.[23] Concerns about personal, sensitive data have intensified in concert with the expansion and power of genomics—a field subject to the powerful logics of "informatic capitalism" and its prolific markets for personal information.[24] Blood and data freely given in the 1950s and 1960s can now potentially be tethered to data-gathering practices that affect a family's access to health care or insurance. This requires that the donors and patients who made some of the very earliest corporeal contributions to genetics and serology are closely protected. Also closed are materials pertaining to those donors who, in the 1950s, explicitly consented for their names to be published in journals. This put out of my reach many sources with potential clues as to who chose to be part of such scientific and medical projects, why they participated, and how they cultivated their relationships with researchers.

Thus, the ease with which names and personal data can today be connected together has created gaps in what historians can learn about the past. In my own research, many of the donors and patients who gave their blood to the transfusion service and to serological genetic research have been hidden twice over. They were obscured by a "big data" enterprise made possible by a vast transfusion bureaucracy that anonymized its contributors and research subjects through the sheer numbers of donors and massive volumes of blood involved. And in addition, many of the paper trails with clues to the experiences and contributions of postwar donors and patients have been partially hidden by new privacy regulations—regulations that have directly responded to the later success and proliferation of the scientific fields they contributed to. The visibilities of the past are dynamic, and are produced by scientific and cultural change; the erasures of mid-century blood genetics have helped me to better understand what I can see and why.

NOTES

1. Sheila Callender, R. R. Race, and Z. V. Paykoç, "Hypersensitivity to Transfused Blood," *British Medical Journal* 2, no. 4411 (1945): 83–84.

2. This was in stark contrast to blood transfusion in interwar Britain: Nicholas Whitfield, "A Genealogy of the Gift: Blood Donation in London, 1921–1946" (PhD diss., University of Cambridge, 2011).

3. The search for and study of blood groups was an international enterprise. By the end of World War II, many countries had transfusion services that yoked together serologists, geneticists, and transfusion specialists, who shared blood and data through international societies, organizations like the Red Cross, and local personal and professional networks.

4. Susan E. Lederer, "Bloodlines: Blood Types, Identity, and Association in Twentieth-Century America," *Journal of the Royal Anthropological Institute* 19 (2013): S118–29.

5. On the many ways that research subjects were racialized by British scientists investigating the genetic complexities of blood: Jenny Bangham, *Blood Relations: Transfusion and the Making of Human Genetics* (Chicago, IL: University of Chicago Press, 2020).

6. On the (partial) invisibility of infrastructures: Susan Leigh Star and Karen Ruhleder, "Steps toward an Ecology of Infrastructure: Design and Access for Large Information Spaces," *Information Systems Research* 7 (1996): 111–34.

7. Callender et al., "Hypersensitivity" (1945); "Miss F. M." was given this identifier in a second paper published the following year.

8. For thoughtful reflections on the demands of knowledge and privacy faced by historians: Susan C. Lawrence, *Privacy and the Past: Research, Law, Archives, Ethics* (New Brunswick, NJ: Rutgers University Press, 2016).

9. This quote comes from some years after the war: Dunsford to Race and Sanger, May 9, 1956, SA/BGU/F.5/3/1, Wellcome Library.

10. Fisher to Race, April 15, 1948, SA/BGU/E.7, Wellcome Library.

11. In another case, the "Kidd" group was named after a child suffering from erythoblastosis fetalis with this novel antigen: "Medical Research Council Progress Report, 1950–1953, of the Blood Group Research Unit," 2–3, FD8/18, National Archives, London.

12. Richard Duffy was another multiple transfused patient—in this case, his name was attached to the blood group revealed during studies of his blood: Stephen Pierce and Marion Reid, *Bloody Brilliant! A History of Blood Groups and Blood Groupers* (Bethesda MD: AABB Press, 2016), 562.

13. On eponymous disease-naming practices in medicine: Andrew J. Hogan, "Medical Eponyms: Patient Advocates, Professional Interests and the Persistence of Honorary Naming," *Social History of Medicine* 29 (2016): 534–56; disease characterization often depended on large numbers of patients over many years; blood group "discovery," by contrast, could often be pinned to a single donor or patient. Thanks to Andrew Hogan for thoughts on this comparison.

14. The need for unambiguous, mobile, and writeable blood group names is consistent with the fact that many such terms were shortened versions of donor names, for example, "Kell" from "Kelleher," and "Ok" from "Kobutso": George Garratty et al., "Terminology for Blood Group Antigens and Genes—Historical Origins and Guidelines in the New Millennium," *Transfusion* 40 (2000): 477–89. See also: Jenny Bangham, "Writing, Printing, Speaking," *British Journal for the History of Science* 47 (2014): 335–61.

15. Serologists recounting the history of their field claim that "Lutheran" was a misspelling of the name "Lutteran" and the result of an incorrectly labelled specimen: Marion E. Reid and Christine Lomas-Francis, *The Blood Group Antigen Facts Book*, Second Edition (San Diego and London: Academic, 2004), p. 193, a story traced to the eminent Patricia Tippett: G. Garratty et al., "Terminology for Blood Group Antigens and Genes," *Transfusion* 40 (2000): 477–89.

16. Steven Pierce, personal communication, October 2020.

17. One apparent exception to this tendency to use personal names for blood groups was "Bombay." But though an important phenotype, it turned out not to represent a new antigen as such, and the term remained in inverted commas in Race and Sanger's textbooks. Nevertheless, provisionally naming a group for an entire city suggests a distinctive shift in granularity when the London-based researchers attended to specimens from donors located overseas.

18. Some antigen-naming narratives can be found in *The Blood Group Antigen Facts Book*; many such narratives are passed informally through successive generations of serologists.

19. Nicholas Whitfield, "Who Is My Donor? The Local Propaganda Techniques of London's Emergency Blood Transfusion Service, 1939–45," *Twentieth Century British History* 24 (2013): 542–72.

20. The "postgenomic era" is a phrase used to denote genetic research after the publication of the "Human Genome Project." The title of this digital collection presumably alluded to efforts to "crack" the genetic code: Simon Chaplin, "Codebreakers: Makers of Modern Genetics," March 1, 2013, http://blog.wellcomelibrary.org/2013/03/codebreakers-makers-of-modern-genetics/. Accessed October 28, 2020.

21. Jenny Reardon et al., "Bermuda 2.0: Reflections from Santa Cruz," *GigaScience* 5 (2016): 1–4.

22. Toni Hardy, private communication with author, June 20, 2016; August 5, 2016.

23. For a cultural history of privacy in Britain: Deborah Cohen, *Family Secrets: Shame and Privacy in Modern Britain* (Oxford; New York: Oxford University Press, 2013).

24. On the landscape of genomic capitalism: Jenny Reardon, *The Postgenomic Condition: Ethics, Justice, Knowledge after the Genome* (Chicago, IL: Chicago University Press, 2017).

16

Invisible Vitality

The Hidden Labours of Seed Banking

Xan Chacko

Since the mid-twentieth century, a global technoscientific enterprise has gained favour as a promising cure to the twinned crises of plant extinction and habitat loss. This endeavour, known as seed banking, takes the seeds of plants that are threatened or valued in their environments and freezes them for use in the indeterminate future.[1] Conservators concerned with the loss of plant species in the wake of the ongoing global extinction event known as the anthropocene, use the technique to "bank" seeds before they are lost forever.[2] Breeders interested in harnessing the diversity of crop types to create new varieties use seed banks as reservoirs for traits and genes they can dip into for particular projects.[3] As sites that capture and embody anxieties about self-preservation, seed banks are also, simultaneously, hyper-visible as beacons of hope.[4] How might such a visible technoscientific endeavour contribute to studies of invisible labour? At the risk of threatening the basis and security placed by governments and international institutions on seed banking, this chapter reveals the project's reliance on a series of layered invisibilities that come together to create a façade of security. Ultimately, I probe the layers of invisible labour, knowledge, and vitality that qualify seeds to be held in a liminal frozen state.

The science behind the freezing of seeds can be traced to a series of experiments conducted some fifty years ago at the newly built Jodrell Laboratory at the Royal Botanic Gardens, Kew. There, a senior chemist and fellow of the Royal Society, Horace T. Brown (1848–1925), collaborated with a young botanist, Fergusson Escombe (1872–1935), to better understand how plant seeds were affected by exposure to extremely low temperatures.[5] Escombe had written a review article in the new journal *Science Progress* on the state of the field of experimentation on seed germability—the ability of a seed to germinate—which he called "vitality."[6] When reporting their findings in the *Proceedings of the Royal Society*, the pair posited that their new experiment, which involved submitting the seeds to 110 consecutive hours of temperatures between –183°C and –193°C (using equipment and "liquid air" generously provided from Professor Dewar's laboratory) conclusively left "no room for doubt": seed vitality was not affected by the exposure to sustained low temperature, and mature plants grown from the thawed seed were equally healthy to those of the controls.[7]

Whether in the Arctic or the Tropics, seed banks share some basic elements. While the exact temperature may vary, most seed banks are cold rooms with shelves of neatly labelled seed packets suspended in a container. There are also exceptions to the cold storage of seeds. Since the advent of agriculture, Indigenous farmers have devised seed-saving techniques using wood, ash, or earthenware to safeguard seeds. For clarity, when I refer to seed banking I mean the technique of storing seeds at below freezing temperatures as a distinctly twentieth-century Western scientific phenomenon. In these banks, seeds are encased in a combination of plastic, metal foil, paper, cloth, or glass, and a hum from the refrigeration is the only sound that fills the space. This vision of quiet dormancy pervades our representations of seed banks but only describes the final moment in a longer and more complex seed-banking cycle. Missing in this perception are the multifarious practices that select, codify, and prepare seeds for their icy future. Also missing are the people who are mutually articulated with the seeds in the contact zones between nature and culture: seed conservation laboratories. A third invisible layer is the history of botanical extraction and circulation under the auspices of "economic botany" that cleared the pathways that are reused today to move seeds around the globe but now under the rubric of "biodiversity conservation."[8] Why are these people and processes omitted from the grand narratives of saving seeds for our survival?

By examining and animating the processes that make seeds ready for storage in the vault, I highlight the subjectivities of the seed scientists. I study the processes and theories used by scientists to turn seeds as specimens into facts through what historian and philosopher Donna Haraway has called "a subject- and object-shaping dance of encounters."[9] Elsewhere, I follow in greater detail the different stages of seed processing from arrival, through cleaning, X-raying, and counting.[10] Paying attention to the oft-erased modalities of care for seeds in their preparation for cryogenic preservation is a feminist act of "attunement."[11] My inquiry reveals the invisible forces and assumptions that, at the same time as creating the seed bank world, belie its security. This chapter builds on ethnographic and archival research conducted at the Millennium Seed Bank in Wakehurst Place, UK, between 2015 and 2017. Using my experience of working as a seed curator, I start by tracing the more obvious forms of invisibility in seed banks and conclude with a discussion of the invisible labour of seed vitality. I argue that to render the labour of reproduction or "vital labour" of the seeds invisible ignores the most crucial keystone on which the whole edifice of the seed bank depends. Could it be that the promise of security is predicated upon an assumption that cold storage is the *only* salient element of seed banking? Once stored, can all seeds be considered saved? This assumption deserves interrogation by considering the diverse practices of care that make seeds ready for banking before they are shelved in the freezer.

AFFECTIVE LABOUR

Before seeds can be placed in the freezer, they must be disciplined into behaving in appreciably generalizable ways.[12] At seed banks this disciplining labour is performed by "seed curators." More than three-quarters of the seed curators I encountered identified as women. Some fresh out of university or on internships, and others well-seasoned veterans with over ten years of experience, these curators brought a mixed set of knowledges and expertise to the job. Seed curators are a class of scientists who share embodied practices with art and natural history conservators, as well as gardeners (Wylie, chapter 19).[13] Invisible technicians of the seed bank world, curators performed a variety of tasks including labelling, cleaning, counting, screening, and viability testing to prepare seeds and create knowledge about the seeds for their futures as valuable "latent life."[14] Each of these processes is also imbued with texture and variability.

Having labelled the unopened seed packets, even temporarily, the curators moved them to a drying room maintained at 15% relative humidity (RH) and 15°C, where the moisture content of the seeds was reduced; too wet and the water inside the seeds would expand on freezing and destroy them from within. Seed collections would sometimes spend months getting dried, which was generally helpful to the curators who were woefully understaffed and often found themselves with a backlog of seeds to be processed through the different stages of seed preparation. The dry room served as a holding space, to which seeds were returned between each stage of preparation. After the seeds were desiccated for nine months, the next stage was cleaning, where the collections were "purified" to a state where only the seeds remain. Seed collections sent to the banks carried not just the seeds but also plant parts like tufts or spikes, termed "debris," which were considered superficial to their capacity to reproduce (figure 16.1). Anthropologist Mary Douglas, writing on the politics and culture of hygiene, contends that "dirt is essentially disorder."[15] The seed bank's rationalization of purity was that pathogens present on and around the seeds in the collection could compromise not just that bunch of seeds but also spread in the bank, infecting other precious, previously banked seeds. Douglas reminds us that, "There is no such thing as absolute dirt: it exists in the eye of the beholder."[16] In this case, separating seed from dirt or "debris" was the creative but undertheorized labour of the seed curator because it depended upon the curator knowing what the seeds *should* look like, what then was the debris, and finally what was the most expedient means of separating one from the other. They might have sacrificed a few seeds by cutting them under the microscope to make visible the inner structures of the seed. A seed curator might have consulted one of the many texts that have accumulated centuries of botanical knowledge to guide this process. However, even once they had identified the reproductive parts of the seed, the means by which the cleaning should be done and when a collection was *clean enough* were left to their discretion.

Seed curators developed the knowledge and mechanical skills required to make these decisions over the course of their employment at the bank; they understood

Figure 16.1. Photograph of a circular metal tray of pine seeds manually being cleaned of debris at the Millennium Seed Bank. The paper-thin seed structures will be kept, while broken pieces and bits of pinecone will be discarded. Photograph by author, February 2016.

their expertise as gained through experiential, iterative, tacit, and haptic processes rather than the purely theoretical. Junior curators might have started with higher educational credentials than senior staff that had come to be seed curators through gardening or horticulture, but their practical and intuitive fluency with seeds could only be learned through doing. In this sense, the skills that were most prized in the seed bank laboratory, while invisible to most, were rendered visible through the valorization of fine motor skills, attention to details, and experience. The open-ended, need-based nature of the cleaning step allowed curators to innovate technological solutions. Using a variety of professional tools such as sieves, bungs, and aspirators, as well as improvised tools such as sticky tape and rubber mats, curators called on their theoretical knowledge, tacit knowledge gained through experience, and took advantage of the collective knowledge of the team with different levels of expertise to devise flowcharts of activity that guided their cleaning process. Ultimately deciding when seeds were "clean enough" was the purview of each individual curator, who had to balance the bureaucratic decision of how much time would be needed to clean a collection further against the overall perceived health, importance, and rarity of the collection. It was often determined to be less expedient to spend additional time cleaning a collection that was very large, or less rare in the collection of bank. Cleaning brought out a tension between the practical aspects of seed banking. On one hand, the bank had limited space, and the cleaning process was meant to reduce the collections to reflect only the most pertinent parts—the seeds. Cleaning was, in practice, less about purity than it was about saving money. If, on the other hand, the bank had unlimited space and resources, it would not have been as important to clean the collections such that only seeds remained. The vault was prime real estate because of limited space and high energy costs to maintain the freezer temperature.

Along with the bags and labels that accompanied the seeds from their original location, all debris was incinerated so that no stray seeds or pathogens could make their way out of the laboratory. This extra seed material had potential value for the understanding of both the environmental and physiological specificities of the seed, as well as its cultural and historical relations with the seed. The preparation of the seeds for their new and indefinitely long life in the bank was thus contingent on a difficult balance between preservation and erasure, both of which were the products of care. Stripping the seeds of any non-seed material that tied them to their originary world not only rendered the origins of the seed invisible but also came with an informational cost. To make the seeds legible in the seed bank thus required a careful erasure of place.

After seeds were sufficiently cleaned of their past connections, the next process of emplacement in the bank world was often an X-ray screening. A subset of the whole collection was scanned to reveal the inner structures of the seeds, which were then quantified based on categories demarcating seed health. When analysing X-ray images, curators categorized all the seeds into four groups: "full," "empty," "part-full," or "infested." While full seeds had bright clear white insides, indicating density, the interior of empty seeds looked black in X-ray images. The other two categories, part-full and infested, were categories that had to capture all the variation between the clearly delineated extremes and depended on the judgement of the curators. While all other parts of the preparation process required the seed curators to work as individuals, in the X-ray screening, the team came together, gathered around the monitor display, and suggested how the seeds seem to each of them.[17] Based on the views of the team, the leader made verbal pronouncements while using her body to indicate the seeds to which the values were being ascribed. This diffused meaning-making process relied on the scientists' ability to translate shapes, shades, and size into the remaining subjective categories of part-full or infested. I found that their pronouncements went further than the descriptive lexicon including affectively charged language such as good/bad, like/dislike, lovely, darling, and wicked. These subjectivities were not just embraced but were translated into numerical data about seed quality that informs the future of their life in the bank. This process, both technical and cultural, drew on the bodies and feelings of the seed scientists.[18] There is no space for the expression of this invisible, affective, and communal labour in the current, prevailing narratives of seed bank worlds. Even less visible is the invisible labour of the seeds.

SEED VITALITY

After screening, the seed collections were counted, dried further, and eventually packaged into the containers where they spend the rest of their future in the bank. Compared with the temperature at which Escombe and Brown held the seeds in their experiment in the 1890s (–193°C), the contemporary practice of seed banking uses a more technically and energetically achievable standard of –20°C. Once seeds were banked, curators periodically performed germination tests by removing individual collections from the freezer, carefully thawing, then growing them out in controlled conditions to test their liveliness—what Escombe called "vitality." While Escombe and Brown concluded from their experiments that freezing renders seeds "absolutely inert," a claim that motivated the endeavour of seed banking, it is the great invisible truth of seed banking that, even in the most pristine frozen conditions, over time, seeds *do* lose their regenerative capacity; as it turns out, seed latent liveliness has a biological clock. The inability of the science of seeds to predict exactly when or why seed death happens is at the heart of the actuarial process of germination testing. Seeds underperforming in their tests jeopardized the security that curators had invested in the whole collection, and often the only recourse left was to propagate the seeds and hope that enough of them grew into plants that would provide new seeds with renewed reproductive vigour. Through propagation, the clock was, in practice, reset. Once new seeds were harvested from the plants grown in the greenhouses of the bank, they formed a new collection that bore the same identifying features as the parent collection—such as name and place of origin—but their claim to a home outside the bank was now once-removed. Products of artificially mediated reproduction, this new collection made of seeds from plants grown from banked seeds also needed to be cleaned, screened, and counted before being banked. The cycle of invisible labour of both plants and curators continued into the next generation.

I spent time with Stacy, who worked in the greenhouses of the Millennium Seed Bank and who has been struggling to get some plants to grow further than a few centimetres, others to flower, and yet others to bear fruit. She suspected that a particular plant from South Africa, which she had been growing in pots since 2006, had not flowered because of a missing symbiotic relationship with a bird or an insect, which she had not been able to recreate. Stacy was growing these plants out from seed because the herbarium specimen that was supplied with the seeds was inadequate for providing a definitive identity for the plants. If/when flowers and fruit emerged, they would be dried and used to complete the sheet identifying the plant. Reanimating Darwin's work after nearly 200 years, Stacy fashioned tools that created affective interspecies interactions to understand what could have been holding these plants in sexual dormancy. Having tinkered, prodded, and nudged the plants in as many ways as she could muster, she resigned herself to the idea that she did not have ultimate control over their reproductive desires. While frustrated at the plants' stubbornness, she was not surprised and asked, "Why should we expect these plants to thrive without their ecological companions?"[19] Analysing Charles Darwin's treatise on orchid fertilization published in 1888, historian Carla Hustak and anthropologist Natasha Myers ask, "What if the topology of insect/orchid encounters were conditioned not just by a calculating economy that aims to maximize fitness but also by an *affective ecology* shaped by pleasure, play, and experimental propositions?"[20] Stacy's frustration with being unable to force the plants to flower in the greenhouses at the bank is a reminder of our patchy understanding of the affective ecologies of reproduction.[21] The invisible labour of plant reproduction evades capture by the standardizing forces of science. The endeavour of seed banking contains within it the conditions of its own destruction because it is unable to acknowledge our still limited control over the invisible vitality of plants.

If Stacy is not able to coax a plant into producing flowers and fruits, its identity will remain unknown. The seeds of this plant are saved in the bank, but what value could they have without a name or the ability to grow to full maturity and reproduce? These problems are by no means the norm, but their existence begs important questions about the outcomes and efficacy of banking seeds away from their natural homes. Simultaneously engaging both natural and laboratory landscapes, Stacy's invisible work also illustrates the flaws in the idea of an epistemic binary between theoreticians and naturalists in biology. A plant fertility expert, Stacy uses theoretical knowledge built from centuries of colonial botany to inspire practical solutions in her greenhouses.

Even if the biology of the seeds cooperates (and this is a big if), I have shown how the labour of care associated with a single seed collection is multiplied through the lifespan of the original seeds and also all of their future descendants. Seed bank managers, aware of the magnitude of labour required to maintain a collection in perpetuity, try to convey these multiplicative factors to their colleagues soliciting funding from the governments, agencies, charities, and corporations: investors who believe in the promise of ensuring plant life for the future through the salvage work of seed banking. Unfortunately, donors are decidedly less attracted to funding the maintenance of existing collections, no matter their intrinsic value.[22] Funders like to attach their support to projects that promise novelty, either by expanding collections to find threatened or hard-to-reach species or searching with neocolonial passions for hitherto unknown but potentially valuable plants in threatened or hard-to-reach places.[23]

Eschewing any objective narrative of purification, thinking with the "dance of encounters" between seed scientists and the seeds shows the complex workings of authority that make specimens into facts in the seed bank world.[24] Highlighting the subjectivities in the care for seeds does not detract from the capacity of seed banking to secure biodiversity. Rather, exploring how human choices affect the results of saving seeds enriches our understanding of the complex invisible processes that scaffold the institution of seed banking.

NOTES

1. Rodney Harrison, "Freezing Seeds and Making Futures: Endangerment, Hope, Security, and Time in Agrobiodiversity Conservation Practices," *Culture, Agriculture, Food and Environment* 39, no. 2 (2017): 80–89, https://doi.org/10.1111/cuag.12096; Sara Peres, "Saving the Gene Pool for the Future: Seed Banks as Archives," *Studies in History and Philosophy of Science Part C: Studies in History and Philosophy of Biological and Biomedical Sciences* 55 (2016): 96–104; Thom Van Dooren, "Banking Seed: Use and Value in the Conservation of Agricultural Diversity," *Science as Culture* 18, no. 4 (2009): 373–95.

2. On the politics of this "salvage" style extraction and accumulation, see Joanna Radin, *Life on Ice: A History of New Uses for Cold Blood* (Chicago, IL: University of Chicago Press, 2017); Kim Tallbear, *Native American DNA: Tribal Belonging and the False Promise of Genetic Science* (Minneapolis: University of Minnesota Press, 2013).

3. Helen A. Curry, "Gene Banks, Seed Libraries, and Vegetable Sanctuaries: The Cultivation and Conservation of Heritage Vegetables in Britain, 1970–1985," *Culture, Agriculture, Food, and Environment* 41, no. 2 (2019): 87–96, https://doi.org/10.1111/cuag.12239.

4. Tracey Heatherington, "Seeds," in *Anthropocene Unseen: A Lexicon*, eds. Celia Howe and Anand Pandian (Santa Barbara, CA: Punctum Books, 2020), 405–9.

5. Horace Tabberer Brown and Fergusson Escombe, "Note on the Influence of Very Low Temperatures on the Germinative Power of Seeds," *Proceedings of the Royal Society of London* 62 (January 1, 1898): 379–87, https://doi.org/10.1098/rspl.1897.0088.

6. Fergusson Escombe, "Germination of Seeds: The Vitality of Dormant and Germinating Seeds," *Science Progress* 6, no. 5 (1897): 585–608, https://www.jstor.org/stable/43414678.

7. Brown and Escombe, "Very Low Temperatures," 163.

8. Xan Sarah Chacko, "Digging Up Colonial Roots: The Less-Known Origins of the Millennium Seed Bank Partnership," *Catalyst: Feminism, Theory, Technoscience* 5, no. 2 (2019): 1–9.

9. Donna Jeanne Haraway, *When Species Meet* (Minneapolis: University of Minnesota Press, 2008), 4.

10. Xan Sarah Chacko, "Creative Practices of Care: The Subjectivity, Agency, and Affective Labor of Preparing Seeds for Long-term Banking," *Culture, Agriculture, Food and Environment* 41, no. 2 (2019): 97–106.

11. For more on how a change of perspective and attention can yield a more nuanced understanding of materials and practices, see Jane Bennett, *Vibrant Matter: A Political Ecology of Things* (Durham, NC: Duke University Press, 2010); Timothy Choy and Jerry Zee, "Condition—Suspension," *Cultural Anthropology* 30, no. 2 (2015): 210–23.

12. On the standardization of the messy present in order to imagine a manageable future, see Susan Leigh Star, "Power, Technology and the Phenomenology of Conventions: On Being Allergic to Onions," in *A Sociology of Monsters: Essays on Power, Technology and Domination*, ed. John Law (London: Routledge, 1990), 26–56.

13. For an example from palaeontology, see Caitlin Donahue Wylie, "'The Artist's Piece Is Already in the Stone': Constructing Creativity in Paleontology Laboratories," *Social Studies of Science* 45, no. 1 (2015): 31–55.

14. Joanna Radin, "Latent Life: Concepts and Practices of Human Tissue Preservation in the International Biological Program," *Social Studies of Science* 43, no. 4 (2013): 484–508.

15. Mary Douglas. *Purity and Danger: An Analysis of Concepts of Pollution and Taboo*, (London: Routledge, 2002), 2.

16. Ibid.

17. No special accreditation is required to use the X-ray machine as part of a seed curation team. However, the training of the scientists to attune their eyes to interpreting the images is a slow incremental process. I asked a curation team leader how long it took to get comfortable with interpreting X-ray images and she said, "Ten years and counting." See Chacko, "Creative Practices of Care," 102–3.

18. Evelyn Fox Keller, *A Feeling for the Organism: The Life and Work of Barbara McClintock* (London: Macmillan, 1984); Carla Hustak and Natasha Myers, "Involutionary Momentum: Affective Ecologies and the Sciences of Plant/Insect Encounters," *differences* 23, no. 3 (2012):

74–118; Eva Hayward, "Fingeryeyes: Impressions of Cup Corals," *Cultural Anthropology* 25, no. 4 (2010): 577–99.

19. Interview with Stacy (name changed to preserve anonymity), February 2016. My decision to anonymize my research subjects also creates invisibility, but it is a required attribute of the permission granted for me to conduct fieldwork by the Internal Review Board at the University of California, Davis in 2015. For more on IRB, see Radin, chapter 20.

20. Hustak and Myers, "Involutionary Momentum," 78. Emphasis in the original.

21. Susannah Chapman and Xan Chacko, "Seed: A Feminist Keyword," *Feminist Anthropology* 3, no. 1 (2021): 1–9.

22. Oreskes, Naomi, "Objectivity or Heroism? On the Invisibility of Women in Science," *Osiris* 11 (1996): 87–113.

23. This section is based on interviews and correspondence with Bruce Pengelly in February 2020. For a longer engagement with funding for seed banks, see Xan Chacko, "Stringing, Breaking, and Remaking the Colonial 'Daisy Chain': From Botanic Garden to Seed Bank," *Catalyst: Feminism, Theory, Technoscience* 8, no. 1 (2022): 1–30.

24. Haraway, *When Species Meet*, 4.

17

Oneida Inscriptions

Judith Kaplan

This chapter is about an archive of Indigenous labour that was almost overlooked. More specifically, it addresses sources for studying the history of efforts to document, analyse, and revitalize the Wisconsin dialect of the Oneida language, a critically endangered member of the Northern Iroquoian group.[1] Through a discussion of efforts to develop and apply a "scientific" orthography—one that details the inscription of inaudible or whispered utterance-final forms in Oneida—I show that the spoken and social environment are essential to the apprehension of meaning, and with it, the research process. In a departure from other "salvage" projects in the early-twentieth-century history of Americanist linguistics, the initial study of Wisconsin Oneida did not rely on recording technologies or extensive fieldwork. Rather, native speakers were encouraged to self-inscribe their language. Through their paid labours, the faintest of sounds came to be seen through writing and translated. Through the precarity of the archive, written records of their voices were almost lost.

THE GRAMMAR OF SCIENCE

Floyd Lounsbury submitted his doctoral dissertation on the subject of Iroquoian morphology to the Faculty of the Graduate School of Yale University in 1949. This was the basis of his subsequent monograph, *Oneida Verb Morphology*, which appeared under the University's imprint in 1953.[2] It was a study of lasting personal significance for Lounsbury, who was immediately awarded a Yale professorship upon its completion. It was also, by all accounts, a scientific success. According to one estimation, "it established the basic framework and terminology that has been applied to the analysis of Iroquoian languages ever since."[3]

Just what did this "framework" look like? The most fundamental problem for Lounsbury at the outset was to identify the relevant units of linguistic analysis. In a departure from the abstract ("morphophonemic") practices of Leonard Bloomfield, Charles Hockett, and others, he advocated for a data-driven approach (a method he called "morpheme alternants").[4] Lounsbury attempted to segment the most basic units of meaning-bearing forms through repeated elicitations that would vary one element (e.g., number) while holding everything else constant (gender, person, and so on). From there, he noted which segments changed, designating the constant a unit. Thus identified, he went on to analyse the formal structure of Oneida in terms of these units, an operation known as "tactics."

The task of analysing Oneida in this way—one that was geared towards transcription rather than the pragmatics of free speech—was especially difficult given the polysynthetic and incorporating nature of the language. As Lounsbury put it, "The place to make a cut between two morphs may be obvious or obscure, and the business of segmenting may be correspondingly easy or difficult according as the data permit of but one manner of interpretation or of several." Oneida language data, it so happens, allows for several such interpretations. He continued, "In languages of the fusional type . . . the place to make the cut may often be decided upon only with difficulty and to some extent arbitrarily, and morphemes in general have many allomorphs. The Iroquoian languages are of this type."[5] What did Lounsbury mean by this? In polysynthetic and incorporating languages like Oneida, nouns and pronoun references are largely folded into the verb itself; prefixes and suffixes abound. For this reason, segmentation appeared difficult—if not arbitrary—to him as a native speaker of English.

One particular challenge in the process was that Oneida characteristically drops utterance-final sounds. Even silences, in other words, can have distinct and literal meanings. For example, compare the following, where /e/ (*to go*) is the root of the verb:

jiten. (*let's (you and I) go, final position*) vs. jitne (*medial position*)
jitow. (*let's (we pl. inclusive) go, final position*) vs. jitwe (*medial position*)[6]

Reflecting on these and related examples, Lounsbury asked, "What then is the locus in the forms, or for the hearer, of the meanin[g] . . . in these examples?" To which he responded:

> It is in the *cumulative implication* of the other morphs. It remains a fact that *the linguist's allocation of a meaning corresponds to only a part of the native hearer's source of derivation of a meaning*, and the correspondence is considerably less close for fusional languages [e.g. Oneida] than for agglutinating languages.[7]

Though not the focus of his study, Lounsbury provided further evidence of the complete loss of the main verb root in utterance-final position. Comparing takná. jih. (*give me the kettle*, imperative, final position) with takná.jyu (*give me the kettle*, imperative, medial position), he demonstrated how the spoken environment might govern morpheme loss in this way.

While these examples may seem quite technical, the fundamental idea is easy to appreciate: it is important to know something about the context in which a given form is spoken in order to apprehend its meaning correctly. This underscores issues of *process* in linguistic data collection and analysis—the transition from orality to textuality—where meaning is just as likely to be lost as it is to be preserved. In this regard, knowing something about the experience of the "native hearer" is necessary to an accurate account of labour relations in the history of linguistic research.

In his 1953 monograph, Lounsbury formally recognized material support from the Wenner-Gren Foundation, the Works Progress Administration, and the University of Wisconsin in his "Acknowledgement." He elaborated further on his debts as a researcher in the book's "Preface" section:

> The materials on which the present volume is based were collected at Oneida, Wisconsin, in 1939–40. I was at that time placed in charge of a project, sponsored by the University of Wisconsin and carried out under the Works Progress Administration, the purpose of which was to employ a small group of bilingual Oneida Indians in the writing of texts in their native language. An initial period of training in phonemic spelling prepared the group for this work. The texts which they wrote and the grammatical and other notes which I accumulated at that time are the basis for the analysis of Oneida morphology presented here. . . . I wish to express gratitude to several of my teachers who have aided me and have been involved in one way or another in this work: to Professor Morris Swadesh . . . Professor Freeman Twaddell . . . Professor Wendell Bennett . . . and to Professor Bernard Bloch. . . . Lastly, to the group of Oneida Indians who cooperated with such interest in our language project I wish to express my appreciation and sincerest thanks.[8]

The experiences of these unnamed Oneida linguists might well have remained invisible to researchers outside of the community. Their autobiographies were recorded in a cache of 167 spiral-bound notebooks that were brought back to the University of Wisconsin-Madison in the midst of World War II. These pencilled inscriptions were eventually consigned to a vast basement storeroom in the Social Sciences Building overlooking Lake Mendota. Forgotten there for some sixty years, the notebooks were recovered only in 1998 by dint of some "anthropological sleuthing" on the part of Professor Emeritus Herb Lewis.[9] Though Lewis had originally been looking for biographical materials on Lounsbury, he immediately recognized how significant the Oneida notebooks would be for cultural anthropology and revitalization efforts. The collection was swiftly processed and transferred to the Wisconsin State Historical Society: copies were "donated" to the Oneida Nation.[10]

A PRECARIOUS ARCHIVE

In one of these notebooks, Andrew Beechtree, a resident at Oneida, recorded his impression of the Works Projects Administration (WPA) intervention in 1941:

> During these lean years of the Great Depression, the whole nation was hard pressed to furnish employment to its millions of unemployed inhabitants. The result was the national, state, local government officials and every other capable organization was thinking up work or occupations of many types and descriptions. It will take no great stretch of imagination to be convinced that the Indian early became part of this great army of unoccupation.

As a member of the "great army of unoccupation" himself, Beechtree was paid for his account as a participant in the Oneida Ethnological Study, which put him to work, like so many others, on public improvement following the Great Depression. This historical study was successor to the Oneida Language and Folklore Project. As Beechtree recalled:

> The project affecting us, the Oneidas, had for its object the recording, for the first time, of the language . . . of the Oneidas, in a methodical or scientific manner. . . . The results from this undertaking were so satisfactory and interesting that a correlated research project, the historical study, was immediately sponsored and approved. This embraces the writing of biography, autobiography, and consulting newspaper and other records.[11]

Participants were paid $45 per month for their labours, which organizers hoped would ultimately contribute to the production of an Oneida dictionary, hymnal, chronicle, and folklore collection—source materials for subsequent linguistic analysis and revitalization initiatives among younger members of the community.[12]

This salary was urgently needed. As others have noted, The Great Depression hit the already vulnerable Wisconsin Oneida community especially hard.[13] Members had been forced to relocate and were then dispossessed of their lands by the Dawes General Allotment Act of 1887. Prior to this piece of legislation, they had shared some 65,436 acres in Wisconsin; by the New Deal era, this had been reduced to less than 800 acres held primarily in individual family allotments. Having been cheated out of their landholdings by exploitative taxation laws, many abandoned agriculture and sought industrial jobs in towns nearby like Green Bay and DePere, or further afield in the major urban centres of the Upper Midwest—Chicago, Detroit, Milwaukee, and Minneapolis.[14] Not only did this accelerate language loss and fragment the community, many of those jobs disappeared with the shock of the Great Depression.[15]

Privileged researchers and research institutions also felt its impact. Despite early financial investments in Americanist linguistics during the mid-1920s, by 1937 all sources of private funding (e.g., Carnegie money) had dried up. The University of Wisconsin was likewise struggling, which motivated an urgent campaign for federal subsidies to help make ends meet. No fewer than twenty proposals were submitted for WPA grants in the year 1938–1939, and nineteen of them were successful. The proposal for a "Phonetic vocabulary of the Oneida Indian language and the accumulation of native folklore" was awarded $10,928 and was set to begin work under the direction of Morris Swadesh in January 1939.[16]

Employment security was something that long eluded Swadesh—a student of Edward Sapir and the controversial pioneer of lexicostatistical glottochronology.[17] His tenure with the University of Wisconsin did not last long, and he was poised to leave for a new job in Mexico before the Language and Folklore Project had even officially commenced. Swadesh tapped Lounsbury, an undergraduate mathematics major who had audited a couple of his classes, to take over its direction in his stead.

The Language and Folklore Project ran for nineteen months, during which time two dozen members of the community took part in a short training session meant to impart the novel orthography. A respite from outdoor labour during the harsh Wisconsin winter, it seems to have generally garnered a positive reaction from those who took part. We know this because of the personal reflections they wrote down in the notebooks that were lost for so long. Oneida community member Guy Elm, for example, wrote the following:

> For the last two years I have been working on the Oneida Language Project, sponsored by the University of Wisconsin here in Oneida. I am very much interested in my work. We are writing all kinds of Indian stories, jokes, and the Oneida history. Someday I hope to see it published in books so that the people can read it and find out for themselves what Oneida people really are—bad or good.[18]

In line with Elm's wishes, his words *were* eventually published, alongside this account from a neighbour who was not employed in the WPA research projects:

> I can't speak my own language, but I can understand it very well and I certainly wish I went to those Oneida language classes when they had them. I think I could have learned very easily, as I do understand the language quite well, but my children kept me from going as they were quite small and I could not leave them so I just missed my chance.[19]

Testimonials like these—misplaced for decades in a basement without climate control—have been used by researchers, teachers, and activists in the years since their rediscovery. Their precarity challenges the idea that textual language is somehow more conservative than spoken language. Put differently, this case study shows that the history of Americanist anthropology involves endangered archives as well as endangered languages.[20] We see research foci, accordingly, moving in and out of the visual field at distinct layers of the WPA-sponsored research. First, pertaining to the outputs of the Language and Folklore Project, the move from spoken to textual language can be described as a process that conceals some content (pragmatic context, nuances of spoken Oneida) in order to reveal other features (morphological units). Lounsbury's definitive study also neglected to give the Oneida linguists who took part the same degree of intellectual recognition accorded to his professors and colleagues. Second, with regard to materials produced by The Oneida Ethnological Study, we are reminded of the fact that textual materials, like oral tradition, can be lost if preservation is not actively valued.

DOCUMENTARY AND ARCHIVAL PROCESSES

How do these reflections on oral and textual culture correspond to the history of descriptive, or documentary, practice in linguistics?[21] Morris Swadesh was well known as an innovator in linguistic method—work he laid out explicitly in correspondence and published papers. A primary consideration, as he made clear in a private exchange with colleague Dell Hymes, was the choice of interlocutor. For instance, in support of Hymes' graduate research on Sinslaw, an Indigenous language of the Pacific Northwest, Swadesh gave the names of potential "informants," noting their ages alongside other socio-linguistic observations. One of them, Billy Dick, age seventy-one, "last summer was still working regularly on the logging boom," Swadesh told Hymes. "He sometimes drinks heavily. His fault is that he tries too hard to form expressions to meet your request."[22] Though full consideration of the pathways and infrastructures by which linguists like Swadesh got to know their interlocutors is beyond the scope of this chapter, this is one shadowed corner of research practice that might be illuminated by making constituent labour practices more readily visible.[23]

But what of the linguist's research design? In the case of the WPA-Oneida research, the very conditions that threatened to extinguish the Wisconsin dialect—forced acculturation to English in government-run schools—facilitated research design, which depended on bilingualism and self-inscription. This followed a violent history: as the University of Wisconsin stated in its 1999 press release on the recovery of the notebooks, "One woman recounted that she had been reprimanded for speaking Oneida in school by having 'a rag tied around my mouth all one day.'"[24] In a departure from traditional linguistic fieldwork, Oneida speakers in these projects interviewed each other directly without the day-to-day intervention of academic researchers.[25] As the *Milwaukee Journal* reported in February 1939, this made the Wisconsin study "the only white collar WPA project among the Indians."

That said, Lounsbury imposed his own selection criteria on the records produced by men like Andrew Beechtree and Guy Elm. Swadesh handed him explicit directives: "Be sure that the native word corresponds with the meaning intended. Use the semantic key, and watch out for ambiguities in the English like these. *back* body part rather than direction, *bark* of tree, *blow* with mouth, *burn* intransitive, *child* young person rather than kinship term, *day* opposite of night rather than abstract measure," and so on.[26] These directives illustrate the enduring point made by theologian Walter Ong that oral "meaning is not assigned but negotiated, and out of a holistic situation in the human life world"—even if such ambiguities are ultimately concealed from readers of the final publication.[27]

Those members of the Oneida Nation who contributed their voices to the projects have eventually been recognized. Indeed, visitors to the Language Archives of the Oneida Nation webpages find those collections organized by contributors' names.[28] Nevertheless, invisibilities persist as a necessary part of the move to text. Scientific collections are often remarkably contingent and fragile in nature, despite claims as to their stability and permanence.[29] We see this in the material history of the Oneida

Ethnological Study notebooks, the misplacement of which likely happened when Lounsbury was abruptly called to serve in World War II.[30]

Such contingency and fragility points to an even more fundamental problem for the language historian and/or the historian of linguistics: the relative conservatism of oral versus written forms of language transmission. In a sense, the basic commitment of these studies—reflecting "salvage anthropology" more generally—was that inscriptions (whether in print, audio, or lately video media) stood a better chance of surviving over time than did oral tradition alone. The archival impulse here was to preserve the ephemeral present for the enduring future by "scientifically" writing it down. And the stakes were undoubtedly high: according to UNESCO standards, the Wisconsin dialect of the Oneida language was "critically endangered" throughout the twentieth century, with just three native speakers still living in 2011, all of them over ninety years old.[31]

This story has a message for anyone who may question the applicability of labour analyses to intellectual activities: it was absolutely clear to participants in the WPA projects that the Oneida linguists and historians were in fact labouring. For example, Swadesh directly expressed his progressive commitments to project member John A. Skenandore in a 1939 letter. Swadesh was just about to leave for Mexico, and was writing to show his appreciation for Skenadore's involvement, emphasizing its value *as labour*:

> Even if I'm not there I still want the project to succeed and I appreciate what you are doing to keep it going in the best way. I particularly realize that the Workers Alliance is necessary to keep the WPA. I know that if it wasn't for the Workers Alliance there would be a big cut in the WPA and the Oneida language project never would have got started. I want to show my appreciation to the members of the project who belong to the Workers Alliance. . . . I want to give you as a present a two-month subscription to the Daily Record. . . . It has news about the labor movement and tells the true facts about what is going on in the world from the worker's point of view. . . . Many newspapers tell things about the WPA to make it seem like the people are only killing time, loafing, wasting money. They try to make people believe that WPA workers are lazy, dumb, useless. They want people to believe that labor unions are vicious and radical and destruction [*sic*] and dishonest. People ought to know the true facts so they can form correct opinions. That's why we ought to read a newspaper that shows the other side.[32]

Not only did Swadesh reveal a keen fundraising acumen with these words, he also defined cultural expression and stewardship as labour in explicit terms. Stadler King, who conducted several interviews as a WPA labourer, gave proud voice to "the other side," shedding light on the Oneida contribution while decentring the role of university researchers and government patrons:

> We have made words for [an] Indian dictionary. We have written stories which we solicited from many of the Oneidas. . . . We have translated all the Indian stories and the words for the dictionary into English. We have written biographies of some of the older people. We have written in Indian and translated some old Indian medicines which they used long ago. All the work has been done by Oneidas under the supervision of some university students.[33]

On this telling, the administrative labour of some unnamed university students was pushed to the background, offsetting the more visible documentary labour of the Oneidas.

Historical understanding of these WPA projects may well have been inverted if the notebooks' pencilled pages had faded or never been found. Part of the significance of this case study therefore lies in its demonstration of the processes that variously render research labour visible or invisible. Still, questions remain as to how historians may interpret the tight bond between endangerment, salvage, and inscription; what negotiations gave rise to the "methodical or scientific" writing system employed in these studies; and what power inheres in the structures of English as a medium of translation and access.

NOTES

1. See the expanded entry for Oneida with data on size and vitality in David Eberhard, Gary Simons, and Charles Fennig, eds., *Ethnologue: Languages of the World*, 24th Edition (Dallas, TX: SIL International, 2021). Online version: http://ethnologue.com, accessed October 20, 2021.

2. Floyd G. Lounsbury, *Oneida Verb Morphology*, Yale University Publications in Anthropology, 48 (New Haven, CT: Yale University Press, 1953).

3. Wallace Chafe and John Justeson, "Floyd Glenn Lounsbury," *Language* 75, no. 3 (1999): 563.

4. On these prominent figures in the history of American structuralism, see Dell Hymes and John Fought, *American Structuralism* (The Hague and New York: Mouton, 1981). For a more selective discussion targeting issues of interest to historians of science, see Greg Radick, "The Unmaking of a Modern Synthesis: Noam Chomsky, Charles Hockett, and the Politics of Behaviorism, 1955–1965," *Isis* 107: 49–73.

5. Lounsbury, *Oneida Verb Morphology*, 12.

6. Lounsbury used a period /./ to symbolize a zero morph, which would not make a sound. The /./ in "jiten." and "jitow." makes the inaudible visible. The striking point about these examples is that the very root of the verb /e/ is dropped when it is at the end of the utterance.

7. Lounsbury, *Oneida Verb Morphology*, 17.

8. Lounsbury, *Oneida Verb Morphology*, 5.

9. "Rediscovered Native History Notebooks Donated to Oneida," University of Wisconsin Press Release, June 1, 1999. University of Wisconsin Archives, Morris Swadesh Faculty File.

10. "Rediscovered," The University's use of the word "donated" suggests that they did not recognize Oneida authorship.

11. Beechtree, cited in Herbert Lewis, "Introduction," in Idem. (ed.), *Oneida Lives: Long-Lost Voices of the Wisconsin Oneidas* (Lincoln: University of Nebraska Press, 2005), xvii.

12. Lounsbury, *Oneida Verb Morphology*. On revitalization in the community, see *The Oneida Speak*, producer, Michelle Danforth, Wisconsin Public Television, Native American Public Telecommunications, Lincoln, NE: Vision Maker Media, 2006. According to the CPI Inflation Calculator of the U.S. Bureau of Labor Statistics, this corresponds to $889.04 in 2021.

13. Jack Campisi and Laurence Hauptman, "Talking Back: The Oneida Language and Folklore Project, 1938–1941," *Proceedings of the American Philosophical Society* 125 (1981): 441.

14. Campisi and Hauptman, "Talking Back," 441; *The Oneida Speak*, Michelle Danforth.

15. Robert Ritzenthaler, "The Oneida Indians of Wisonsin," *Bulletin of the Public Museum of the City of Milwaukee* 19 (1950): 14.

16. Campisi and Hauptman, "Talking Back," 443.

17. For Swadesh's biography, see Dell Hymes, "Morris Swadesh: From the First Yale School to World Prehistory," in *The Origin and Diversification of Language*,ed. Joel Sherzer (Chicago, IL: University of Chicago Press, 1967). On his politics, see David Price, *Threatening Anthropology: McCarthyism and the FBI's Surveillance of Activist Anthropologists* (Durham, NC & London: Duke University Press, 2004), chapter 5. On this controversial methodology, see Judith Kaplan, "From Lexicostatistics to Lexomics: Basic Vocabulary and the Study of Language Prehistory," *Osiris* 32 (2017): 202–23.

18. Guy Elm, "Report on Economic Conditions before Relief Came into Effect in Oneida," in *Oneida Lives*, 82.

19. Mrs. James Skenandore, "Struggling in the Depression," *Oneida Lives*, 131.

20. See Judith Kaplan and Rebecca Lemov, "Archiving Endangerment, Endangered Archives: Sound Collection, Archiving and Analysis in Americanist Linguistics and Anthropology," *Technology and Culture* 60 (2019): 161–87.

21. On the rise of documentary linguistics, see work by Andrea Berez, for example, her co-authored chapter with Ryan Henke, "Language Archiving," in Kenneth Rehg and Lyle Campbell, *The Oxford Handbook of Endangered Languages Online* (Oxford: Oxford University Press, 2018): 1–26.

22. Swadesh to Hymes, 1/2/1954. APS Ms. Coll. Dell H. Hymes Papers No. 55 Series I, Morris Swadesh correspondence.

23. The role of the Bible translation organization, SIL International, is particularly salient in this regard. See the special issue of *Language*, September 2009, on this topic.

24. "Rediscovered Native History Notebooks Donated to Oneida," University of Wisconsin Press Release, June 1, 1999. University of Wisconsin Archives, Morris Swadesh Faculty File.

25. Lewis here reports on the details: "Bilingual and literate Oneidas were invited to apply to work on the project, and about twenty were selected for preliminary instruction and testing. Eventually, two women and nine men were employed on the project for eighteen months." (Lewis, "Introduction," *Oneida Lives*, xxiii.)

26. APS Ms. Coll. Floyd Lounsbury Papers No. 95, Folder 15: Morris Swadesh correspondence. The underlined words are part of Swadesh's 200-word list of "core vocabulary" that he used in his lexicostatistical method. See Morris Swadesh, "Lexicostatistic Dating of Prehistoric Ethnic Contacts," *Proceedings of the American Philosophical Society*, 96 (1952): 452–63.

27. Walter Ong, "Before Textuality: Orality and Interpretation," *Oral Tradition* 3 (1988): 267.

28. See https://oneida-nsn.gov/Language/Archives/, accessed September 1, 2015. Also see the entries in Karin Michelson, Norma Kennedy, and Mercy Doxtator, eds., *Glimpses of Oneida Life* (Toronto, Buffalo, London: University of Toronto Press, 2016).

29. Rebecca Lemov, *Database of Dreams: The Lost Quest to Catalog Humanity* (New Haven, CT: Yale University Press, 2015).

30. "Rediscovered." There are interesting resonances here with the case studies by Jenny Bangham (chapter 15) and Lara Keuck (chapter 21) in this volume.

31. See http://uwdc.library.wisc.edu/collections/Oneida/about accessed September 1, 2015.

32. Swadesh to Skenandore, April 4, 1939. APS Ms. Coll. Floyd Lounsbury Papers No. 95, Folder 15: Morris Swadesh correspondence.

33. King, cited in Lewis, "Introduction," *Oneida Lives*, xxxiv.

18

Making Movement Matter

Whitney Laemmli

In 1981, the American folklorist Alan Lomax described his dream of a "great library of the visual arts, where all important cinematic documents would be stored, catalogued, and analyzed." This "temple of knowledge," he argued, would "cost no more than an atomic submarine, but its influence would far outrun the famed library of Alexandria or, indeed, all the libraries that ever existed, since it would preserve a living, moving record of all human behavior."[1]

Existing archives, Lomax mused, were filled to bursting with novelists' early manuscripts, documents detailing the machinations of political leaders, and the scribblings of scientists. The picture of history these texts left, however, was fundamentally incomplete, as it contained no trace of the embodied experiences of everyday life. How did a young woman in Montana walk down a street in 1917? How did a middle-aged man in Auckland greet a friend in the early nineteenth century? What physical techniques did Gambian groundnut farmers use to extract their harvest in the 1940s? These movements and gestures, Lomax contended, were just as important to the historical record as their textual counterparts but had long been neglected.

Lomax hoped to help rectify the situation, and, in many ways, he was well-qualified to do so. By the time Lomax turned his attention to the body's movements, he was almost certainly the most famous folklorist in the United States, known for his role in the "discovery" of musical legends like Jelly Roll Morton, Lead Belly, Woody Guthrie, and Muddy Waters. Born in Texas in 1915, Lomax spent his childhood accompanying his father, John, himself a folklorist, on his trips to record the songs of prisoners and sharecroppers in the rural South. After completing his undergraduate education at the University of Texas and at Harvard, Alan participated in his father's work more fully, joining him on expeditions for the Library of Congress, and co-authoring two books, *American Ballads and Folk Songs* (1934) and *Negro Folk Songs as Sung by Lead Belly* (1936). Alan also began embarking on his own expeditions, including a trip with Zora Neale Hurston in the summer of 1935. In 1937, he was appointed the inaugural "Assistant in Charge" of the Archive of Folk Song at the Library of Congress. As his influence grew, Lomax hosted popular television and radio programs, produced dozens of folk records, recorded oral histories with jazz musicians, and helped pioneer the "man on the street" interview. Whether interviewing ordinary Americans in the wake of the attack on Pearl Harbor or recording the songs of southern sharecroppers, Lomax saw his folkloric career as a form of political work, making the unseen seen, the silent heard, the intangible felt.[2]

It was thus with similar aims that, beginning in the 1960s, Lomax broadened his focus from the capture of sound to the recording of movement. He focused particularly on dance, arguing, among other things, that dance needed to be taken seriously as a critical form of cultural labour. And while a full realization of his imagined "temple of knowledge" might have remained out of reach, Lomax was determined to make a start, setting out to gather at least one filmed sample of dance from every culture on the planet. In terms of sheer volume, he was remarkably successful: between approximately 1965 and 1985, Lomax and his collaborators—dance experts Irmgard Bartenieff and Forrestine Paulay—obtained more than 250,000 feet of dance footage from nearly 2,000 cultural groups. They also analysed the films in great detail, confident that dance was an untapped resource for charting humanity's past and shaping its future. They called the project "Choreometrics."[3]

This chapter explores Choreometrics, focusing first on Lomax's lofty vision for a new kind of repository that would make the work performed by dance and dancers visible in new ways. In the process, he hoped to reshape both academic thought and the prospects for sustaining diverse cultures across the globe. Second, the chapter reveals the often hidden labour involved in Choreometrics' own creation—the work of the postal workers, administrators, publishers, movement experts, graduate student coders, statisticians, film-makers, and dancers upon which the construction of Lomax's temple depended. In doing so, the chapter uncovers a number of ironic contradictions at the heart of Lomax's quest and argues that reckoning with these paradoxes of visibility is crucial for engaging with Choreometrics and its descendants today.

BUILDING THE TEMPLE

The scale of Lomax's ambitions for Choreometrics may have been unique, but he was not alone in his newfound interest in the moving body. By the 1960s, a growing contingent of anthropologists had begun to argue that language alone could not possibly capture the functioning of a culture in all its complexity, contending that other sensory and corporeal experiences were equally worthy of study.[4] This attention was particularly acute in the new subfield of "kinesics." Led largely by the anthropologist Ray Birdwhistell, those working in kinesics sought to call attention to the messages communicated by the physical body, positing that non-verbal behaviour was patterned and analysable in the same way as spoken language.[5]

Lomax met Birdwhistell in 1961 when he enrolled in a special seminar Birdwhistell held on the fundamentals of kinesic research; Lomax credited the course with inspiring his new interest in dance.[6] But while the metrics for evaluating music—tone, timbre, harmony, and so on—were second nature to Lomax, he was unaware of a similarly structured system for physical movement. Again, Birdwhistell provided guidance, connecting Lomax with Irmgard Bartenieff and Forrestine Paulay, former dancers with expertise in Labanotation, a system of movement notation developed by the German choreographer Rudolf Laban in the 1920s.

Bartenieff and Paulay were enthusiastic about the project and its goals, but—despite the massive scale of the planned endeavor—Choreometrics' limited resources meant that they could only be paid for part-time work.[7] Nevertheless, their contributions were significant: not only did they provide technical expertise on dance; they also collaborated on nearly all other aspects of project development.

Before any analysis could begin, however, the Choreometrics team needed data. Lots of data. Using George Murdock's Human Relations Area Files at Yale University as a guide, Lomax made an initial list of approximately 1,900 distinct cultural and ethnic groups for analysis: the team set out to obtain at least one dance sample from each.[8] With countless phone calls and hundreds of hours of research, they reached out to anyone and everyone that might have filmed dance, from anthropologist Margaret Mead to kuru-researcher Carleton Gadjusek, to a retired vascular surgeon, adventurers' clubs, and the U.S. military. The team even unsuccessfully attempted to locate rumoured lost footage made by Roy Disney and Orson Welles. The process was long and sometimes contentious: letters negotiating fees, rental arrangements, and postage fill nearly a dozen boxes of the Lomax archive. While some institutions—such as the Institut für den Wissenschaftlichen Film in Göttingen—were eager to participate, others—including the Russian cultural ministry—required delicate diplomatic negotiations. Wealthy adventurers with outsized egos were often the most difficult to work with, requiring handling with "kid gloves."[9] This collecting enterprise, however, was only the beginning. Once the films arrived at Lomax's Columbia University offices, Lomax, Bartenieff, and Paulay began to watch and code them, looking for the "principles that unify and differentiate the movement styles of the species."[10]

Lomax, Bartenieff, and Paulay briefly considered using traditional Labanotation to make and record these observations but soon decided it was not well-suited to their needs.[11] Labanotation was enormously detail-oriented, designed to record choreographic works in their entirety for future performances or preservation.[12] The Choreometrics team, however, was more interested in the overall characteristics of movement. As such, as they made a first pass at the films, they agreed to simply write down the movement elements that jumped out especially forcefully, ultimately agreeing on a list of approximately eighty-five key traits and setting them down in the official "Choreometric Coding Book."[13]

The "Coding Book" was then shared with a larger group of previously untrained raters—mostly graduate students from a variety of disciplines—enlisted to code the remaining dance samples. Once a dance was fully coded, technicians transferred the data from coding sheets to punch cards and fed them into Columbia University's IBM 7094 computer. There, with the help of Columbia University statisticians, Lomax, Bartenieff, and Paulay began to search for meaningful patterns.

MOVEMENT REVEALED

Lomax, of course, had undertaken the project with a number of objectives in mind. His first goal was to amass evidence that cultural practices like dance were not mere window-dressing on the human experience but rather crucial to human survival. While non-human animals depended on genetic change to produce new adaptive behaviours, humans, Lomax argued, passed on knowledge about how to thrive in varied environments through symbolic cultural codes.[14] In this view, music, dance, art, and literature all had direct relationships to the fulfilment of both basic human needs and species-level survival.

Lomax further theorized that dance functioned as a particularly important storehouse of collective knowledge about adaptive bodily practices. Drawing on the Choreometrics data, he observed that the movement patterns that characterized a given society's dance also appeared in its repertory of everyday and working movements. He noted, for example, that the stooped posture and "deep shoulder rotation" characteristic of West African dance mirrored the widespread use of the short-handled grubbing hoe in agriculture and that the distinctive hand movements of carnival dancers in a champagne-producing French village had parallels in "the vintners, pacing the aisles of bottles, rotating two bottles at a time."[15] To Lomax, this suggested that dance was not simply a matter of individual artistic creativity but rather a tool for preserving and communicating the forms of movement that allowed a community to flourish. Lomax also linked dance style and climatic conditions, pointing out, for example, that Eskimo dance prominently featured explosive elements, as such movements represented "one effective way to generate heat in the extreme cold."[16] Community dances in tropical climates, on the other hand, taught gardeners the "graduated and flowing" habitus necessary for agriculture, just as dances in northern Europe and Asia schooled mountain hunters in the rigid body postures they assumed when poised to strike. In essence, the Choreometrics team contended that dance enshrined the movement patterns "of maximal importance to the actual physical survival of the culture," preserving ways of moving that were "as necessary as breath and food."[17]

Given these apparent links between dance and work, movement and environment, Lomax was all the more troubled that this kind of cultural labour was so rarely the subject of academic study. In his critiques, he echoed the more recent work of dance scholars like Lynn Brooks Matluck, who has pointed out that dance long suffered from the curse of "double invisibility": overlooked first because it was relatively difficult to record, and second, because it was so often associated with women.[18] Lomax anticipated that Choreometrics would alter this state of affairs—not only transforming the scholarly picture of *how* societies functioned but of *whose* labour ensured their continued existence.[19] Without dance, he insisted, many societies would not long endure.

Lomax's second goal for Choreometrics was about visibility in a broader sense, rooted in his alarm about the flattening effects of a global media environment increasingly dominated by the cultural products of the United States. Lomax bemoaned in particular the way in which television "impose[d] its US mainstream cultural tyrannies everywhere" and warned of an insidious process of worldwide homogenization he called the "greying of culture." "With every passing month," he warned, "we are being moulded and remade by what we are allowed to see and hear."[20] He believed, even more alarmingly, that this shaping touched even the physical body, resulting in a kind of creeping somatic standardization. Suddenly, he noted, the "confining and stiff-waisted European" habitus seemed to be almost everywhere, the "head-back, chest-out, erect posture of the North European elite . . . held up for universal admiration as the only way for a *real* human being to carry himself." From soldiers to schoolchildren, Western ways of moving proliferated like invasive species, with similarly "ridiculous and unfortunate consequences."[21]

Lomax's hope was that Choreometrics could help reverse this process.[22] Drawing on their massive stores of dance footage, Lomax, Bartenieff, and Paulay produced four documentaries that explained their work and highlighted a diverse array of movement styles (*Dance & Human History* [1974], *Step Style* [1977], *Palm Play* [1977], and *The Longest Trail* [1984]). All were intended for widespread distribution: Lomax anticipated that they would be shown both on American public television and in farther-flung places where minority or "traditional" cultures seemed threatened.

He also planned to publish a detailed handbook that would allow *any* reader to analyse dance in the same way the Choreometric specialists did, breaking it down into its constituent parts and connecting its characteristics to specific histories of work, environment, and culture. He was adamant that the system be "understandable to anyone who wanted to use it—whether sociologist, filmmaker, or schoolboy; whether American, African, or Polynesian."[23] Lomax hypothesized, moreover, that once these new lay-experts were trained to analyse dance, their day-to-day experience of non-dance movement would also transform. Newly attuned to the variety of human movement styles, novel scientific understandings would demolish old prejudices.[24] No longer would the "shuffling" movements of African Americans connote laziness; instead, they would tell a story about climatic adaptation and agricultural technology. Indeed, as Choreometrics-trained observers moved through a city, they would encounter hard evidence about the long course of human history in the body of every person they passed. A trip to the grocery store might teach as much as an afternoon at a natural history museum. As Lomax put it, the experience would be akin to "looking through a microscope or underwater for the first time."[25]

Ultimately, he hoped that these new ways of seeing would ripple outward globally. In a draft book manuscript about the project, Lomax contended, "Once these cultural traditions of movement style become visible, members of all the varied human traditions, whether they be viewers or program makers, film-makers or film goers, can no longer easily be shamed or enticed out of their birthright." They would, instead, be able to "build upon their inherited visible culture, cope creatively with the media and participate in developing the multi-channel, multi-cultural civilization that a healthy human future demands."[26] It would be a total "recalibration" of the human perceptual apparatus.[27]

VISIBILITY'S PARADOXES

Lomax's global dreams were, however, at odds with more local realities. On the most basic level, Lomax's ambitions were frustrated by both a lack of funding and on-the-ground issues that limited film distribution. While the documentaries Choreometrics produced were aired on U.S. television and screened at academic conferences, the American Museum of Natural History, and the Museum of Modern Art, there is little evidence that they ever became widely available across the globe. Lomax's book manuscript, provisionally titled *Dancing: A World Ethnography of Dance and Movement Styles* and envisioned as a chimera of academic tome and do-it-yourself movement instruction manual went unpublished. Crushed under the weight of its own ambition—and more than a thousand pages of data, photographs, and coding sheets—it simply grew too large for any publisher to take on.[28] Focused on remaking the world writ large, Lomax seemed to ignore the less glamorous, practical work required to realize his ambition.

As a result, instead of moving across time and space, the Choreometrics footage ended up sitting statically, quietly populating the shelves at the Library of Congress. When I first visited the Lomax collections in 2015, the archivist told me that not a single person had touched the Choreometrics collections in the years he had overseen them. Recently, however, there has been a renaissance of interest in Choreometrics, as scholars and activists have embarked on a new set of efforts to digitize Lomax's dance films and make them publicly available online.[29]

In many ways, these new digitization ventures represent a fulfilment of Lomax's long-ago vision, a second chance to make this wild eruption of stomps, leaps, and turns visible once more. Choreometrics' potential resurrection, however, should also prompt a re-evaluation of its practices, including a re-examination of the hidden labour that produced the films upon which the project depended.

Throughout Choreometrics' history, Lomax remained stubbornly unwilling to recognize the ways in which his distributed band of filmmakers had shaped his underlying dance "data." When he began collecting, Lomax recognized that much of the footage he was gathering was not produced for the purposes of scholarly analysis. But while acknowledging in passing that an inexperienced or biased cameraman might shape the data in minor ways, Lomax contended that movement style was so profoundly entrenched in the body that its basic elements could not help but emerge.[30] This methodological naiveté was not uncommon for the time, but it is striking in the context of Lomax's ongoing criticism of most commercial filmmaking. While decrying the work of Disney, U.S. military filmmakers, and the gentleman adventurers who disturbed the peace of remote tribes, Lomax relied upon their labour—and upon their directorial judgement. As much as he may have wanted to undermine global media hegemony, he could not extricate his work from its systems of production.[31]

It is important, therefore, that those who might encounter the Choreometrics footage today or in the future—watching, perhaps in awe, as past movements stream across some as-yet-undetermined internet portal—have some awareness of the processes that brought it into being. To ignore the work of hundreds of filmmakers, whose eyes and hands chose some images and foreclosed others, is also to allow the films to be read as transparent representations of static cultures, rather than as documents produced through complicated relations of power at particular historical moments. Indeed, even in the 1970s, some criticized Choreometrics for seeming to offer a version of movement's history uninflected by change and creolization, in which dancers and their traditions were trapped in amber, rather than evolving dynamically over time.[32]

In fact, the dancers themselves represent one final lacuna haunting the Choreometric project. They are, paradoxically, both the *most* and the *least* visible actors in its story. When watching one of the films, they are undeniably present: it is they who compel the viewer's gaze and their actions that are the object of study. Information about individual dancers—their lives, training, motivations, or aesthetic preferences—however, is conspicuously wanting.[33] How did they understand the role of dance in their own histories or communities? Why did they allow themselves to be filmed? What did they hope to get out of the process? What, even more simply, were their names? In focusing only on the images of the body, rather than the experiences or subjectivity of the dancer, Lomax provided no answer to these questions.

In part, this absence is a result of Lomax's desire to illuminate the basic movement patterns of large groups, rather than the particularities of individual styles. He had long believed that the field of folklore properly concerned the art and wisdom produced by communities and criticized American folklorists' "individualistic" reticence to focus on this collective knowledge.[34] This orientation—towards the group, away from the individual—was firmly embedded in the system of recording and analysis the Choreometrics team utilized. As a result, at least in the printed texts Lomax produced, there is very little sense of what these dances—or dancers—looked like. As Lomax himself acknowledged, excluded from the study's purview were "the sequences of movements, the gestures, the costumes, the dramas, the themes, the functions, [and] the contexts in which particular dance sequences acquire their meanings."[35]

Thus, even as Choreometrics sought to elevate dancers' work, proclaiming its worth as a subject of study and preservation, the dancers themselves disappeared. This elision is particularly ironic given Lomax's loud criticism of those who viewed dance as an unthinking, pleasurable pastime, rather than as a form of true labour, both physical and intellectual. Indeed, a tension sat at the heart of the project between two different notions of the nature of dance: one that held it up as a form of highly skilled, individual expression with an infinite degree of variation, and another that suggested it was merely a collection of basic cultural vernaculars, mimicked with varying degrees of perfection by largely anonymous practitioners. In the end, Lomax's view seemed to tend towards the latter. Even within the walls of his new temple, dancers remained mere receptacles of knowledge, rather than skilled knowledge-producers.

Tellingly, Lomax did ultimately thank the "explorer film-makers of many countries [for] generously sharing their hard-won findings with us," but not the dancers whose groaning muscles, sweating brows, and aesthetic prowess made their films possible.[36] The reasons for the absence are manifold and perhaps not uncommon: gender, race, political and economic power, as well as the prevailing tendency to denigrate bodily labour. Lomax's own inability to recognize these contradictions speaks to how deeply embedded some of these prejudices are, but it also attests to the ongoing importance of the collective endeavour represented by this volume. Particularly at a moment when new forms of representation, data collection, and archive production are proliferating, it is vital to remember how the tools we use to make phenomena visible continue to constrain what we can see.

NOTES

1. Alan Lomax, draft manuscript for "Dancing: A World Ethnography of Dance and Movement Styles," c. 1981, Library of Congress, Alan Lomax Collection, Folder 4/18–01.

2. For a biographical treatment of Lomax, see: John Szwed, *Alan Lomax: The Man Who Recorded the World* (New York: Penguin, 2010).

3. For a fuller treatment of Choreometrics, see: Whitney E. Laemmli, "The Living Record: Alan Lomax and the World Archive of Movement," *History of the Human Sciences* 31, no. 5 (December 2018): 23–51.

4. For a history of visual anthropology, for example, see: Anna Grimshaw, "Visual Anthropology," in *A New History of Anthropology*, ed. Henrika Kuklick (Oxford: Blackwell Publishing, 2008).

5. Ray L. Birdwhistell, *Introduction to Kinesics: An Annotation System for Analysis of Body Motion and Gesture* (Louisville, KY: University of Louisville Press, 1952).

6. Ray L. Birdwhistell, Letter to Alan Lomax, February 21, 1961, Library of Congress, Alan Lomax Collection, Folder 1–1/01.

7. Because he lacked a permanent institutional affiliation, Lomax often scrambled for funding, cobbling together a living from contract work for the Library of Congress, the sales of his popular books and records, and small pockets of grant money from institutions like the American Council of Learned Societies. Institutionally, Choreometrics was supported by Columbia University's Department of Anthropology as well as its Bureau for Applied Social Research, though its financial investment remained limited. The project later received grants from the Rockefeller, Ford, and Wenner-Gren Foundations as well as the National Institute of Mental Health. Paulay remained part of Choreometrics throughout its life, while Bartenieff's involvement ended in the late 1960s.

8. On the creation of the HRAF, see: Rebecca Lemov, "Filing the Total Human: Anthropological Archives from 1928 to 1963," in *Social Knowledge in the Making*, eds. Charles Camic, Neil Gross, and Michèle Lamont (Chicago, IL: University of Chicago Press, 2001), 119–50.

9. Margaret Bach, "Letter to Alan Lomax," September 4, 1972, Library of Congress, Alan Lomax Collection, Folder 9.2–3/01.

10. Lomax, "Dancing."

11. Alan Lomax, "Brief Progress Report: Cantometrics-Choreometrics Projects," *Yearbook of the International Folk Music Council* 4 (1972): 142–45.

12. For more on Labanotation, see: Whitney E. Laemmli, "Paper Dancers: Art as Information in Twentieth Century America," *Information and Culture* 52, no. 1 (January 2017): 1–30.

13. Lomax, "Dancing."

14. Lomax, "Dancing."

15. Ibid.

16. Ibid.

17. Alan Lomax, Irmgard Bartenieff, and Forrestine Paulay, "Dance Style and Culture," in *Folk Song Style and Culture*, ed. Alan Lomax (Washington, DC: American Association for the Advancement of Science, 1968).

18. Lynn Brooks Matluck, *Women's Work: Making Dance in Europe Before 1800* (Madison: University of Wisconsin Press, 2007).

19. Lomax's aims resonate with those of the feminist anthropologists chronicled in Alison Wylie, "Doing Science as a Feminist: The Engendering of Archeology," in *Feminism in Twentieth-Century Science, Technology, and Medicine*, eds. Angela Creager, Elizabeth Lunbeck, and Londa Schiebinger (Chicago, IL: The University of Chicago Press, 2001), 23–45.

20. Lomax, "Dancing."

21. Ibid.

22. Like many contemporaries, Lomax held a loosely cybernetic view of culture, and he understood dance as a crucial element in its regulation. On cybernetics in the social sciences, see, for example: Steve J. Heims, *The Cybernetics Group* (Cambridge, MA: MIT Press, 1991).

23. Alan Lomax, "Choreometrics and Ethnographic Filmmaking," *Filmmakers Newsletter*, February 1971, 24.

24. "Choreometrics—Groundwork. Progress Report—Undated," Library of Congress, Alan Lomax Collection, Box 9.1–1/02.

25. Lomax, "Dancing."

26. Lomax, "Dancing."

27. Alan Lomax, Irmgard Bartenieff, and Forrestine Paulay, "A Handbook for the Analysis of Dance," 1970, Library of Congress, Alan Lomax Collection, Folder 4–11/05.

28. "Facts on Dancing: A World Ethnography of Dance and Movement Style," c. 1980, Library of Congress, Alan Lomax Collection, Folder 39.04.01.

29. See, most notably, the work of the Association for Cultural Equity and the "Re-Imaging and Re-Imagining Choreometrics" project: http://www.reimaginechoreometrics.com, accessed April 1, 2021.

30. Alan Lomax, "Report: Sources of Films," Library of Congress, Alan Lomax Collection, Box 9.1/01–07. As Anna Grimshaw has pointed out, this uncritical view was shared by a number of early ethnographic filmmakers. See: Grimshaw, 2008.

31. As Jonathan Sterne has noted regarding nineteenth-century recordings of Native American music, the "work of anthropological cultural stewardship coincided with the decimation that necessitated the stewardship in the first place." Jonathan Sterne, *The Audible Past: Cultural Origins of Sound Reproduction* (Durham, NC: Duke University Press, 2003), 332.

32. See, for example: Joann W. Kealiinohomoku, "Caveat on Causes and Correlations," *CORD News* 6, no. 2 (1974): 20–24; Drid Williams, "Choreometrics Discussion," *CORD News* 6, no. 2 (1974): 25–29.

33. Scholars in folklore and sound studies are paying increasing attention to these kinds of omissions. See: Stephen Wade, *The Beautiful Music All around Us: Field Recordings and the American Experience* (Champaign: University of Illinois Press, 2012).

34. Alan Lomax, "Notes," n.d., Library of Congress, Alan Lomax Collection, Box 13/01–02.

35. Lomax, "Dancing."

36. Ibid.

19

Invisibility as a Mechanism of Social Ordering

How Scientists and Technicians Divide Power

Caitlin Donahue Wylie

Invisibility often indicates systemic oppression. This is true for the unrecognized labour of enslaved people who built the physical and social infrastructure of the United States (including the university where I work). Within the world of science, we also see oppression at work on the Indigenous people who guided European naturalists to collect crucial specimens (and food), and the women who have contributed to science as "computers," bubble chamber operators, technicians, and researchers in their own right yet who have not received the same renown—or even acknowledgement—as their male counterparts.[1] Here, invisibility reflects a society's divisions of power. However, invisibility can also sometimes *empower*. Laura Stark (chapter 6) illustrates a fascinating case of U.S. medical research volunteers who strove for invisibility as a way to legitimate their service. Alexandra Noi (chapter 5) documents the cases of ex-prisoners who chose to work on scientific expeditions to remote Siberian forests because of the freedom that this afforded from Soviet state surveillance. In some situations, then, freedom from scrutiny, surveillance, and documentation can allow people to structure their own behaviour, work, and knowledge. In this chapter, my study of workers in vertebrate palaeontology laboratories illustrates three usually invisible processes: defining power relations (among scientific workers), preparing scientific evidence (fossils), and learning how to study people (my journey as an ethnographer).

Invisibility—like all power relations—is dynamic. It is a process. Sociologist John Law reminds us that a single stable "social order" does not exist; there is only the ongoing, shifting activity of "social ordering."[2] Workers negotiate their labour relations in every interaction; sometimes the invisible are in charge, while in the next conversation they may be marginalized. Invisibility is therefore not always a social problem, though it is always a revealing sign of how people construct and maintain their relationships and hierarchies. As such, it is worthy of study and (sometimes) activism. Ways to investigate invisibility without assuming oppression include asking crucial questions such as *who* is invisible, *to whom*, *in what ways*, and *in what contexts*. These questions remind us, as researchers, whose perspective we are perhaps inadvertently taking—most likely that of the powerful visible.

I began my PhD thesis as a study of what I thought was a modern example of "the invisible technician" who laboured in natural philosopher Robert Boyle's seventeenth-century laboratory.[3] Boyle obscured these skilled workers and their techniques in reports of his research, portraying them only as low-status servants who were interchangeable and thus irrelevant to the outcomes of the experiments they performed. This portrayal bolstered the credibility of the evidence and knowledge that Boyle wrote about as natural and real, not as the carefully crafted achievements of skilled individuals. I saw a parallel with today's fossil preparators, who chip the rock off vertebrate fossils and thus literally shape the specimens that are the basis of our knowledge about evolution and Earth history.[4] As Xan Chacko explains (chapter 16), seed conservators do similar specimen-defining work with little documentation of their careful decisions about how to prepare seeds for deep-freezing. Likewise, Margaret Bruchac (chapter 4) highlights the challenges faced by historians attempting to figure out how someone made an object (wampum belts, in her case) when that work has no textual documentation.

Despite its importance, supposedly "technical" work in science is often omitted from written records, due perhaps to its apparent simplicity, low status, or its threat to the credibility of evidence and knowledge. Accordingly, scientists rarely describe how a fossil is prepared for study—or by whom—in publications or specimen records. They do not list preparators as co-authors, though they sometimes thank preparators in papers' acknowledgements. Preparators themselves rarely publish. There are no protocols or authoritative manuals about how to prepare fossils; instead, preparators draw from and create a variety of methods and tools, thereby tailoring their techniques to reveal, repair, and reconstruct each fossil.[5] It seemed to me, based on my undergraduate job as a fossil preparator, that preparators' work, expertise, and identities were being left out of published reports of science, to the detriment of preparators as well as fossil-based knowledge. I rushed to the rescue, to observe preparators in their labs, interview them, and tell the world what they do.

But fieldwork held surprises, as it so often does. I saw no scientist–tyrants screaming instructions at cowering, docile preparators but the opposite: preparators were controlling the physical space of the lab and the practical decision-making space of fossil preparation. Some museums even feature glass-walled fossil preparation labs in which staff and volunteer preparators are on display while working, as the very visible public faces of palaeontology.[6] In the lab, glass-walled or not, preparators made decisions. As an example of a typical interaction, I witnessed scientist Henry bringing a slab of rock containing tiny mammal fossils to the preparation lab. He told preparator Kevin that he wanted to know about the "feasibility" of "breaking this apart . . . to get little bones out."[7] This discussion could be interpreted as an expert scientist assigning a task to a less-qualified, lower-status technician. However,

Henry was *asking* Kevin about "feasibility"—he wanted Kevin's opinion. Only when Kevin said that it was possible did Henry's question turn into a work request. Then Kevin said thoughtfully, "A little acid?" and went off to try dissolving the rock with acid. He was thinking out loud rather than asking permission, and Henry gave no instructions. Deciding how to prepare the specimen—including which tools and materials to use, which tasks to do first, and even crucial judgements of what is fossil and what is rock—is the domain of the preparator.

As a result of preparators' ownership of the tasks of choosing and applying preparation techniques, scientists tend to act more like visitors to the lab and seekers of information than overbearing micromanagers. This relationship resembles other categories of low-status workers who assert power in unexpected and sometimes invisible ways, such as supposedly unskilled factory workers who know how to operate machines that their supervisors do not, coal miners who collectively slow down production to punish interfering supervisors, and slaughterhouse workers who de facto manage themselves because supervisors avoid the rooms housing the "dirty" work.[8] Histories of labour thus have much to teach us about the perspective-dependent nature of invisibility.

In addition to controlling their methods and work practices, preparators also have remarkable control over their community. For example, they organize their field by training new staff members and volunteers informally on the job, deciding who works on which fossils, and defining desired skills through their recently established society, the Association for Materials and Methods in Palaeontology.[9] They thus have "craft control," unlike many technicians who must follow strict directions and who are hired and managed by people who are not technicians.[10] For example, though they are excluded from scientists' publications, many preparators build community by giving conference talks about best practices (both at scientific societies and their own preparation-specific meetings) and by visiting other labs to observe different ways of working or to lead training workshops. A few preparators even publish papers about methods and training strategies in online forums, self-organized conference proceedings, and, occasionally, scientific journals.[11] These workers are not oppressed, I realized. Actually, they are thriving under their own leadership, both in the lab and as a profession.

I argue that this stringent division of labour creates a sheltered space for preparators to define their own methods and identity.[12] From the outside, it seems that scientists are more powerful, with their higher status and salaries, academic degrees, and published knowledge claims. And it is true that scientists are the founders and funders of fossil preparation labs, and preparators sometimes resent scientists' power. But from the perspective of everyday work, as well as the physical shape of the fossils, preparators are in charge. Seen from their view, the lab is not a place of oppression but rather of divided labour, with scientists most often writing papers and grants in their offices while preparators work with specimens and tools in the lab. Sociologist Andrew Abbott describes this construction of divided labour as a process of groups claiming "jurisdiction" over different problems, tasks, and skills, and thereby defining each group's professional identity and domain of power.[13]

The perceived divide between the jurisdictions of scientists and preparators is deepened by each group's ignorance about how to do the other's tasks. Scientists and preparators are therefore not high- and low-status members of the same group; instead, they see themselves as separate work communities. Other skilled technicians, such as in medicine and computing, likewise hold separate knowledge and identities from their bosses.[14] In sociologist Pierre Bourdieu's terms, because scientists and preparators have distinct capital (skills and expertise), doxa (knowledge and beliefs), and habitus (practices), they are separate fields.[15] This perceived distinction is both a cause and a result of severe conflicts that I witnessed in rare situations of jurisdiction violations, that is, when preparators do research or researchers prepare fossils. For example, one preparator described research as high-status work that was reserved for—and ruthlessly protected by—scientists:

> We started doing experimentation on adhesives and consolidants . . . comparing things. We started setting experiments and that was too scientific, too much like research for our boss and he put the kibosh on it. In fact, he stole our glues! . . . It was basically anything that was not simply removing rock from bones and sticking [bones] back together and putting them on a shelf was frowned upon by our boss.

This preparator suspected that this scientist-boss considers preparators' research a threat to his own status.

Likewise, preparators are offended when scientists prepare fossils, because that is not their jurisdiction. For example, when a scientist tried to prepare a fossil and broke it in the process, preparator Brent criticized him both for insulting the fossil's preparator, Jane, and for interfering in the preparators' domain: "So he broke [Jane's] specimen. God . . . He just has no respect for anyone else. This is [Jane's] effort. . . . He should just let her do the prep. Back away from the fossil" (figure 19.1). Tellingly, Brent ascribed ownership of both the fossil and the work of preparing it to the preparator, not to the scientist who had collected the fossil and would publish about it and whose grants helped fund the preparator's job. However, even if invisibility in print empowers preparators in the lab, their lower status still means that scientists' decisions override theirs, such as this scientist's decision to prepare (and damage) a fossil despite the preparators' disapproval. Because scientists have higher status, they are likely to have the power to confiscate glues or even fire preparators to prevent boundary crossing. In comparison, preparators can only complain and resist, as Brent and Jane did and as preparators in one museum did by installing locks on the lab's specimen cabinets to prevent a rogue scientist from stealthily preparing the fossils. The divisions between scientists and preparators are thus locally obvious and fiercely enforced, though with unequal consequences due to scientists' higher status.

The strictness of these boundaries both explains and is enacted by preparators' absence from scientists' papers. Omitting preparators from publications serves to set them aside from the research community, but this omission is also justified by the belief that preparators and scientists belong to separate fields.[16] Assigning invisibility is thus a dynamic and continuous method of defining groups. For the same reasons,

Figure 19.1. This fossil fish was broken by a scientist and later repaired by a preparator. As the glue reuniting the two halves of the specimen dried, the preparator tried to protect the specimen with warning notes, signalling her power over the fossil and the lab space despite her lack of power to prevent the scientist from damaging the fossil in the first place. Photograph by author.

scientists are invisible in preparators' everyday work. Both groups rely on each other, but they see their work as complementary rather than overlapping. Law explains that this "deletion" of groups is necessary to create and represent a social structure within an organization, that is, the process of social ordering.[17] He ascribes the concept of deletion to symbolic interactionists, who study local, microsocial interactions as the basis of social order.[18] While ordering a society, Law argues, people and things must be deleted, meaning excluded or omitted, to reduce complex reality to manageable patterns and to create social meaning and structures. Deletion, as a way of simplifying and focusing to represent a certain social order, is not necessarily bad. However, in his ethnography of a synchrotron laboratory, Law observes how workers "delete" other groups based on status in addition to group identity: "Outsiders tend to delete the work—and particularly the heroism—that is involved in the efforts of others. And they tend, in particular, to delete the work of subordinates."[19] Thus, problems arise when ordering includes "ranking" people and things by status and value, and accordingly deleting the unvalued as a form of "disenfranchisement" and "silence."[20] The power of deciding whom and what to delete is the site of inequality and unfairness. Law observes that, like preparators, synchrotron technicians feel underappreciated but also enjoy moments of craft control allowed by their deletion: "[Technicians] tell of autonomy, of being left to get on with a responsible job like running the machine overnight. They don't necessarily mind being ignored."[21] We see here the striking complexities of the multiple effects and perspectives of social ordering, which workers are always simultaneously enacting, resisting, negotiating, and constructing.

Without the sense of their fields' separateness and the resulting invisibilities of outsider groups, I surmise, the fossil laboratory community would look very different. If scientists had to describe fossil preparation methods in a publication, for example, then they would pay more attention to preparators' decisions and actions. They would stop deleting the work and the workers. They would probably become more involved in the everyday life of the lab, by supervising preparators' work and by suggesting or perhaps even requiring certain methods, despite their limited knowledge of preparation techniques and skills. They might begin hiring preparators themselves; most likely based on applicants' scientific credentials more than the hands-on experience and skills that preparator's value. Because preparators' work would be published in their papers, scientists would be more invested in it. This could detract from preparators' autonomy, by limiting their power over their work and their community. Sociologists Susan Leigh Star and Anselm Strauss argue that making workers and their work visible can cause "the eradication of discretion from skilled workers" as a result of increased supervision or newly standardized tasks.[22] Thus, preparators might be *less* powerful when visible, if by becoming part of scientific papers they then lose their power to design methods, train new preparators, and manage their workplace.

This case shows the complexities and multidimensionality of invisibility. Social order is not always as it appears. Workers may be oppressed, overlooked, and unappreciated for reasons of social status, class, gender, race, ethnicity, and many other factors. On the other hand, they may be empowered craft workers whom we are viewing from the perspective of a separate field that does not include them and from which they distinguish themselves. Workers also oscillate between these extremes and most often operate somewhere on a spectrum between "oppressed" and "empowered." The concept of invisibility is not static or universal, though it is a widespread way of differentiating groups and creating a multi-group social order based on local priorities. Invisibility is selective and purposefully so, for reasons such as those described by Star and Strauss: "Visibility can mean legitimacy, rescue from obscurity or other aspects of exploitation. On the other [hand], visibility can create reification of work, opportunities for surveillance, or come to increase group communication and process burdens."[23] It is our mission as researchers to understand the ongoing construction of social relations, and therefore the multiple roles that invisibility can play. We can do this by reflecting on our own perspectives and assumptions and by actively striving to access a variety of points of view, especially those of the seemingly least powerful people in a community.

NOTES

1. On the labour of enslaved people, see William C. Allen, "History of Slave Laborers in the Construction of the United States Capitol" (Washington, D.C., 2005), https://emancipation.dc.gov/sites/default/files/dc/sites/emancipation/publication/attachments/History_of

_Slave_Laborers_in_the_Construction_of_the_US_Capitol.pdf; "President's Commission on Slavery and the University" (University of Virginia, 2018), https://slavery.virginia.edu. On Indigenous guides to European naturalists, see Charles Darwin, *A Naturalist's Voyage Round the World: Journal of Researches into the Natural History and Geology of the Countries Visited during the Voyage Round the World of H.M.S. Beagle under the Command of Captain Fitz Roy, R.N.* (London: John Murray, 1913); James Delbourgo, *Collecting the World: Hans Sloane and the Origins of the British Museum* (Cambridge, MA: Harvard University Press, 2017). On women's unrecognized labour in science, see Margot Lee Shetterly, *Hidden Figures: The American Dream and the Untold Story of the Black Women Mathematicians Who Helped Win the Space Race* (New York, NY: William Morrow and Company, Inc., 2016); Peter Galison, *Image and Logic: A Material Culture of Microphysics* (Chicago, IL: University of Chicago Press, 1997), 199–200; Sharon Traweek, *Beamtimes and Lifetimes: The World of High Energy Physicists* (Cambridge, MA: Harvard University Press, 1988), 28–29; Denise Kiernan, *The Girls of Atomic City* (New York: Simon and Schuster, 2013); Mar Hicks, *Programmed Inequality: How Britain Discarded Women Technologists and Lost Its Edge in Computing* (Cambridge, MA: MIT Press, 2017); Dava Sobel, *The Glass Universe: How the Ladies of the Harvard Observatory Took the Measure of the Stars* (New York, NY: Viking, 2016); Naomi Oreskes, "Objectivity or Heroism? On the Invisibility of Women in Science," *Osiris* 11 (1996): 87–113; Margaret Rossiter, *Women Scientists in America* (Baltimore, MD: Johns Hopkins University Press, 1982).

2. John Law, *Organizing Modernity: Social Ordering and Social Theory* (Cambridge, MA: Blackwell, 1994), 1–2.

3. Steven Shapin, "The Invisible Technician," *American Scientist* 77, no. 6 (1989): 554–63; Steven Shapin, *A Social History of Truth: Civility and Science in Seventeenth-Century England* (Chicago, IL: University of Chicago Press, 1994).

4. Caitlin Donahue Wylie, "Overcoming the Underdetermination of Specimens," *Biology & Philosophy* 34, no. 24 (2019): 1–18.

5. Caitlin Donahue Wylie, "Preparation in Action: Paleontological Skill and the Role of the Fossil Preparator," in *Methods in Fossil Preparation: Proceedings of the First Annual Fossil Preparation and Collections Symposium*, eds. Matthew A. Brown, John F. Kane, and William G. Parker, 2009, 3–12; Caitlin Donahue Wylie, " 'The Artist's Piece Is Already in the Stone': Constructing Creativity in Paleontology Laboratories," *Social Studies of Science* 45, no. 1 (2015): 31–55; Caitlin Donahue Wylie, "Trust in Technicians in Paleontology Laboratories," *Science, Technology, and Human Values* 43, no. 2 (2018): 324–48; Caitlin Donahue Wylie, *Preparing Dinosaurs: The Work behind the Scenes* (Cambridge, MA: MIT Press, 2021).

6. Caitlin Donahue Wylie, "Glass-Boxing Science: Laboratory Work on Display in Museums," *Science, Technology, and Human Values* 45, no. 4 (2020): 618–35.

7. All names are pseudonyms.

8. Kenneth Kusterer, *Know-How on the Job: The Important Working Knowledge of "Unskilled" Workers* (Boulder, CO: Westview Press, 1978); Michael Yarrow, "The Labor Process in Coal Mining: Struggle for Control," in *Case Studies on the Labor Process*, ed. Andrew Zimbalist (New York: Monthly Review Press, 1979), 170–92; Timothy Pachirat, *Every Twelve Seconds: Industrialized Slaughter and the Politics of Sight* (New Haven, CT: Yale University Press, 2011).

9. Matthew Brown et al., "The Smithsonian Institution's Exhibit Fossil Preparation Lab Volunteer Training Programme, Part II: Training and Evaluating Student Preparators," *Geological Curator* 9 (2010): 179–86; Steven Jabo et al., "The Smithsonian Institution's Exhibit Fossil Preparation Lab Volunteer Training Programme, Part I: Design and Recruitment,"

Geological Curator 9 (2010): 169–78; "About Us," Association for Materials and Methods in Paleontology, 2014, https://paleomethods.org/About-Us-2; Wylie, *Preparing Dinosaurs.*

10. Jeffrey Keefe and Denise Potosky, "Technical Dissonance: Conflicting Portraits of Technicians," in *Between Craft and Science: Technical Work in U.S. Settings*, eds. Stephen R. Barley and Julian E. Orr (Ithaca and London: Cornell University Press, 1997), 78–79.

11. As an example of an online forum, see the Society of Vertebrate Paleontology's Preparators' Resources website, https://vertpaleo.org/preparators-resources-2/, accessed August 12, 2021. As an example of preparator-organized conference proceedings, see Matthew A. Brown, John F. Kane, and William G. Parker, eds., *Methods in Fossil Preparation: Proceedings of the First Annual Fossil Preparation and Collections Symposium* (2009). As an example of a published paper by preparators, see Constance Van Beek and Matthew A. Brown, "Three Dimensional Preparation of a Late Cretaceous Sturgeon from Montana: A Case Study," *Geological Curator* 9, no. 3 (2010): 149–53.

12. Caitlin Donahue Wylie, "The Plurality of Assumptions about Fossils and Time," *History and Philosophy of the Life Sciences* 41, no. 21 (2019): 1–21.

13. Andrew Abbott, *The System of Professions: An Essay on the Division of Expert Labor* (Chicago, IL: University of Chicago Press, 1988).

14. Stephen R. Barley, Beth A. Bechky, and Bonalyn J. Nelsen, "What Do Technicians Mean When They Talk about Professionalism? An Ethnography of Speaking," *Research in the Sociology of Organizations* 47 (2016): 125–60.

15. Pierre Bourdieu, *Sociology in Question* (London: Sage Publications Ltd, 1993), 72–73; Aaron L. Panofsky, "Field Analysis and Interdisciplinary Science: Scientific Capital Exchange in Behavior Genetics," *Minerva* 49, no. 3 (2011): 295–316; Wylie, "The Plurality of Assumptions about Fossils and Time."

16. Wylie, "The Plurality of Assumptions about Fossils and Time"; Wylie, "Overcoming the Underdetermination of Specimens."

17. Law, *Organizing Modernity.*

18. See, for example, Susan Leigh Star, "Craft vs. Commodity, Mess vs. Transcendence: How the Right Tools Became the Wrong One in the Case of Taxidermy and Natural History," in *The Right Tools for the Job: At Work in the Twentieth-Century Life Sciences*, eds. Adele Clarke and Joan H. Fujimura (Princeton, NJ: Princeton University Press, 1992), 257–86.

19. Law, *Organizing Modernity*, 131.

20. Law, *Organizing Modernity*, 113, 116, 132.

21. Law, *Organizing Modernity*, 133.

22. Susan Leigh Star and Anselm Strauss, "Layers of Silence, Arenas of Voice: The Ecology of Visible and Invisible Work," *Computer Supported Cooperative Work* 8 (1999): 20–21.

23. Star and Strauss, "Layers of Silence," 9–10.

IV

PRACTICE

Commentary: Teaching Practices with Invisible Labour

Judith Kaplan

The chapters in this section invert our analytic focus on invisibility and labour, subjecting historical scholarship and STS to the kinds of questions about power, people, and processes carried through the rest of the volume. Like playing an open hand of cards when learning a new game, it is our hope that by laying on the page some of the routinely invisible elements of our own practices, these chapters will be particularly useful in teaching research methods and ethics. Having taught a research-oriented undergraduate seminar several times under the rubric of "invisible labour," I offer a few thoughts on this aspiration in what follows. While the labours of the classroom are often relegated to a domain that is less than fully visible to those wielding power over academic careers, teaching is what many of us do for most of our time. Moreover, as most teachers will attest, students themselves do much to illuminate and question the objectives and methods of disciplined research, so I want to begin this chapter by acknowledging my profound debts to students at the University of Pennsylvania.

I contend that invisible labour is a compelling analytic focus for engaging newer scholars in the disciplined study of science as a social and cultural activity. This gives teachers and students the opportunity to ask: why might some stories be under-represented in the historiography of science, where do we find them, and how might our accounts contribute to future efforts to understand what science, technology, and medicine mean?

In addition to shining a light on contributions that may have been marginalized in science—past and present—contributors to this section of the volume maintain that invisible labour can help us better situate ourselves with respect to the problems we think and write about.

When I look back over the genesis of my own research program in the history of linguistics, for instance, I find that the meta-disciplinary question I keep circling back to (on what grounds has linguistics been identified with the study of nature versus culture?) derives from my mother's own scholarly pivot from literature to comparative linguistics. Growing up, I was exposed to the idea that "science" could liberate one's mind from narrow social scripts, and this has undoubtedly shaped the way I look at modern developments in the language sciences. My mother's career in linguistics has also made it relatively easy for me to access sources and interlocutors. Just as other pieces in the volume encourage us to pay attention to paratextual genres that attend the research publication—acknowledgements and footnotes, for example—the point here is that we might do well to look closely at author biographies in the history of science and STS.[1]

Take the authors of this book, for example. Readers will note right off the bat that the majority are women, drawing attention to the genealogy of the project in feminist science studies (see Clarke, Commentary: People). Then, there are issues of geographic and linguistic focus to consider. Among the contributors to this volume, it may be said that some have chosen research problems that are relatively close to home—as my own trajectory attests, for instance. In other cases, a desire to resist the expectation of cultural fluency can motivate studies further afield: one participant in the workshops leading up to this volume spoke of a reluctance to be put inside a research box limited to their country of origin. Third, we might look at the volume as a record of professional networks past and present: the editors actively contemplated the affordances and limitations of our institutional and collegial associations in our efforts to assemble a wide range of case studies. This meant working together and across the disciplinary boundaries imposed by our training, an experience that highlighted for us the efforts that many colleagues make to embrace interdisciplinarity in the history, sociology, and anthropology of science. As a larger group, we have had to commit a great deal of energy to making ourselves understood—glossing insider terms, simplifying and justifying narrative conventions, bringing each other up to speed on literatures and debates. These efforts are another reason for hoping that these chapters will be well suited to classroom use.

My experiences teaching with case studies of invisible labour lie, unsurprisingly, behind a number of these assertions about recovery and self-reflexivity. The utility of the theme for discovering new things about the world was easy to convey—my students were keen to elevate marginalized experiences and contributions to science.[2] The practical constraints of the semester schedule also helped us focus on and situate our individual experiences as having epistemic value. It is no easy task to identify, research, and write something up on a semester schedule, making local or personal topics an ideal choice for logistical reasons and for the cultivation of self-reflexivity in problem choice and research design.

These ideas and constraints yielded a number of rich and well-crafted essays.[3] For example, one student carried out a study of child language brokers in U.S. health care. In framing the project, he drew upon personal memories of translating mail, answering calls, scheduling doctor appointments, attending parent–teacher conferences, and disputing legal matters for his parents, who were refugees from Vietnam. He conducted a number of interviews with members of a group of first-generation and low-income Asian students in Philadelphia, counter-balancing an extensive literature on the informal labour of Spanish-speaking youth in American health services. His paper pointed up the necessarily fragmented nature of American health care delivery, as well as the inversions such translational labour can initiate within traditional family hierarchies. Crucially, he gave voice to this invisible labour force through the inclusion of his personal testimony and extensive interview notes with peer translators.

Another student, prompted by reports that disability statistics would be largely cut out of the 2018 Colombian census as an administrative cost-saving measure, conducted a qualitative analysis of those who provide care to people with disabilities in her hometown. This population was poised to become doubly invisible from a governmental point of view. Working through the social media groups of disability advocates and with legal records, she found numerous reasons why caregivers might prefer invisibility, something she interpreted as an obstacle to the fuller inclusion of people with disabilities in Colombian society. Like classmates who targeted issues in pharmacology, law, and engineering, considering public health from the standpoint of invisible labour helped her to cultivate a critical perspective on the profession she was stepping into upon graduation.[4]

A third student used extensive archival and interview materials to call attention to computing and data infrastructures at Penn. Her research pointed to a split between the rhetoric and practice of computer and information sciences on campus, and, following an era of centralized campus computing, followed maintenance workers and contracts out of the university into outsourced labour arrangements and external computing services. Focusing on the School of Arts and Sciences, the administrative unit that embraced our class, she documented resource disparities across departments and analysed the complicated dynamics of training a workforce required to have both technical expertise and service-oriented flexibility. The project advanced awareness of the hidden social and environmental costs of our own research, teaching and learning in the College—concerns that only escalated with the move to remote operations in March 2020.

These projects all featured original research, and they each added nuance to my understanding of the development of techno-scientific systems. My students taught me to pay closer attention to the temporal dynamics of medical translation, the ways in which family structure conjugates with labour hierarchies, the religious motivations for *not* seeking compensation (see Stark, chapter 6), and the hidden labour enabling my own efforts with respect to writing this comment (I am thinking here of the need to look up student records, to access remote library services, and to use university-sponsored cloud storage). These studies demonstrate the cultivation of self-reflexivity at personal, professional, and institutional levels.

In sum, the students in my classes embraced the productivity of invisible labour for talking about the logics and logistics of archival research. They reflected actively on the claims to authority made on behalf of textual evidence versus oral history. And they brought an impressive array of language skills to the table in their efforts to think globally about invisible labour in scientific practice.

Reading the chapters drawn together in this section against the student projects just described, another layer of questions comes into view. We are called upon to ask: is the professional mandate to carry out original research ever at odds with the rights and interests of historical subjects and their communities? Why should historical visibility be a coveted goal, and on what terms? It is difficult but meaningful work to slow down and circle back in classes designed to empower young scholars to forge ahead.

We believe that the chapters gathered up here will help students wrestle with shifting dynamics of problem choice, evidentiary practice, and narrative position. Describing experiences that date back to the early phases of their careers, the contributors compel us to ask several fundamental questions about research practice: what—or who—might be a source? What implicit values underpin the research questions we ask? What efforts, exchanges, and infrastructures make our research possible? And what perspectives do we indulge or exclude when we practice the craft of argumentation?

One prominent theme of the chapters in this group is that "knowledge work" is iterative in nature—it necessitates re-visiting, re-telling, re-writing, and re-presenting events, experiences, and epistemic negotiations. In Joanna Radin's piece, for example, this manifests in anthropologist Jonathan Friedlaender's self-awareness of "becoming historical." When Friedlaender was prompted to review his life's work to archive it, he highlighted the ethical assumption that preservation is an inherent good. Radin uses this as a wedge to pry open and extend critiques of the universalizing tendency in modern science: not only do historians need to consider the gaps that emerge between groups surveyed cross-sectionally, they must also consider how their own seemingly self-evident values change over time. In a complementary move, Lara Keuck shows how archivists, for example, invest historical visibility with inherent value, setting the parameters of our own source selection and subsequent interpretive labour.[5] The iterative theme here plays out in her chapter as meanings are assigned and re-assigned to patient records being moved from one institutional locus to another.

For Boris Jardine and Alexandra Widmer, revisiting heretofore unpublished aspects of their own work offers the opportunity to reflect on the range of labours that go into object display and fieldwork, respectively. Both of their chapters suggest that the "finished product"—whether that be a conference talk, research publication, or gallery installation—is often too tidy. To return, in other words, allows Jardine and Widmer to consider an inverted relationship between acknowledgements and the research product, which allows us to better understand the inner mechanics of STS (broadly construed) in turn. Yet again, the theme of iterative practice can be traced through Michaela Spencer's account of epistemic translation. Not only does her narrative reveal the layers of mediation that come through direct quotations, fieldwork notes, and subsequent rounds of analysis, she argues that re-telling stories is a way to promote understanding among potential members of an inclusive audience. Children learn languages through purposive interactions and repetition—the same is true of building consensus across epistemic traditions. This kind of practice can only expand the range of possible meanings associated with humanistic studies of science, technology, and medicine.

With Rosanna Dent's chapter, there is the opportunity to think about how repetition and return scale up and across generations. She describes how anthropologist Nancy Flowers returned to the site of her graduate field research after a period of twelve years to find a different set of political opportunities. Dent also reflects on how the experiences of past generations have determined her own, "I joined the long line of scholars that the village has hosted," she writes of her travels to Pimentel Barbosa (Mato Grosso, Brazil), "carrying with me images and publications that invited village residents to comment on those who had come before me." Her research is thoroughly engaged with the inter-generational (i.e., temporal) definition of field sites, thus it shines a light on relationships between teachers and students.[6]

There are several potential points to recommend an iterative approach to the study of research methods in the history and sociology of science. For one thing, our students can question us. As bell hooks points out, critical thinking calls for "radical openness" and "does not simply place demands on students, it also requires teachers to show by example that learning in action means that not all of us can be right all the time, and that the shape of knowledge is constantly changing."[7] Writing and re-writing—a formalized practice of "hypergraphia" (see Keuck, chapter 21)—simultaneously allow us to diagnose implicit biases in the field and to give narrative structure to lives and human experiences in the history of science. Resonating with the chapters presented here, one colleague in anthropology asks students in her research seminars to consider the following:

> What led you to the question that is central to your research? What do you understand about the history of the question, and the history of the site(s) where you will explore the question? In what ways are you implicated within, or complicit with, the practices your research seeks to address? How will you identify methodological strategies that take these implications and complicities into account? To whom (or what) is your research directed, and what is at stake for them? What forms of accountability will you develop throughout your research and writing process? What gives you the right to write, and what responsibilities accompany that right?[8]

As these questions make clear, no researcher labours in isolation. This brings me to my second point about the chapters in "Practices": they share a common interest in the co-constitution of knowledge, one that has the potential to pluralize notions of authorship in history and STS. In some cases, the recognition of co-constitution is bound up with disciplinary identity: collaborative scholarship and the joint production of meaning are all routinely emphasized in oral history (Radin, chapter 20), public history (Jardine, chapter 22), and anthropology (Widmer, chapter 23; Spencer, chapter 25). More specifically, Radin raises the right of the interviewee, or narrator, to edit their own story. This reflects an evolving set of foundational beliefs within the field of oral history that has shifted from a positivist emphasis on oral testimony as data to a post-positivist inquiry into the workings of memory and subjectivity. In the current iteration of its "principles and best practices," practitioners avow "not only respect for narrators and their communities, but also the importance of being attentive to those who are especially vulnerable," reemphasizing the "dynamic, collaborative relationship between interviewer and narrator with a commitment to ongoing participation and engagement," as well as "sensitivity to differences in power, constraints, interests and expectations."[9] What might it mean to extrapolate this spirit of co-authorship and respect to historical actors who are no longer living, known only through their paper or material remains? Keuck invites this question asking, "How do historians' attitudes towards sources differ from those of archivists?" Not only does her analysis recognize that archivists are co-producers of knowledge, Keuck also calls attention to the institutional parameters of history by setting up the comparison. It is always helpful to ask, at every possible career stage, how access to the sources we end up deciding are "relevant" has been facilitated or limited by a wide range of potential co-authors and disciplinary precedents.

Indeed, the co-constitution of meaning is often achieved through research infrastructures that are not readily visible in scholarly or scientific outputs. Widmer and Dent make this clear in their reflections on the hospitality extended to researchers engaging in fieldwork. Widmer argues that hospitality is one way that Indigenous communities inhabiting restrictive power matrices can assert sovereignty over land, labour, materials, and knowledge. Her analysis encourages readers to think about what benefits might have accrued to the citizens of Vanuatu through their support of her own fieldwork experience with young children. In a complementary way, Dent points out that leaders in the Xavante community have built "social infrastructures to mediate and distribute the work and compensation associated with research," whether biomedical, geographical, or social-scientific in nature. She helpfully defines a research "infrastructure" as something that endures and "undergirds the creation and movement of the materials of science," and emphasizes that this infrastructure can be social as well as material. Taken together, these chapters can foster classroom conversations about what might support the production and circulation of those sources on which historical and STS scholarship depends, on the researcher's position within a "field" of inquiry, and on the potential ways of *repaying* (not just acknowledging) our debts to actors and their communities.[10] How might we think about the distribution and reciprocity of benefits—the "balancing entry"—when it comes to classroom projects?

It is important to note that the "co-constitution of knowledge" is open-ended, not finished. This point comes through powerfully in the chapters by Spencer and Jardine, who are both especially sensitive to the open-endedness of meanings put out into the world, to the continual re-negotiation of meaning in under-determined spaces. While much of Spencer's chapter concerns the breach of communication across epistemic traditions, the effort to cultivate shared understandings ultimately casts a hopeful eye to the future. As Joy Bulkanhawuy states in her contribution to Spencer's project, "We as a Yolŋu people, are already strong. We know where to find everything and can show this to the non-Indigenous people, so we can work together. But this work will come out from cultural awareness and from Yolŋu people's knowledge of our djalkarri-ŋur (foundations)." What meanings might be made through subsequent readings of these words? Similarly, Jardine emphasizes the disconnect between an "installation shot" of a gallery and the "often intense work of interpretation that visitors put into interpretation—suppressing or bringing into the space their own expertise, hopes, fears, prejudices, knowledge, profound and everyday experiences." Thinking about these reflections in tandem, one takeaway from the study of research methods and ethics may well be an enhanced sense of responsibility to (unanticipated) uses and re-uses of our work. This is a common point of critique in STS and historical scholarship—consider accounts of nuclear weapons, artificial intelligence, and genome sequencing—but something that should be taken on board self-reflexively as well.[11] These conversations can, and probably should, be bound up with the concrete consideration of open-access licensing as student research nears the point of publication.

It is worthwhile asking what these grounded reflections on the theme of co-constitution might have to do with a critical and self-reflexive attitude towards discovery procedures in the humanistic study of science, technology, and medicine. Moreover, it is important to recognize that many who study the history of science and STS in school may go on to completely unrelated professional lives: what could be the significance of these chapters for them, for you? Tracing the under-appreciated rise of professional "alternatives" to the professoriate in 2011, Anthony Grafton and Jim Grossman insisted that advisors and departments must

> make clear to all students that they will enjoy . . . unequivocal support, whether they seek to teach at college or university level, join a nonprofit agency, or head off into business or government. We teach our students to question received ideas and to criticize inherited terminologies and obsolete assumptions. It's past time that we began applying those lessons ourselves.[12]

So what might be the payoff of thinking with invisible labour, and how might this shake out, in particular, for those not contemplating history of science or STS as a career?

First, and foremost, I believe our theme encourages a healthy appreciation for the various conditions and labours that make pursuing such questions possible: from land acknowledgements and systemic issues of educational access to recognizing the "reproductive labour" of those working in, say, food service on campuses and the "background work" that might go into syllabus design. As our disciplinary network evolves and adapts to broader institutional and market pressures, "public engagement" is increasingly valued. This desideratum is consistent with the focus of this section on collaborative knowledge work, which tends to foreground the "range of individuals who contribute to and are impacted by the fundamentally human enterprise of science."[13] It amounts to a call for a new kind of "scientific literacy": whereas an earlier generation contemplated how much science education might be needed for citizenship in a "science-oriented social order," the goal here is to link learning about science with social responsibility and community development.[14]

Second, the emphasis on co-construction running through these chapters underscores the point that knowledge is a resource, extracted and exchanged like other goods and materials. What do historians, sociologists, and anthropologists of science stand to gain from the privilege of presenting a given narrative, and what do we owe in return?

Finally, several of the chapters in this section present a challenge to the fundamental logic of discovery that usually motivates humanistic studies of science in the first place, including studies of invisible labour. This motivation is evidenced, for example, in Steven Shapin's insistence that "the price of . . . continued invisibility is an impoverished understanding of the nature of scientific practice."[15] It is reinforced by the guidelines for authors and reviewers provided by the editors of *Isis*, the oldest English-language journal in our field, who "encourage authors to meaningfully consider non-elite actors, especially those whose roles have previously been obscured by historians."[16] While the recent adoption of these guidelines (in 2021) suggests that teaching with "invisible labour" may be an accurate reflection of contemporary disciplinary standards, the chapters in this section make an even deeper point. They show that we should contend with the fact that sometimes it is not our place to render people, powers, and processes visible; that the clock might run out on our professional and historiographic values; and that preservation does not necessarily entail permanence.

In the final weeks of my classes on invisible labour, I pivoted to questions about the publication process—offering up draft manuscripts of my own and comments from reviewers and editors as a way of introducing students to the "next steps" they might consider taking with their projects. The peer-review process, which depends on a huge amount of necessarily hidden effort, is an immensely significant way in which knowledge in our field(s) is certified and made fit for public consumption. In recent years, journals and publishers have partnered with services like Publons to recognize the considerable amount of time and effort that goes into this disciplinary activity.[17] At the same time, authors trust their reviewers to catch any omissions and call out potential blind-spots, and there is the feeling that a piece of work is "done" when these queries have all been resolved. By contrast, in the preparation of this volume, most of our peer reviews were rendered face-to-face—through workshops and several rounds of editorial discussion. Although this has been, on the whole, an incredibly rich and collegial practice, it is a departure from the way in which authority and credibility are usually achieved—a departure that has occasioned some anxiety about the final stages of manuscript preparation. Given the opportunity to teach with invisible labour again, I would contrast the experience of preparing this volume with the method of "double-blind" peer review.[18] Rather than distributing trust in the findings and interpretations presented across a network of anonymous readers, the authors in this section invite you to trust them through deeply personal accounts of their research: the pathways they have taken, the support they have enjoyed, the obstacles they have faced, the ethical concerns they have had, and the limits to what they feel they can say.

NOTES

1. On the various functions of paratext, see, for example: Anthony Grafton, *The Footnote: A Curious History* (Cambridge, MA: Harvard University Press, 1999); Annemarie Mol, *The Body Multiple: Ontology in Medical Practice* (Durham, NC: Duke University

Press, 2002); Emily Callici, "On Acknowledgements," The American Historical Review 125 (2020): 126–31.

2. The importance of this approach was made forcefully by members of the *Annales* school, students of history "from below," and more recently by ancient historians engaging with these traditions in social history. See, for example, Kim Bowes, "Introduction: Inventing Roman Peasants," in idem, *The Roman Peasant Project 2009-2014: Excavating the Roman Rural Poor* (Philadelphia: University of Pennsylvania Press, 2020), 1–13 and references therein.

3. Student information is private. If readers would like to reach out to the authors of these projects, they should contact Judith Kaplan.

4. Personal communication. On the unbounded training of those who study the history of science but do not enter into the academic profession, see Lynn Nyhart, "The Shape of the History of Science Profession, 2038: A Prospective Retrospective," *Isis* 104, no. 1 (2013): 131–39.

5. See Michelle Caswell, "Introduction: Silence, Agency, and the Social Life of Records," in idem, *Archiving the Unspeakable: Silence, Memory, and the Photographic Record in Cambodia* (Madison: University of Wisconsin Press, 2014), 6–7.

6. See Rosanna Dent's essay, "Whose Home Is the Field?," Isis 113, no. 1 (2022): 137–43.

7. bell hooks, *Teaching Critical Thinking: Practical Wisdom* (New York; London: Routledge, 2010), 10.

8. Deborah Thomas, "What Constitutes Fieldwork? Or, How Might We Thinking about Ethnography in a Decolonizing Anthropology?" Presentation for the workshop "What Is a Field?" held at the University of Pennsylvania, April 10, 2021.

9. OHA, https://www.oralhistory.org/principles-and-best-practices-revised-2018/#Introduction, accessed August 6, 2021; see also Rebecca Sharpless, "The History of Oral History," in *Handbook of Oral History*, eds. Thomas Charlton, Lois Myers, and Rebecca Sharpless (Lanham, MD: AltaMira, 2006).

10. For anthropological reflection on the latter of these, see: Elizabeth Povinelli, *Labor's Lot: The Power, History, and Culture of Aboriginal Action* (Chicago, IL: University of Chicago Press, 1983); Marilyn Strathern, *Kinship, Law, and the Unexpected: Relatives Are Always a Surprise* (New York: Cambridge University Press, 2005).

11. See, for example, Ulrich Beck, *Risk Society: Towards a New Modernity*, trans. Mark Ritter (London: Sage, 1992); Lydia H. Liu, "The Future of the Unconscious," in idem, *The Freudian Robot* (Chicago, IL: University of Chicago Press), 249–66; Alexandra Luccioni and Yoshua Bengio, "On the Morality of Artificial Intelligence," *IEEE Technology and Science Magazine* 38 (2020): 16–25; Jenny Reardon, "Commentary: Ends Everlasting," *BJHS Themes* 4 (2019): 283–91. For an example of a recent public initiative, see The Montreal Declaration for a Responsible Development of Artificial Intelligence, https://www.montrealdeclaration-responsibleai.com/the-declaration, accessed September 30, 2021.

12. Anthony Grafton and Jim Grossman, "No More Plan B," *Chronicle of Higher Education* 58, no. 8 (2011).

13. Nyhart, "The Shape of the History of Science Profession," 137; *Isis* guidelines, https://www.journals.uchicago.edu/journals/isis/referees, accessed August 27, 2021.

14. For examples, see: Charles L. Koelsche and Ashley G. Morgan, Jr., *Scientific Literacy in the Sixties* (Athens: University of Georgia Press, 1964); Wolff-Michael Roth and Angela Calabrese, *Rethinking Scientific Literacy* (New York: Routledge, 2004).

15. Steven Shapin, "The Invisible Technician," *American Scientist* 77 (1989): 563; see also Susan Lindee's discussion of accuracy in her commentary for this volume.

16. "Guidelines for Manuscript Authors," https://www.journals.uchicago.edu/journals/isis/instruct, accessed August 2, 2021.

17. See, for example, https://publons.com/in/wiley/, accessed September 15, 2021.

18. The issues of trust here are reminiscent of the system of witness testimony and personal recommendation that contoured early modern scientific letters. See, for example, Lorraine Daston, "Marvelous Facts and Miraculous Evidence in Early Modern Europe," *Critical Inquiry* 18 (1991): 93–124.

20

Collecting Human Subjects

Ethics and the Archive[1]

Joanna Radin

Anthropological collectors have long engaged in "salvage"—the attempt to metaphorically freeze those artefacts, traditions, and languages in danger of disappearing.[2] During the Cold War era, justifications for salvage were re-articulated as new techniques emerged that changed the kinds of materials that could be collected and maintained. The metaphor of freezing had become a reality in practice: New access to technologies of mobile and long-term cold storage—including mechanical refrigeration, dry ice, and liquid nitrogen—supported the accumulation and preservation of thousands of vials of blood extracted from members of human communities around the globe.

Under the auspices of the International Biological Program (IBP), a large-scale effort to establish global baselines of biological variation, circa 1964–1974, a number of biological anthropologists and human geneticists emphasized the importance of salvaging blood samples from Indigenous people, whose survival they considered to be endangered by the corrosive forces of modernity.[3] In the late 1950s through the 1970s, it was only possible to discern a small number of meaningful markers in the molecular analysis of blood, such as abnormal haemoglobins, blood groups, and a few proteins. Almost immediately, however, research using blood collected from East African communities demonstrated an evolutionary relationship between the presence of the gene for sickle cell anaemia and resistance to malaria.[4] This finding led scientists to believe that studying the blood of the members of Indigenous groups, whom they characterized as portals to the past, would yield many similarly powerful examples of natural selection in humans.[5] Though this particular prediction was not borne out, blood did end up being filled with biomedical potential. By the 1970s, two scientists—Carleton Gajdusek and Baruch Blumberg—had been awarded Nobel prizes for research using tissues collected from Indigenous peoples living in the Western Pacific.[6]

In the lab, freezers filled with this blood became archives, making such bits of bodies available for repeated analyses.[7] By the 1990s, scientists were thawing these old, cold blood samples in order to mine them for the human DNA latent within. This practice, which became known as "freezer anthropology," gained momentum amid controversy surrounding the demise of the Human Genome Diversity Project (hereafter referred to as the Diversity Project).[8]

In many ways, the Diversity Project had been a successor to the human-centred blood collection projects of the IBP in that its primary goal was to sample and archive human variations in the genomic age.[9] It even involved some of the same scientists and their students. As in the IBP, the focus was on collecting genetic material from Indigenous communities. In the intervening decades, however, the practice of salvaging and preserving Indigenous blood went from being regarded as a relatively uncontroversial technological breakthrough to the focus of an emerging critical conversation about the ethics of human subjects research and the nature of belonging.[10] More recently, it has led to a movement in which Indigenous peoples become the stewards of their own biobanks.[11]

The shifting ethical and epistemological status of these frozen human remains has led historians to inquire about the circumstances of their original procurement. Many of the scientists who participated in what one referred to as the "heyday" of anthropological blood collection have retired or have passed away, though their frozen blood samples sometimes endure.[12] In certain cases, even their students are reaching the end of their careers. While some of these veteran collectors have been acutely sensitive to their legacies and have carefully curated their personal papers, others do not have a sense of themselves as "historical subjects" whose stories might be lost. As historians turn our attention to these scientists, we may have the opportunity to interview practitioners. In the process, we may also find ourselves in the position of turning human subjects researchers into the subjects of historical research. In this sense, I have struggled with the feeling that I, myself, have engaged in a form of salvage.[13]

This chapter focuses on the collection practices of one such individual, biological anthropologist Jonathan Friedlaender, with whom I conducted and published an oral history.[14] The oral history described transformations in the field of biological anthropology over the past forty years, as well as Friedlaender's involvement in developments in biomedical research ethics. In the process of participating in the oral history project, Friedlaender began to salvage materials from his own career, assembling an archive that would ultimately be deposited at the American Philosophical Society in Philadelphia.

By putting the circumstances that led to the creation of Friedlaender's archive of frozen blood alongside an account of the circumstances that led him to create an archive of his career, it becomes possible to simultaneously consider the value of viewing certain forms of historical and scientific practice in terms of the ethics of collection creation, use, re-use, as well as repatriation and refusal. Specifically, I use Friedlaender's experience of "becoming historical" to reflect on the ethical consequences of an assumption held by many historians as well as scientists: that collections should endure indefinitely.[15] In an era in which concerns about research involving human subjects are transforming how we make knowledge about our species, it has never been more important for historians to think reflexively about the assumptions that guide their own research enterprises.[16]

COLLECTING BLOOD

Jonathan Friedlaender was a PhD student in biological anthropology at Harvard when he took his first blood samples in 1966. He had travelled to the Western Pacific as part of an IBP-affiliated initiative known as the Harvard Solomon Islands Project.[17] The Harvard Solomon Islands Project involved several seasons of fieldwork, including the collection of thousands of blood samples and other kinds of biomedical data. The idea that various communities of Solomon Islanders were undergoing dramatic changes due to globalization motivated anthropologists to make collections that would enable them to obtain a snapshot of such populations at a specific moment in time. The organizers of the Harvard Solomon Islands Project agreed that it was urgent to salvage as much blood as possible from these unique populations and that this blood should be preserved indefinitely for future, as-yet-unknown, uses.

Friedlaender would recall that, during that field season, "I learned [how to take blood samples] using humans as practice. I don't know exactly what would be acceptable behaviour in the U.S. now, but I assume you have to be certified. That was just not even an issue then."[18] He soon became skilled enough at the practice to incorporate it into a study of human variation on the island of Bougainville.[19] Over the next several decades he repeatedly returned to the Western Pacific and accumulated thousands more frozen blood samples, which he kept for many years at minus 20 degrees Celsius in his lab.

Along the way, Friedlaender also found himself at the centre of controversy. In the early 1990s, he was involved in highly politicized debates associated with the Diversity Project about who could use Indigenous blood and how. Friedlaender was also wrongfully accused of attempting to patent genes extracted from the blood of Indigenous people. "This was simply not true," he stated, but "world-wide, the prospect of collecting cell lines [blood cells that have been altered to allow them to produce limitless amounts of DNA] for the prospective Human Genome Diversity Project (or any other sort of study) became a *cause celebre.*"[20] The limitations of "informed consent"—the idea that the subjects of research should understand the nature of their participation in research—was formulated as a central ethical problem for the practice of human biology.[21]

Those who sought to accumulate Indigenous blood—in both the 1960s and the 1990s—assumed that it could and should remain in the freezer indefinitely, a thought that has, more recently, gone to the heart of concerns about whether those who agree to participate in human subjects research can ever be reasonably informed about the purposes to which their data will be used in the future. Such uses are often unrelated to those that justified their initial collection.[22]

BECOMING HISTORICAL

In the postscript to his oral history, Friedlaender began by noting that he "was flattered and pleased when Joanna Radin first approached me" about the project.[23] However, when first asked to answer a young historian's questions, he was not immediately responsive to inquiries and, even then, expressed doubt that he had much to share. Initially, he communicated only through an intermediary, Muriel Kirkpatrick, the curator of the Temple Anthropology Lab. This lab, located in the basement of the building that houses Anthropology, as well as several other social science departments, might be more accurately described as a repository for collections—of bones, manuscripts, and cultural objects—which had become historical. Though no blood samples were ever maintained in this space, Kirkpatrick was ultimately able to locate the documentation that supported the studies Friedlaender participated in during the 1960s and 1970s, including the Harvard Solomon Islands Project. This documentation filled three tall metal filing cabinets, stored in a back corner of the basement lab. Their contents were tantamount to a "lost" archive: a unique set of evidence upon which a historian might begin to reconstruct some of the practices involved with turning Indigenous blood into anthropological resources in the mid-twentieth century.

It was also an ethically problematic archive. In addition to the detailed genealogical and anthropological information Friedlaender had collected over the course of the Harvard Solomon Islands Project and his own subsequent research, there were hundreds of photographs. Long forgotten, this was a collection that, from the University's perspective, constituted "protected human subjects material," which includes identifying information about participants in biomedical or social science research. In a powerful sense, the project of investigating the salvage of Indigenous blood during the Cold War reanimated aspects of the lives of those who had participated in their acquisition.

By the time the old filing cabinets had been recovered, Friedlaender had warmed to the idea of becoming a historical subject. He soon began the project of "collecting himself": gathering up the manuscripts, photos, and ephemera that had survived several marriages, moves, retirement, and illness.[24] Friedlaender was suffering from a serious illness, which—he would later intimate—heightened his sense of urgency in salvaging his potential historical legacy.

Prior to proceeding with the oral history, Friedlaender and I discussed the risks involved in this form of research. Friedlaender was required to complete a consent form in order to proceed with the oral history—making him a human subject, as well as someone who was of interest to a historian because of his own work with human subjects. The process of obtaining informed consent was one that had become familiar to him from his more recent efforts in collecting blood from Melanesian research subjects in the 1990s and early 2000s. He had himself drafted a number of such forms, which, incidentally, he contributed to his personal archive. In some cases, these informed consent documents were written in multiple languages and read aloud to those who were not literate. In Friedlaender's case, the primary risks involved with participating in the oral history as stated on the consent form were that there could be no guarantee of anonymity—the very purpose of the project ran against that norm—and that he might experience emotional distress from discussing issues from his past. It did not explicitly include the taken-for-granted-by-the-historian likelihood that the information he shared might be used by people other than myself, and for purposes other than that for which it was originally collected.

ETHICS OF THE ARCHIVE

By the time the oral history was finished and had been published by Friedlaender in December 2008, it became necessary to find an institutional home for the personal archive he had assembled. It felt inappropriate that I should retain the archival materials as part of my "personal" collection, nor did it occur to either of us that I would destroy them. The assumption that the materials should be preserved is a value promoted by professional historians and archivists, but which is in tension with emergent norms of human subjects research, which can include the discarding of research materials at the conclusion of the study.[25] In this instance, I privileged my status as a historian over that of my status as a human subjects researcher.

Friedlaender's initial response to the suggestion that we find an archive for his personal collection was a mixture of humility and surprise. Would anyone really find a use for this stuff? As a reflex, but without much reflexivity, I replied that we could not know for sure, but that unique collections—like his blood samples—warranted preservation. I did not view myself as an agent of coercion, but in retrospect, I had indeed encouraged him to embrace a sensibility in which salvaging his letters and his thoughts for the sake of posterity was an unambiguous good.

As I indicated before, one of the most significant ethical issues to emerge during Friedlaender's career has been whether or not materials he and his colleagues accumulated can be re-used for purposes other than that for which they were initially collected. What has made this problem so intractable, despite a great deal of critical scholarship on issues of property and informed consent regarding the use of human remains, is that there are no universal or definitive answers.[26] When I invited Friedlaender to take on the role of historical subject, I was, in a sense, asking him to consider if he would want the material traces of his career to be used for purposes other than that for which I was collecting them.

Though historians of science and medicine do not often explicitly consider their work as part of the enterprise that guides biomedical research, some Institutional Review Boards (IRBs) in American universities may require historians who wish to conduct oral history interviews to undergo approval processes that are similar to those from whom we wish to learn. From the perspective of the IRB, we are all engaged in "human subjects research." At the same time, even though their orientations towards such research may be quite different, scientists and historians alike are aware that the IRB process fails to account for a vast range of ethical aspects of research and, in some cases, can generate new forms of harm or blockages of knowledge.[27]

In 2011, U.S. federal officials proposed overhauling rules that would affect historians' access to certain collections of archived materials, including those that document human subjects research.[28] While many historians do excellent work without archives by relying on published books and journals, the archive has been a central component of historical practice at least since the nineteenth century.[29] The proposed change to the rules affecting access to biomedically relevant research led historians of human subjects research to consider the extent to which their reliance on new uses for old materials—the bread and butter of their archive-dependent knowledge practices—represented what one anthropologist of science has referred to as an "ethical plateau," a point of conflict in which previously taken-for-granted assumptions and practices must be carefully attended to.[30]

In pointing this out, I am not claiming that the power relationships between a historian and a scientist-as-historical-subject are or have ever been equivalent to that between a scientist and their human "research subject," especially when that subject does not have shared access to the resources of the research university. What I am arguing is that the political and cultural changes that have led these relationships to be placed side-by-side in frameworks like those provided by the IRB reveal otherwise obscured ethical concerns. For example, when I chose, as the subject of my research, scientists who themselves did research on humans, I entered into an ethically ambiguous relationship with both the scientists *and* their research subjects. In other words, the IRB makes research subjects out of people with an extremely broad range of relationships to power and therefore very different kinds of vulnerabilities and interests.

Nevertheless, being placed into the position of "human subjects researcher" can be valuable in that it provides a vantage point from which to re-assess assumptions about how people and things become subjects and, sometimes, objects of knowledge. I chose to conceive of my encounter with Friedlaender as one of collaboration, a decision facilitated both intellectually and socially by our shared location in American research institutions. This collaboration was also supported by his sense of my skill and trustworthiness—a complex intuition in which the official IRB consent document may well have been the least important factor.[31] One consequence of the way in which our "research" relationship proceeded was that when he subsequently decided that he did not want certain things he had told me or given me to become historical, I honoured his wishes by redacting them. This is the same strategy that stewards of now old blood have turned to when people or their kin have requested the return or destruction of frozen blood. Such reversals have not come without tensions and, in some cases, legal action.[32] And they highlight the utmost importance of the personal, affective relationship of trust between all parties.[33]

I am suggesting, then, that the examination of assumptions about salvage and preservation as they pertain to collections of archived manuscript material may help prepare those who study human subjects as well as the history of human subjects research for the need to become more comfortable with evanescence, to reject completeness, and to accept that people have a right to resist becoming historical. My own thinking along those lines has continued to be informed by Indigenous and Black scholars who have articulated the political and epistemological stakes of refusal as a strategy for resisting the hegemonic impulses of extractive research enterprises.[34]

The practice of history—be it based on written documents or bodily substances—is one of time travel. It requires the mingling of the past in the present with an eye towards the future. It is a form of knowledge production that builds narratives out of the lives and actions of others. This does not excuse the historian from responsibility to those actors—be they dead for centuries, frozen in pieces, or very much alive.[35] There are, to cite an important volume on the ethics of the archive, "no innocent deposits."[36] Since Friedlaender's oral history was published, other scientists have asked me if they should be collecting their own personal papers. My answer is no longer the automatic "yes," that salvage instinct of my discipline. Rather, it marks an opportunity to begin a conversation about the ethical and temporal uncertainties inherent to the archive.

As a historian of human subjects research, I have an indisputable intellectual dependency on the past. However, among the most important lessons I have learned from its study is that there may also be certain histories that cannot or should not be told. Indeed, there may be more to be gained from understanding why and when that is the case, than seeking to keep the possibility open of telling them at any cost. This is an ethical project that exceeds the authority and the responsibility of the IRB, and it is one in which historians, archivists, and scientists—as both human subjects researchers and human subjects—should now find themselves collectively and vigorously engaged.

NOTES

1. A version of this chapter was first published as Joanna Radin, "Collecting Human Subjects: Ethics and the Archive in the History of Science and the Historical Life Sciences." *Curator: The Museum Journal* 57, no. 2 (2014): 249–58. A fuller discussion can be found in Joanna Radin, *Life on Ice: A History of New Uses for Cold Blood* (Chicago, IL: University of Chicago Press, 2017).

2. Jacob W. Gruber, "Ethnographic Salvage and the Shaping of Anthropology," *American Anthropologist* 72 (1970): 1289–99.

3. Ricardo Santos, "Indigenous Peoples, Postcolonial Contexts, and Genomic Research in the Late Twentieth Century: A View from Amazonia (1960–2000)," *Critique of Anthropology* 22 (2002): 81–104; Joanna Radin, "Latent life: Concepts and Practices of Human Tissue Preservation in the International Biological Program," *Social Studies of Science* 43 (2013): 483–508.

4. Anthony C. Allison, "Protection Afforded by Sickle-Cell Trait against Subtertian Malarial Infection," *British Medical Journal* 1 (1954): 290–94.

5. M. Susan Lindee, 2004. "Voices of the Dead: James Neel's Amerindian Studies," in *Lost Paradises and the Ethics of Research and Publication*, eds. Francisco M. Salzano and Magdalena Hurtado (New York: Oxford University Press, 2004).

6. Baruch Blumberg, *Hepatitis B: The Hunt for a Killer Virus* (Princeton, NJ: Princeton University Press, 2002); Gajdusek's work is described in Warwick Anderson, *Collectors of*

Lost Souls: Turning Kuru Scientists into Whitemen (Baltimore, MD: Johns Hopkins University Press, 2008).

7. Joanna Radin, "Standardizing Variation: Creating Human Blood Serum Reference Banks at the World Health Organization, 1958–1970." *History and Philosophy of the Life Sciences, Section C* (2014).

8. Jenny Reardon, *Race to the Finish: Identity and Governance in an Age of Genomics.* (Princeton, NJ: Princeton University Press, 2005); Joanna Radin, "Latent life."

9. Santos, "Indigenous," 81–104.

10. Emma Kowal, Joanna Radin, and Jenny Reardon, "Indigenous Body Parts, Mutating Temporalities, and the Half-Lives of Postcolonial Technoscience," *Social Studies of Science* 43 (2013). Kim TallBear, *Native American DNA: Tribal Belonging and the False Promise of Genetic Science* (Minneapolis: University of Minnesota Press, 2013).

11. Katrina G. Claw, Matthew Z. Anderson, Rene L. Begay, Krystal S. Tsosie, Keolu Fox, and Nanibaa' A. Garrison, "A Framework for Enhancing Ethical Genomic Research with Indigenous Communities," *Nature Communications* 9, no. 1 (2018): 1–7; Krystal S. Tsosie, Joseph M. Yracheta, Jessica Kolopenuk, and Rick W. A. Smith. "Letter to the Editor: Indigenous Data Sovereignties and Data Sharing in Biological Anthropology," *American Journal of Physical Anthropology* 1 (2020): 4.

12. D. Andrew Merriwether, "Freezer Anthropology: New Uses for Old Blood," *Philosophical Transactions: Biological Sciences* 354 (1999): 121–9.

13. Rosanna Dent, chapter 24; Miranda Johnson, "Making a Treaty Archive: Indigenous Rights on the Canadian Development Frontier," in *Law, Memory, Violence: Uncovering the Counter-Archive*, eds. Stewart Motha and Honni van Rijswijk (London: Routledge, 2016), 195–214; Tom Özden-Schilling, "Technopolitics in the Archive: Sovereignty, Research, and Everyday Life." *History and Theory* 59, no. 3 (2020): 394–402.

14. Jonathan S. Friedlaender and Joanna Radin. *From Anthropometry to Genomics: Reflections of a Pacific Fieldworker.* iUniverse.com Press, 2009.

15. Even though they often do not, as discussed in the papers collected in Boris Jardine, Emma Kowal, and Jenny Bangham for the special issue "How Collections End: Objects, Meaning and Loss in Laboratories and Museums," *BJHS Themes* 4 (2019): 1–27.

16. Claire Waterton, "Experimenting with the Archive: STS-ers as Analysis and Co-constructors of Databases with Other Archival Forms," *Science, Technology and Human Values* 35 (2010): 648.

17. Jonathan S. Friedlaender, William White Howells, and John G. Rhoads, *The Solomon Islands Project: A Long-Term Study of Health, Human Biology, and Culture Change* (Oxford and New York: Clarenden Press; Oxford University Press, 1987).

18. Friedlaender and Radin. *From Anthropometry*, 163.

19. Jonathan S. Friedlaender. *Patterns of Human Variation: The Demography, Genetics, and Phenetics of Bougainville Islanders* (Cambridge, MA: Harvard University Press, 1975).

20. Friedlaender and Radin. *From Anthropometry*, 163.

21. Jonathan S. Friedlaender, "Genes, People, and Property: Furor Erupts over Genetic Research on Indigenous Groups," *Cultural Survival Quarterly* 20, no. 2 (1996); Jonathan S. Friedlaender, "Commentary: Changing Standards of Informed Consent: Raising the Bar," in *Biological Anthropology and Ethics: From Repatriation to Genetic Identity*, ed. Trudy Tuner (Albany: State University of New York Press, 2005); Jenny Reardon, "The Human Genome Diversity Project: A Case Study in Co-production," *Social Studies of Science* 31 (2001): 357–88; Laura Stark, *Behind Closed Doors: IRBs and the Making of Ethical Research* (Chicago, IL: University of Chicago Press, 2011); Joanna Radin, "Ethics in Human Biology: A Historical Perspective on Present Challenges," *Annual Review of Anthropology* 47 (2018): 263–78.

22. Kowal, Radin, and Reardon, "Indigenous Body Parts."

23. Friedlaender and Radin, *From Anthropometry*, 225.

24. Michael Lynch, "Archives in Formation: Privileged Spaces, Popular Archives and Paper Trails," *History of the Human Sciences* 12 (1999): 65–87.

25. Claire Waterton, "Experimenting with the Archive: STS-ers as Analysis and Co-constructors of Databses with Other Archival Forms," *Science, Technology and Human Values* 35 (2010): 645–76.

26. Jenny Reardon and Kim Tallbear, "Your DNA Is Our History," *Current Anthropology* 53 (2012): 233–45; Henry T. Greely, "Informed Consent, Stored Tissue Samples, and the Human Genome Diversity Project: Protecting the Rights of Research Participants," In *Stored Tissue Samples: Ethical, Legal, and Policy Implications*, ed. R. Weir (Iowa City: University of Iowa Press, 1998); Susan C. Lawrence, "Beyond the Grave—The Use and Meaning of Human Body Parts: A Historical Introduction." In *Stored Tissue Samples: Ethical, Legal, and Policy Implications*, ed. Robert F. Weir (Iowa City: University of Iowa Press, 1998); Michelle M. Mello and Leslie E. Wolf, "The Havasupai Indian Tribe Case—Lessons for Research Involving Stored Biological Samples," *New England Journal of Medicine* 363 (2010): 204–7; M. Susan Lindee, "The Repatriation of Atomic Bomb Victim Body Parts to Japan: Natural Objects and Diplomacy," *Osiris* 13 (1998): 376–409.

27. Stark, *Behind Closed Doors*.

28. Patricia Cohen, "Questioning Privacy Protections in Research," *New York Times*, October 23, 2011.

29. Carolyn Steedman, *Dust: The Archive and Cultural History* (New Brunswick, NJ: Rutgers University Press, 2002); Arlette Farge, *The Allure of the Archives* (New Haven: Yale University Press, 1989/2013).

30. Michael M. J. Fischer, *Emergent Forms of Life and the Anthropological Voice* (Durham, NC: Duke University Press, 2003).

31. Harriet Washington, *Carte Blanch: The Erosion of Medical Consent* (New York: Columbia University Press, 2021).

32. Kowal, Radin, and Reardon. "Indigenous Body Parts."; Jennifer Couzin-Frankel, "Researchers to Return Blood Samples to the Yanomamo," *Science* 328 (2010): 1218.

33. Emma Kowal, "Orphan DNA: Indigenous Samples, Ethical Biovalue and Postcolonial Science," *Social Studies of Science* 43 (2013).

34. Linda Tuhiwai Smith. *Decolonizing Methodologies: Research and Indigenous Peoples*. Third Edition (London: Zed Books Ltd., 2021); Audra Simpson, "On Ethnographic Refusal: Indigeneity, "Voice" and Colonial Citizenship." *Junctures: The Journal for Thematic Dialogue* 9 (2007); Ruha Benjamin, "Informed Refusal: Toward a Justice-Based Bioethics," *Science, Technology, & Human Values* 41, no. 6 (2016): 967–90; Kim Tallbear, "Beyond the Life/Not-Life Binary: A Feminist-Indigenous Reading of Cryopreservation, Interspecies Thinking, and the New Materialisms," *Cryopolitics: Frozen Life in a Melting World* (Cambridge, MA: MIT Press, 2017): 179–202.

35. Lauren Kassell, "Inscribed, Coded, Archived: Digitizing Early Modern Medical Casebooks," *Journal for the History of Knowledge* 2, no. 1 (2021).

36. Richard J. Cox, *No Innocent Deposits: Forming Archives by Rethinking Appraisal* (Lanham, MD: The Sacrecrow Press, 2004).

21

Locating Sources, Situating Psychiatry, Complicating Categories

A Journey through Three German Archives

Lara Keuck

This chapter is the third version of a paper on my historical practice and the challenges I faced when searching for and working with psychiatric sources. As such, it is not only a reflexive piece on how social encounters and legal regulations concerning access to sources (in)forms history writing but also on the continuous work of situating my perspective. To make this visible, I have called attention to my wording and thinking at the various stages of writing and revising. Instead of "updating" my reflections on the archival work that I pursued seven years ago, I chose to flag the layers of this chapter and add new concluding remarks. This represents another shift in perspective that situates my earlier reflections, and it allows me to revisit them *after* having worked with the sources in my own historical accounts—and after engaging with the other contributions in this book.

DRAFTS AND REVISIONS

The first version of this chapter was an unpublished essay that I wrote in September 2014 after a research trip to Southern Germany.[1] I was searching for records to compare and contextualize the diagnostic procedures that were used to identify the peculiarity of cases of Alzheimer's disease. The sources that I was able to access, including patient files, diagnosis books, unpublished correspondence and typescripts, were helpful in giving me a better understanding of the making and use of the diagnosis of Alzheimer's disease as an exploratory category in the 1900s. However, this archive trip also led me to think about the practices that historians use and what these practices imply for historiography: what makes something a source? What does it imply to treat something or someone as a source? And how does my view on the sources differ or coincide with that of the people who decided to keep and care for old materials in an institutionalized or private archive?[2]

The second version of this chapter, "Thinking with Gatekeepers," was written for the *(In)visible Labour* conference that Jenny Bangham and Judy Kaplan organized in 2015 and published in a MPIWG Preprint in 2016.[3] I chose the title to stress that we gain more than bureaucratic access to historical materials through our interactions with archivists. Their very approaches to the sources that they are responsible for can provide windows through which we can see the meanings and functions that these materials can have. Gatekeepers of historical sources (be they professional archivists or not) and the ways in which they treat their archives impact how historical research is done, but they are often invisible or encoded in acknowledgement sections and footnotes.

My attempt was to understand the epistemic implications of the accessibility to sources on my history writing. I therefore discuss the location of patient records and related early twentieth-century psychiatric sources in three very different archives—a university archive, a hospital, and a private collection. These locations shaped the extent to which I was provided access and freedom to see, use, and reproduce documents. I reflect on how the sources were presented to me: as historical research materials, as documents of human lives, or as relics of a great man's discovery. These perspectives had practical implications for how I used these sources, while also becoming themselves topics of my historical research. The following version of this chapter is a slightly edited version of the second and adds a new conclusion that reflects on how gatekeepers' perspectives informed my own history writing.

SOURCES AND GATEKEEPERS

In November 1901, the German psychiatrist Alzheimer encountered a severely confused fifty-one-year-old female patient, Auguste Deter. After her death five years later, he examined her brain and presented his clinical and histopathological analysis under the heading of "a peculiar disease" at a conference in Tübingen.[4] In 1910, Emil Kraepelin, the director of the *Königlich Psychiatrische Universitätsklinik* in Munich where Alzheimer then worked, recounted Alzheimer's case story in the eighth edition of his influential psychiatry textbook, where he labelled the disease with the term "Alzheimers Krankheit."[5] Kraepelin's textbook and Alzheimer's conference paper are published sources; the patient file is kept in the public archive of the city of Frankfurt (Main). Excerpts of the file, including a photograph of the woman that was taken in the psychiatric clinic, can nowadays even be found in Wikipedia under the patient's full name, Auguste Deter.[6] However, other sources that could enrich the history of Alzheimer's disease are less easily accessible. Some are kept in private estates. Some are lost. Some are available but reproduction is forbidden.

For instance, the patient files of the Munich psychiatric clinic where Emil Kraepelin and Alois Alzheimer worked from 1904 onwards are missing. However, the archive holds the original diagnosis books [*Diagnosenbücher*] in which the hospital staff recorded the names, ages, incoming and outgoing dates, and clinical diagnoses of all patients. Additionally, there are books in which doctors summarized patient profiles with a one-page clinical record per person, so-called epicritical reports [*Epikrisenblätter*]. When visiting the Munich clinic, I was allowed to view the diagnosis books and epicritical reports for my historical research, but when I asked if I could scan them, I was informed that medical confidentiality continued to exist after the death of a patient. Since the documents contained both personal data as well as diagnoses, their reproduction was not possible.[7] This influenced my work with the sources in several respects; since I could not visually reproduce the documents that I studied, I described them in as much detail as possible in my field notes. This made me very attentive to material details: which coloured pens were used, who produced the paper, and how diagnoses were corrected and amended. In fact, the limitation turned out to be epistemologically productive for me. The sensitivity of information about psychiatric patients also provided me with questions about my own practices, such as whether to anonymize my sources, and if so, how.

I encountered a very different situation in Heidelberg, where Kraepelin and Alzheimer had worked in the psychiatric clinic in 1903. Whereas in Munich, the historical medical documents have remained in the psychiatric clinic, in Heidelberg, the old patient files were relocated to the University archive in the 1990s. Here, I could ask for copies of everything, even of complete patient files. Moreover, the archivist explained to me that to correctly quote a patient file in my work, I should include the last name and first name of the respective patient—or alternatively, a reference to the file number. When I asked whether it was sensitive information to provide the full name, the archivist replied that she did not want to impede historical research and that even the strictest regulations of the federal archive allow for publication of personal information 110 years after birth.[8] To this archivist, the fact that the patient files were kept and carefully maintained made them, besides all other possible meanings, historical sources, embedded in the institutional setting of a university archive. If I wanted to work with these sources, I had to treat them according to given disciplinary standards and legal and institutional regulations concerning historical research in German archives. With the transfer of the patient records from the clinic to the university archive, their status apparently changed from being primarily confidential medical documents to becoming primarily research sources for historians. When the institutional setting and authority (clinic or archive) that was responsible for regulating the use of the documents changed so did the rules according to which historians of medicine were allowed to access and work with patient records.

The more I was faced with different stances towards access and reproduction in my search for sources, the more I thought about how the archivists and other gatekeepers needed to be part of a story of the history of a disease. I discuss three possible approaches to treat historical psychiatric materials in the following: as documents of human lives, as research materials, and as treasured relics. All of these relate to different reasons why historical psychiatric objects have been kept and protected.

HUMAN LIVES

When I searched the archives for patient files and epicritical reports, I searched them with this question in mind: what writing and diagnostic practices were applied to single out relevant from irrelevant facts? I found many traces in the files, but I also found myself constantly replaying a related question about my own methodology: what do I include as relevant information and what do I leave out?

Many epicritical reports contain information about disrupted families, suicide attempts, and precarious life situations. These might be presented as causes or symptoms for a mental disease, but they also mirror human tragedies. This becomes particularly evident in the patient files. The files comprise a variety of different documents, including transcripts of doctor-patient interviews; intelligence inquiry forms; photographs; medical examinations; requests of other hospitals or asylums to lend the file; and letters, drawings, and sometimes, poems of the patient. The diagnostic purpose of extensive patient interviews and the storage of letters at the time of their recording and writing is why these documents have been preserved. The question is: how to read these documents today? This relates to the methodological challenges of writing the history of medicine from below, but it also pertains to the epistemological double task that I set out at the beginning of this chapter: to understand the diagnostic reasoning of the time I am studying and to find my own stance in approaching the patient records and describing how they reflect the making and working of psychiatric knowledge.

This starts with the question of quoting the names of the patients or not. In most medical reports and histories of medicine, patient names are abbreviated or pseudonymized. In this chapter, I have used the full name of the patient, Auguste Deter, who became known as the first case of Alzheimer's disease. I decided to do so, because her name was already publicly known and because I think that giving her full name instead of only writing about "the first case" or abbreviating her name, puts more emphasis on the fact that she was not only the material for Alzheimer's discovery but actually performed her own agency in the diagnostic process. A cursory Google search confirms that quotes from an interview in her patient file—most prominently *Ich habe mich sozusagen verloren* ("I have lost myself")—are recurrently used to characterize the disease in diverse popular and academic contexts.[9] Against this background, it might even be considered as a question of authorship to give the credentials to Auguste Deter.

The practice of abbreviating the patient's name ensures medical confidentiality. But confidentiality not only protects sensitive data; it also defines what should remain hidden from the public world in the first place. Maybe my struggle in finding a good way to deal with the names of patients, but also of doctors, hospital staff, and gatekeepers, is most of all a reflection of how difficult it is to find an appropriate way to write not only about the making and working of psychiatric knowledge but about the humans involved in and affected by this endeavour.

RESEARCH MATERIALS

One approach to the patient files is to regard them as a collated, fragmented archive of (an episode in) a patient's life. Their archival function is reflected in their materiality; the files contain papers, documents, letters, tables, and requests that were perforated with a small hole in the upper left corner. A small piece of twine was pulled through them and the paperboard document folder to hold the file together. Different colours and layouts of the document folder covers indicate that archived files were sometimes transferred into new folders if the patient was admitted again in later years. Some of the folders bear several archival signatures, indicating that the re-identification of a once admitted person was an issue. An archivist alerted me to the fact that the continuous updating of files was facilitated by the *badische Bindheftung* technique. You can easily loosen the twine to add more pages or to browse the file like a book. The small hole in the upper corner allows paper of very different formats and margins to be added. This binding technique is known from notary documents: a twine is pulled through the pages that comprise the legal contract and is then sealed. In the case of a patient's file, the sealing never took place. Requests and replies were continuously added to the file, not only if the patient had long left the Heidelberg psychiatric clinic but also if the patient had died long ago. In a way, the files continued a life of their own as research materials for psychiatrists and historians.

When I searched the patient files and read the diagnosis books, I found evidence indicating that my practices of historical inquiry and some of the practices of the psychiatric clinic were very similar. The patient files contained underlining in red- and blue-coloured pencil, exclamation marks and dashes on the margins, signalling the different doctors' evaluations of important facts, and reflecting the collective authorship of a patient record. Also, the diagnosis books included notes and signs in different handwritings and colours. I could identify patterns: red circles indicated cases of dementia praecox, orange circles cases of manic-depressive confusion, blue circles cases of paralysis, violet hooks cases of alcoholism, a reversed hook in pencil indicated cases of epilepsy.[10] I noticed this while I was myself counting and grouping diagnoses. For my historical analysis of the establishment and use of the diagnostic category of Alzheimer's disease, I searched for entries of this diagnosis as well as of related ones such as arteriosclerosis, organic brain disease or senile dementia. The doctors' practice of compiling statistics and re-evaluating diagnoses a century ago and my practice of reconstructing their diagnostic and classificatory reasoning looked quite alike. We were both reusing archived paperwork as research materials.

TREASURED RELICS

The role of archiving and reassessing diagnoses is but one example of the various uses of history in psychiatry. Another one is memorializing founder figures. In 1995, the pharmaceutical company Eli Lilly bought a house in the old part of Marktbreit, a small village in Bavaria on the river Main. Konrad and Ulrike Maurer had identified the house as Alois Alzheimer's birthplace, had initiated its transformation into a museum, and the sponsorship of that process by Eli Lilly.[11] A dedicated local tour guide who showed me the museum recounted how Mr. and Mrs. Maurer got in touch with the heirs of Alzheimer and collected all sorts of items that related to him, ranging from his old microscope to the embroidery work of his sister. Why were these objects deemed relevant to preservation and exhibition? In the context of the museum, this question is framed against the background of the aims and scope of exhibiting and memorializing the life and work of Alzheimer. For the Maurers, Alzheimer's work was important for psychiatry but was, moreover, a cultural achievement that needed to be recognized.[12] Their long-term vision for the museum was to have it nominated as an UNESCO cultural heritage site.

Konrad Maurer had worked as professor of psychiatry first in Würzburg, a city near Marktbreit, and then in Frankfurt from 1993 until his retirement in 2009. In Frankfurt, he continued his search for the remains of Alzheimer's work and Auguste Deter's life. In 1997, Maurer and two colleagues published in the "medical history" section of *The Lancet* that they found the original patient file of the first case of Alzheimer's disease.[13] In 1999, Maurer published together with his wife Ulrike a popular science book on Alzheimer and the patient file of "Auguste D"—to whom they never refer fully by name but for whom they give many biographical details that extend the medical case report.[14] When I asked Mr. and Mrs. Maurer why they thought that the case of Auguste Deter was so important, they replied that this was self-evident: Alzheimer's disease had become the infamous disease of our time, and this was the first case. Konrad Maurer referred to his most cherished material—Alzheimer's original histological slices of Auguste Deter's brain, given as a gift—as a *relic*.[15] He made clear that he did not use the religious term as a metaphor. Rather, he explained to me that these slides hold the "embodied" disease.[16] Maurer's approach to sources as treasured relics provided me with yet another perspective on historical psychiatric materials. The idea that a histopathological slide is a relic provides a lens to scrutinize the ways in which the factuality of bodily remains has been used as evidence for a continuous history of a real disease.

The story narrated by my visits to the Maurers and the museum in Marktbreit is about the exceptional role of Alzheimer and of the first case of the disease that was named after him. The exceptional status of Alzheimer is framed by the conception of the gatekeepers for whom their collection, museum, and publications on Alzheimer are a life's task. I met and interviewed the Maurers at their home in Frankfurt, where Konrad Maurer keeps a private collection of historical psychiatric resources. These resources and the ones kept in Marktbreit are not freely accessible. They are (literally) treasured. At first, I thought that the most interesting part of my meeting with the Maurers would be to look at their private archive. But while interviewing them, I realized that more than the historical material itself, I was interested in their view of the material. The reasons why the Maurers kept and treasured these objects could tell me something about the remaking of the foundational role of Alzheimer's first case after its rediscovery in the mid-1990s. My trip to Frankfurt, which had begun as a search for archival material, ended with a piece of oral history being my main source. Gatekeepers do not only regulate accessibility to historical sources; their views on the material they govern can be sources of inspiration of their own.

COMPLICATING CATEGORIES

My project on the history of Alzheimer's disease resulted in a series of articles that complicated categories in one way or another. For instance, I used the diagnostic books to historicize the use of psychiatric categories in the early twentieth century, showing how Alzheimer's disease was neither defined nor discovered, but *explored* as a working title for "interesting cases." The epistemic function of these cases—in particular, the description of Auguste Deter's disease course—changed throughout the past century. The "interesting case" became the "prototypical case." In accordance with this shift, the "provenance" of the cases as patients *of* Alois Alzheimer was presented as an epistemic argument in the late modern scientific debate on what should count as Alzheimer's disease.[17] Thinking with the gatekeepers about what the location of sources reflect about the situatedness of psychiatry informed my analyses and contextualizations of the history of a psychiatric category.

The perspective of complicating categories expands, however, beyond my case study. When this chapter is read in conjunction with other chapters in this book, other complicated categories are foregrounded. Along with Blacker, chapter 10, and Dent, chapter 24, I highlight the meaning of authorities and authority in processes of governing knowledge. The conditions of memorializing and silencing or reactivating functions of archives and museums become an object of inquiry when put in dialogue with the case studies and analyses of Jardine, chapter 22; Bruchac, chapter 4; and Kaplan, chapter 17. When read together with Burton, chapter 7, the focus shifts to how the often invisible labour of the many hands involved in making, storing, and (re-)using of paper work has not only shaped medical ways of knowing in past times but also how historians of science and medicine approach their questions today. In this latter context, this chapter is itself evidence of my approach to complicate clear-cut categorizations, or put differently, to show the complications of demarcations between medicine and the history of medicine, between hospital records and historical sources, between the perspectives of actors and analysts. We are trained to distinguish, in our analyses and bibliographies, between sources and secondary literature. However, especially when writing contemporary history, there is often no temporal gap between the two genres: sources are influenced by the same *Zeitgeist* as the frameworks we use to analyse them. Is my work, then, a history of Alzheimer's disease or a history of histories of Alzheimer's disease? It can, and perhaps even needs to, be read as both.

NOTES

1. Special thanks go to Sabrina Zinke, Norbert Müller, Konrad Maurer, Ulrike Maurer, and the staff of the *Heidelberger Universitätsarchiv* and of the *Klinik für Psychiatrie und Psychotherapie der Ludwig-Maximilians-Universität München.*

2. Susan Lindee's critical but appreciative approach to gatekeepers and their sources and narratives inspired me to write the first version of this chapter: Susan Lindee, *Moments of Truth in Genetic Medicine* (Baltimore: Johns Hopkins University Press, 2005), 231–6. For the second version, further inspiration was drawn from literature on the use of paper work and patient files in the history of medicine and psychiatry. In order to cohere with the word limit of this third version, I kept only the references to literature necessary for understanding my arguments, and references to primary sources. For the list of references of the second version of this chapter, see Lara Keuck, "Thinking with Gatekeepers. An Essay on Psychiatric Sources," in *Invisibility and Labour in the Human Sciences*, eds. Jenny Bangham and Judy Kaplan (Berlin: Preprint of the Max Planck Institute for the History of Science 484, 2016), 114–15.

3. Keuck, "Thinking with Gatekeepers," 107–115.

4. Alois Alzheimer, "Über eine eigenartige Erkrankung der Hirnrinde," *Allgemeine Zeitschrift für Psychiatrie und Psychisch-Gerichtliche Medicin* 64 (1907): 146.

5. Emil Kraepelin, *Psychiatrie. Ein Lehrbuch für Studierende und Ärzte. Achte, vollständig überarbeitete Auflage, II. Band, Klinische Psychiatrie, I. Teil* (Leipzig, Germany: Verlag von Johann Ambrosius Barth, 1910), 624.

6. The Wikipedia entry for "Auguste D" was created on March 14, 2007 and was immediately discussed as a candidate for "speedy deletion as an article about a real person that does not credibly indicate the importance or significance of the subject." Yet, it remained in the Wikipedia and expanded quickly. The photograph was added ten days after the creation of the website; the heading of the entry was changed to include the woman's full last name in May 2010, see: "Auguste Deter: Revision History," https://en.wikipedia.org/w/index.php?title=Auguste_Deter&action=history, accessed August 20, 2021.

7. Norbert Müller, email to author, August 26, 2014. However, the museum of the psychiatric hospital in Munich used to display the epicritical report (including the biographical information about the patient) of one of the early cases of "Alzheimer's disease," and reproduced it in one of their publications, which was offered to me as a present when being admitted to the clinic archive: Hanns Hippius, Hans-Jürgen Möller, Norbert Müller, and Gabriele Neundörfer, *Die Psychiatrische Klinik der Universität München 1904–2004* (Heidelberg, Germany: Springer Medizin Verlag, 2005).

8. Sabrina Zinke, personal communication with the author, September 2, 2014; the quotation rules of the federal archive are statutorily regulated, see http://www.bundesarchiv.de/bundesarchiv/rechtsgrundlagen/bundesarchivgesetz/index.html.de, §5 (2), accessed January 12, 2016.

9. The patient file was translated and partially published first in Konrad Maurer, Stephan Volk, and Hector Gerbaldo, "Auguste D and Alzheimer's Disease," *The Lancet* 349 (May 24, 1997): 1546–49.

10. Psychiatrische Klinik der Ludwig-Maximilians-Universität München (LMU-P) Diagnosenbücher Männer/Frauen, 1904–1912.

11. See also Peter Whitehouse, Konrad Maurer, and Jesse F. Ballenger, eds., *Concepts of Alzheimer Disease: Biological, Clinical, and Cultural Perspectives* (Baltimore, MD: Johns Hopkins University Press, 2000), xii–xiv.

12. Interview with Konrad and Ulrike Maurer in Frankfurt/Main on July 30, 2014.

13. Maurer, Volk, and Gerbaldo, "Auguste D and Alzheimer's Disease."

14. Konrad Maurer and Ulrike Maurer, "Alzheimer: Das Leben eines Arztes und die Karriere einer Krankheit" (München, Germany: Piper Verlag, 1999).

15. Interview with Konrad Maurer on July 30, 2014. Maurer's comparison of the histological specimen with a religious relic echoes nicely with how Rudolf Virchow treated his collection of pathological preparations as "secular relics": Angela Matyssek, "Die Wissenschaft als Religion, das Präparat als Reliquie. Rudolf Virchow und das Pathologische Museum der Friedrich-Wilhelms-Universität zu Berlin," in *Sammeln als Wissen. Das Sammeln und seine wissenschaftsgeschichtliche Bedeutung*, eds. Anke te Heesen and Emma C. Spary (Göttingen: Wallstein Verlag, 2001), 142–68.

16. Konrad Maurer used the English term *embodiment* in our German interview.

17. See Lara Keuck, "Diagnosing Alzheimer's Disease in Kraepelin's Clinic, 1909–1912," *History of the Human Sciences* 31 (2018): 42–64; Lara Keuck, "History as a Biomedical Matter: Recent Reassessments of the First Cases of Alzheimer's Disease," *History and Philosophy of the Life Sciences* 10 (2018). https://doi.org/10.1007/s40656-017-0177-7.

22

Turing, or An Exhibition Should Not Mean but Be

Boris Jardine

INSTALLATION SHOT

All is calm in the gallery. The artefacts, attached to walls and perching on plinths, are at rest. The floors are clean. There are no visitors, but the lights are on. In fact, the lighting is perfect, because this is an "installation shot"—a specific act of documentation, developed in the early-twentieth century as a technique of recording the finished form of an exhibition (figure 22.1).[1] Like any other visual genre it has its own rules: the space is empty of people; sight-lines are clear and the composition is geometrically elegant. The installation shot records the way in which artworks and artefacts are hung or placed, their relation to labels, and, to some extent, the experience of the show itself. It is a way of creating a document of something that is bound to change: while artefacts stay more or less the same over time, galleries and exhibitions do not.

But there is another view of the exhibition—impossible to photograph but perhaps possible to imagine—as a site of intense and dynamic forces. Exhibitions exist at the meeting point of two different kinds of labour: that of the exhibition visitors and exhibition developers. Each exhibition is conceived, designed, and installed by a large number of people with a range of professional identities. In the installation shot we are not witness to this work—often spanning many years—which includes curatorial conception and planning, administration and management, conservation, graphic and set design, security, and maintenance. Exhibition development occurs within complex institutional hierarchies, and draws on the sedimented labour of previous curators, conservators, administrators, and many others.[2] Once opened to the public, an exhibition is then visited by people with differing demands, preconceptions, and reasons for being there. Each visitor *works* to construct their own interpretation of material, and this may or may not align

Figure 22.1. An early and influential 'installation shot'. This the so-called 'Armory Show' (1913), which was decisive in the introduction of continental modernism to America; the full title was 'International Exhibition of Modern Art'; the first venue was New York City's 69th Regiment Armory, hence the colloquial name. Public domain.

with the intentions of those who put the show together. So just as the installation shot is a crucial tool for recording the layout of exhibitions that would otherwise be lost to history, it also conceals the labour that goes into the creation and experience of exhibitions.

It might be contended that, as the finished form of an exhibition is both the intention of exhibition creators and the idealized viewing space, the installation shot does record quite a large amount of these forms of labour. Yet the finished form itself is only one stage in a process that has generated many other possible forms—rejected objects, discarded narratives, graphic and set redesign. And the installation shot also suggests a fixity of meaning that is at odds with the often intense work that visitors put into interpretation—supressing or bringing into the space their own expertise, hopes, fears, prejudices, knowledge, and profound and everyday experiences.

Installation shots, then, reveal and conceal at the same time. Yet the work that they do—stabilizing an inherently dynamic and complex process—is an important starting point because of the yet-unrealized contribution that exhibitions can make to science studies. Exhibitions at once contribute to and complicate notions of communication, celebration, and narrative. What visions of science do we wish to present, and to whom? This question, somewhat baldly stated, is absolutely central to exhibition design but can often be submerged in the dense critical and theoretical reflection explicit or implicit in the history, philosophy and sociology of science. I propose that the *event* of an exhibition and the invisible labour of creators and audiences offer us a model for reflecting on our own (largely written) practice as commentators on modern science.

CODEBREAKER

The reasons that exhibitions have not been integrated into science studies are not hard to find. Exhibitions are first and foremost events to which visitors go, and are therefore not easy to place in historical argument. Yet even the notion that an exhibition is an event is not as straightforward as it sounds. To attend an exhibition is to place oneself in a physical relation to objects that have been preserved for a variety of reasons and in various states of (dis)repair. Curators and other museum workers attempt to give meaning to these objects through labelling and spatial arrangement.

Here it will help to introduce examples, and these will all come from the 2012 exhibition *Codebreaker: Alan Turing's Life and Legacy*, with which I was involved as curatorial assistant to the lead curator, David Rooney. This exhibition celebrated the 100th anniversary of the birth of computer pioneer Alan Turing (1912–1954), and occupied a medium-sized, roughly square gallery space near the entrance to the Science Museum, London.

One important feature of this space was that it has three points of entry, all of which had to be considered as possible "starting points" for any given visitor entering the exhibition (figure 22.2). Turing's life, therefore, could not be told in a strictly (or at least not solely) linear fashion. Already we can glimpse the kind of labour that is central to exhibition-making but which has only weak analogies in historical writing: the entire labour of exhibition design, label writing and object selection was geared towards making a virtue of this necessity, and therefore rendering what might have been a problem invisible to all but the most perceptive of visitors. How would anyone wishing to "cite" or discuss the Turing exhibition contend with—or even notice—the problem of multiple-ingress?

To begin at the three beginnings, then, Rooney—working closely with the design company Nissen Richards—resolved to construct a "main line" through the exhibition, which would start with an introductory panel at the entrance nearest to the Museum's own main entrance. From here visitors would move to a pre-history of computing, then to a text and set-dressing display giving information on Turing's early life. This was followed by a large section dealing with Turing's work in World War II, which was concerned with codebreaking, and in particular the application of mechanical techniques to this problem. The next section occupied another potential point of entry, and here we decided to place some highly personal documents from Turing's youth alongside artefacts relating to his later studies on artificial intelligence. A visitor beginning here would meet the young, speculative Turing and then be able to move either to the wartime or postwar work. Rounding the corner of the exhibition (roughly top-centre of the plan in figure 22.2) three large artefacts formed a startling juxtaposition: the "Pilot ACE" computer, finished in 1950; a piece of the

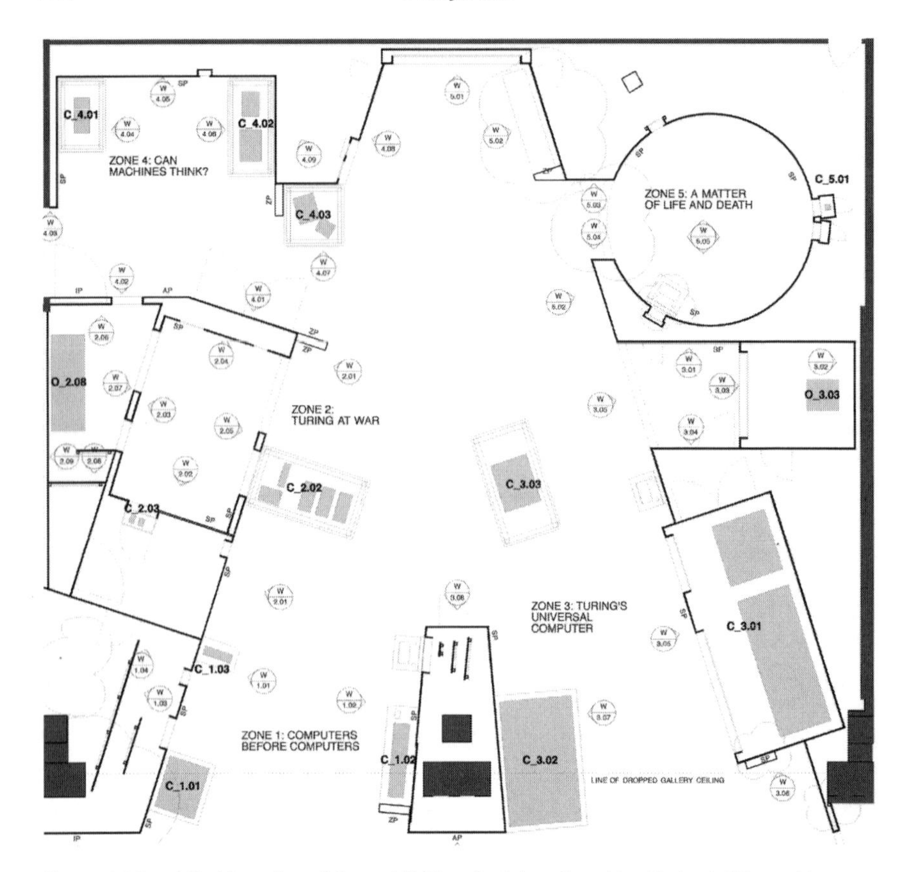

Figure 22.2. Working plan of the exhibition Codebreaker: Alan Turing's Life and Legacy, which ran at the Science Museum, London, June 2012–October 2013. Design by Nissen Richards studio, London, May 2012. The main entrance to the show is from the bottom right (into "Zone 3: Turing's Universal Computer"), but the chronological beginning is at the bottom left (at "Zone 1: Computers before Computers"); another entrance is at the top left, into "Zone 4: Can Machines Think?" The show concludes in the circular room at the top right, "Zone 5: A Matter of Life and Death." Courtesy of Nissen Richards Studio.

1954 wreckage of a Comet jet aircraft; and a model of vitamin B12 by Dorothy Hodgkin, *c.*1957.

Pilot ACE was in fact the only physical artefact directly related to Turing's work. Rooney's label text explains its significance:

> This was one of the first electronic "universal" computers. Its fundamental design was by Alan Turing, who wrote the specification in 1945 while working at the government's National Physical Laboratory. It was completed in 1950.
>
> Turing's idea was to build a large computer, to be known as the Automatic Computing Engine or ACE. But slow progress, coupled with changes in project direction

imposed on Turing, left him deeply frustrated, and he quit in 1948. This small-scale trial version, called Pilot ACE, was completed in his absence.

The vitamin model and section of plane wreckage gave physical form to the uses of the Pilot ACE—in performing crystallographic calculations, and mathematical analysis relating to structural failures that resulted in a series of air crashes in the 1950s.

Pilot ACE itself was positioned at the "exit" to the exhibition, though it was in fact not the conclusion of the narrative—in this way visitors who came in at this point would first see the highlight of the show.

Owing to the lack of a culmination to the "walkthrough" of the exhibition, the story's dramatic or narrative end was told in a separate small circular chamber, constructed for the exhibition (top right in figure 22.2). Here a projector showed slides from the extensive archives relating to Turing's final work, on the biochemical problem of "morphogenesis." A panel on the wall of the chamber showed a reproduction of Turing's death certificate, revealing that he had died from cyanide poisoning. To this day there is no fully satisfactory explanation of Turing's death, but a year prior he had been forced to undergo a course of "chemical castration" after being found guilty of "gross indecency" for homosexual acts. A vial of oestrogen, the chemical used in the process, was displayed with little comment.[3]

Organizing an exhibition, even of this relatively small size, is a herculean endeavour, involving years of planning, in particular around the complex issues of object selection, narrative development, loans in, conservation, publicity and internal communication. Central to the curatorial process is the selection of objects which are themselves the legacy of many years of prior curatorial work—all collections are, after all, a mix of planning and contingency.[4] Then, the exhibition has to be explained, justified and critiqued within the museum, and feedback is regularly given to the major "stakeholders," including external funders (here, Google), and in this case Turing's surviving relatives. Relationships with lenders are complex and varied. For *Codebreaker* we were seeking artefacts from the National Archives (these came with limitations on light exposure that proved, in the end, too stringent); from relatives of Turing's childhood friend Christopher Morcom (the latter having died of tuberculosis, causing Turing to write an extraordinary letter to Morcom's mother, which we included in the exhibition); from Bletchley Park; and, less obviously, from Mick Jagger—who owns an Enigma Machine, and whose involvement in the exhibition we frankly sought for reasons of publicity.

I make no claim that any *special* kind of labour was involved in striking up and managing these relationships. Some negotiations were difficult, some handled better than others, and naturally for larger exhibitions this kind of list could be multiplied many times over. My purpose is rather to show the range of *kinds* of work that go into exhibition development. At every stage of exhibition development, the intended visitor experience is examined and refined. One seldom acknowledged point that emerges from this is that an exhibition is "live" or "active"—that is, communicating with a wide range of people—from the moment it is included on a forward planning document. Certainly, most people would agree that *Codebreaker* meant the most to the most people during the time it was open to the public—but prior developmental stages must also be recognized as exhibitionary acts.

Over the last thirty years or so, the emphasis in museum studies literature has been focused on the visitor experience of exhibitions, and has emphasized the various kinds of presuppositions visitors bring with them, the point being that the "meaning" of an exhibition is co-constituted between the finished form of the exhibition and the visitors themselves.[5] While this has had hugely beneficial effects within exhibition design, it has also, in my view, led to an imbalance in the understanding of what an exhibition *is*, and a neglect of the various kinds of labour that go into exhibition production.

One reason to insist on this point in relation to *Codebreaker* is that there were features of the exhibition development that were invisible in a very special sense. One of these was a series of interviews conducted by myself and David Rooney, convened by the Science Museum's Head of Audience Research and Advocacy, Annika Joy, with a group of over-fifties LGBT+ participants, all members of the charity Opening Doors London, which specifically advocates for and serves that community.[6] Our intention in conducting the interviews was to gain an understanding of the experience of homosexuality in mid-century Britain—but we had no intention to use any of the material gathered in these interviews in the exhibition itself. Nor was it our desire to explain or even communicate the details of *Codebreaker* to a specific interest-group, although naturally Turing's life story and the work we were doing did come up during the interviews.

The content of these interviews was and remains permanently embargoed, so it would not be appropriate for me to go into details of the exchanges that took place. I found these interviews deeply moving, and at times the conversation was challenging: there was, understandably, some wariness about our intentions in conducting them. At times the experiences and identities of the interviewers (myself and David Rooney) were also under scrutiny.

What can I say about this process if I cannot quote from the interviews themselves? Opening Doors London's own history and purpose are instructive here. Citing positive changes in social attitudes and (especially) legislation in post-war Britain, Opening Doors London draws on research showing that

> older LGBT+ people—many of them members of first post-1967 generation—carried with them the real benefits of the changes since the 1970s; but also the scars of a painful history, which in turn were accentuated by all the differences that bisected an increasingly diverse world: of class, gender, race and ethnicity, geography, access, and sexual needs and desires.[7]

The group with whom we talked, then, were uniquely placed to give us insights not only into gay life in mid-century Britain, but specifically into difficulties encountered in living through periods of persecution and liberation, and into the ways in which history can play both a positive and negative role in present experience.

Certainly the conversations we had with members of Opening Doors London conveyed a powerful sense of ambivalence, operating at many levels: the "bad old days" could be set against our own more liberal times; but specific acts of solidarity and forms of community were seen to have had greater meaning in the past than they could do now. Personal journeys of increasing confidence, self-awareness, love and sexuality were understood in relation to broad concepts of justice, societal change and the relation of the individual to the state. Although many participants did know each other prior to our meetings, evidently much was discussed that had not been shared previously.

The process was integral to our curatorial work on *Codebreaker*, and specifically to Rooney's exhibition text: it gave him confidence to talk about homosexual experience in the 1950s because he had heard first-hand testimony about this experience. The interviews had not settled any matters of fact, but had deepened our understanding of this historical episode and allowed us to ask the right questions: in what ways could homosexuality be experienced as liberating in mid-century Britain? what was the lived experience of state persecution after World War II? how does sexuality relate to intellectual work, career path, and professional identity? *Codebreaker* talked frankly about the sense of oppression of those years not because of anything Turing said about his own experience, but because of what we came to know about the range of possible experiences. Here is the "story panel" text which introduced the section of *Codebreaker* dealing with Turing's last years:

> Until the late 1960s, most homosexual acts were illegal. Many people lived in constant fear of being caught by police, prosecuted and either imprisoned or fined.
>
> Lives were routinely destroyed in the aftermath of such humiliating events. People were often sacked from their jobs and ostracised by families, friends and the wider community. Some felt suicide was the only option. In parallel, scientists and doctors were experimenting with ways to "cure" gay people or remove their sexual urges.
>
> In 1952, Turing was arrested for a sexual relationship with a man, and sentenced to a one-year course of female hormone treatment. At the time he had been advising the government on secret codebreaking projects, but his security clearance was revoked and he was later placed under surveillance.
>
> Two years after his arrest, in 1954, Turing was found dead at his Wilmslow home. The official verdict was suicide from cyanide poisoning, the coroner believing "his mind had become unbalanced."

This was the visible product of our curatorial labour. The second paragraph in particular speaks with a confidence only possible in light of the testimony we received. But what of the question of visibility itself? One way of thinking about the significance of this process is to consider it as a form of invisible labour, specifically a form of invisible labour concerned with invisibility. Our interviews often covered questions of visibility in relation to "hidden" homosexual experience, and the kind of work that went into avoiding persecution. Yet our intention in the exhibition was precisely to draw on these interviews in order to make visible certain aspects of Turing's lived experience.

Why then did we decide to keep the interviews themselves private? There were two main reasons, one having to do with the exhibition, and the other with the discussions themselves. As mentioned, the exhibition faced a number of challenges, some having to do with the space, others having to do with the subject matter—i.e. how to tell the intellectual biography of someone whose main professional identity was as a mathematician through artefacts? I explore the specific answer to this question below. In terms of narrative voice, however, we decided that incorporating the voices of the older LGBT+ participants would introduce a separate and distracting layer to what was already a complex story.

We were also mindful of the space in which the discussions were to take place. We wanted to create an environment that would be conducive to free discussion, and so we announced at the very beginning of the first session that the interviews would not be quoted or used in our work. This itself presented certain difficulties: what, we were asked, was the purpose of these discussions? No one present had known Turing, or anyone connected to him, so what insight could participants shed on the matter? We insisted that it was our intention to hear the testimony of participants for its own sake, and to inform ourselves about something beyond our own social experience. In journalistic terms—and without wishing at all to downplay the significance of the process—we were engaged in "background" research.

Here is a clear point of intersection between exhibition work and writing about science. The printed page can convey the notion that all evidence that could be of use has been presented to the reader, so that they might be able themselves to reconstruct the episode under discussion. What could be the benefit of "holding back" anything discovered during research? Yet issues of narrative structure, selectivity and standpoint occur in both forms of communication. The lesson of *Codebreaker* is that the overall sense or tone of a piece of writing must be determined by much more than the evidence documented in footnotes. What museum theorists call "atmosphere" is also essential in directing (rather than determining) visitor experience.[8] I propose that arranging and managing the atmosphere of an exhibition and of a text is as much a matter of seeking to broaden one's own (curatorial, historical) standpoint as it is about specific items chosen for display and specific facts conveyed through prose.[9]

MATERIAL BIOGRAPHY

At the outset I suggested that we have not yet allowed exhibitions to inform the work of historians, philosophers, and sociologists of science. One reason for this is that it is hard to say just how exhibitions *mean*. The physical comportment of a visitor in a specific space, in relation to other visitors and an array of artefacts, contributes to the interpretation of any one artefact.[10] If this were not the case, exhibitions would simply be pointless. Yet there is no easy way of accessing specific experiences within exhibitions, let alone aggregating them or incorporating them into historical arguments. Exhibitions are also heavily reflexive: because an exhibition cannot be reduced to its propositional content, and because it simultaneously draws on historical literature that evidently *can* be reduced to its propositional content, exhibitions always need to present an argument for their own physical form.

This was especially clear in the case of *Codebreaker*. There are virtually no surviving three-dimensional artefacts made by Alan Turing or directly relating to his life. There is a large archive at King's College, Cambridge, and there are some personal documents in private hands and official ones at the National Archives. Yet the Science Museum was tasked simultaneously with providing some kind of explication of Turing's scientific work, and incorporating artefacts either loaned in, or sourced from its own collections. Here Rooney's somewhat unusual institutional position proved essential: Rooney was trained as a historian of horology, and was at the time of *Codebreaker* responsible for the Science Museum's collections of planes, cars, bicycles and related artefacts. The story told in *Codebreaker* was primarily about the social and cultural significance of technology. This was, in effect, a contextual history of computing, through which the life story of Turing was threaded. As Rooney later put it,

> How can an exhibition for the widest possible audience cover the most complicated science and technology, and offer a biographical account that does not shy away from the heartbreaking circumstances of its subject's life and legacy?[11]

Because we were aware that Turing himself might become subordinate to the artefacts of this history—including mechanical calculators, ciphering machines, a robotic tortoise, and the Pilot ACE (Automatic Computing Engine) computer—we made the image of Turing central to the exhibition, first by using a theatrical "stage-set" constructed from photographs of Turing as a device at crucial locations, and second by reproducing an engaging photograph of him at the focal point of the room (figure 22.3). These were not subtle techniques, but in a room populated by engineered metal and heat-mouldable plastic, they had the dramatic effect of bringing one single person to life.

Still, this does not offer very much to the historian who would like to use or draw on the insights of the exhibition. The lessons of *Codebreaker* are, I think, not strictly biographical. Rather they are about the role of biography in historical argument. To offer a comparison: the "Armory Show" (figure 22.1) was not significant in the annals of modernism for any specific proposition made in the accompanying text. Rather, the Armory Show acted as a public statement that the Continental avant-garde had a major role to play in American aesthetics. Evidently, the analogy is not perfect: no one would dispute Turing's significance for the history of computing, and in fact his own prominence dramatically obscures the work of many others.[12] Instead, the importance of *Codebreaker*, I believe, lies in questions of identity and standpoint. Too often biography is the invisible motor of historical argument, and when the lives of actors are cherry-picked for anecdotal support, they contribute to arguments that could be made without them. In the process, another mistake is made, in which the complexities of lived experience itself are subordinated to intellectual achievement.[13]

Figure 22.3. Installation shot of the exhibition *Codebreaker: Alan Turing's Life and Legacy*. The display case on the left shows a salvaged fragment of a crashed Comet jet; a crystallographic model can be seen straight ahead; the Pilot ACE computer is in the recessed display case to the right. Turing can be seen, seated, in the large photograph at the far end of the gallery. Image courtesy of David Lambert.

Turing's biography is undoubtedly one of intense suffering and joy. We know from the pioneering work of Andrew Hodges that it is not possible to separate Turing's effeminacy and enthusiasm from his highly productive time at King's, and his homosexuality from the tolerant atmosphere of Bletchley Park.[14] Turing's belief in the persistence of the soul was not, in his own mind, separate from his speculations about artificial intelligence. To feel confident in telling the story of Turing's last years, we knew it was necessary to seek out and hear the testimony of people who had experienced oppression in mid-century Britain owing to their sexuality. Yet in the popular imagination, Turing's achievement is one of the coldest rationality: the creation of the mathematical justification for modern digital computing.

In the spirit of Donna Haraway's claim that we need "feminist figures of humanity," *Codebreaker* offered Turing as an embodied, queer, loving, feeling man who was instrumental in precisely the universalizing essentialism that we equate with rational modernity.[15] Making visible the connection between people, artefacts, and historical episodes is possible in an exhibition precisely because it is an experience and not a set of propositions. The combination of background research and respect for the standpoint of participants/visitors/readers (as well as historical actors), the life told through physical artefacts (including images and documents *as* physical artefacts, with their own materiality and "stage presence"), the emphasis on the contingency of lived experience and its destabilizing relation to rational modernity, these are all moves that can be made in exhibition design that can provide lessons and models for historical writing.

NOTES

1. A useful history of the installation shot is given in Julie Sheldon, "Picturing the Installation Shot," in *A Companion to Modern Art*, ed. Pam Meecham, (Hoboken, NJ: Wiley, 2017), 127–44.

2. See the chapters by Keuck and Radin, this volume, for more on the ways in which collections are amassed, maintained, and given (many) meaning(s).

3. It was instructive to spend time in this chamber and observe visitors. The vial of oestrogen was a representative one, and the death certificate was clearly a reproduction. Yet I heard a number of people remark that these were "the actual" artefacts surrounding Turing's death; I did not interpret this as a mistake.

4. See Lara Keuck's and Joanna Radin's contributions to this volume.

5. For a useful summary, see Sharon Macdonald, ed., *A Companion to Museum Studies* (Malden, MA; Oxford; and Carlton, Victoria, Australia: Blackwell Publishing, 2006), Part IV, "Visitors, Learning, Interacting."

6. See https://www.openingdoorslondon.org.uk (accessed October 14, 2021).

7. Jeffrey Weeks, "Opening Doors London: A Short History," available online at https://www.openingdoorslondon.org.uk/who-we-are (accessed October 14, 2021).

8. Regan Forrest, "Museum Atmospherics: The Role of the Exhibition Environment in the Visitor Experience," *Visitor Studies* 16 (2013), 201–16.

9. See Alison Wylie, "Why Standpoint Matters," in *Science and Other Cultures: Issues in Philosophies of Science and Technology*, eds. Robert Figueroa and Sandra G. Harding (London; New York: Routledge, 2003), 26–48.

10. Reflexive appreciation of this point is relatively recent; for an insightful treatment of the history of display technique within the Science Museum, see Andrew Nahum, "Exhibiting Science: Changing Conceptions of Science Museum Display," in *Science for the Nation: Perspectives on the History of the Science Museum*, ed. Peter J. T. Morris (Houndmills, Basingstoke: Palgrave Macmillan, 2010), 176–93.

11. See https://www.davidrooney.uk/exhibitions/codebreaker (accessed October 14, 2021).

12. For one of the countless examples, see Andrew Baker and Kirstin Sibley, *E A Newman and Pilot ACE: Turing's Legacy* (London: Invisible Numbers, 2017), available online at https://issuu.com/invisiblenumbers/docs/pilot_ace_newspaper; see also http://www.invisiblenumbers.co.uk/publications.html (both accessed October 14, 2021).

13. For a more fully developed version of this argument, see Thomas Söderqvist, "The Seven Sisters: Subgenres of 'Bioi' of Contemporary Life Scientists," *Journal of the History of Biology* 44 (2011): 633–50.

14. Andrew Hodges, *Alan Turing: The Enigma* (London: Burnett Books, 1983).

15. Donna Haraway, "Ecce Homo, Ain't (Ar'n't) I a Woman, and Inappropriate/d Others: The Human in a Post-Humanist Landscape," in *Feminists Theorize the Political*, eds. Joan Scott and Judith Butler (New York: Routledge, 1992): 87–101.

23

Invisibilities of Care

Reproductive Labour and Indigenous Hospitalities in Post/Colonial Fieldwork

Alexandra Widmer

"Why should we bring children into the world only to work for the white man?"[1] people in Vanuatu[2] asked Cambridge anthropologist W. H. R. Rivers in the early twentieth century.[3] Rivers sought to explain the rapid population decline in this southwestern Pacific Island country in terms of the psychological effects of colonialism as well as other factors, especially infectious diseases brought by Europeans. This contrasts dramatically with the demographic analysis a century later, which emphasized a looming strain on resources attributed to a rapidly growing youth population with high rates of teenage pregnancy.

In my own research, I am interested in situating and historicizing narratives that quantify, explain, and seek to manage populations in the southwestern Pacific Islands between these two moments. These narratives often assign the stakes and responsibility for birth rates to women's agency and responsibility. This naturalizes their role in reproduction and skews the place of broader social issues that shape reproduction. I have also conducted research on the changes in women's lives between these two historical moments, especially with respect to knowledge, social networks, and care practices of birth, pregnancy, and raising small children.

It has been my privilege to do this work, analysing missionary, British, and French documents about the colonial administration of marriage practices and the medical training of Indigenous men and women as physicians and nurses in colonial and mission institutions. I interviewed retired nurse–midwives who had trained in the country's largest mission hospital and women who had given birth at the same mission hospital in the 1950s and 1960. In Pango village, just outside the capital of Port Vila, Vanuatu, I learned about massage healers' care during pregnancy and young women's experiences of birth at the government hospital in 2010 and their experiences of caring for small children through ethnographic research and interviews. When I reflect on the invisible work that made this research possible, I feel humbled by and grateful for the kindness of ni-Vanuatu (citizens of Vanuatu) who not only contributed their knowledge and skills but also their hospitality during my fieldwork. As well, there was the work of Vanuatu Cultural Centre and those who approved my research permit under the Vanuatu Cultural Research Policy. Furthermore, since this research and writing overlap with the growth of my two daughters, I am grateful to the many people (including their father) who enabled my research by contributing to their care.

The previous paragraph may sound like a normal research acknowledgement—detailing the care of the researcher and their family and the debts to research participants and to a host organization and community. But rather than sequestering the discussion of such reproductive labour and hospitality in a separate section, this chapter examines such activities directly. I focus on the necessities of care activities during fieldwork—especially fieldwork carried out by researchers with children—care activities that are only partially rendered visible (if at all) in formal publications. As historian Emily Callaci points out about "acknowledgements" sections, they often carry information about the "great deal of invisible, uncompensated work" on which scholarly institutions are built.[4] By unpacking some of these—at least those that pertain to my own fieldwork—my analytical goal is to highlight the aspects of the entangled experiential and conceptual terrain of "reproductive labour" and "hospitalities," in order to reflect on the multiple dimensions of labour, work, and care that are necessary and then rendered invisible (or partially visible) in the scientific knowledge published afterwards. My hope is that scholars and students analysing the histories of the sciences—especially those that involve relationships between those of settler ancestry and Indigenous people—recognize the multiple kinds of invisible care labour involved in making knowledge claims.

I focus on my own experiences of care my family and I received during a field research trip to Port Vila and Pango Village, when I was accompanied by my small children and their non-anthropologist father.[5] I use two experiences to frame my analysis: The first is the help I needed for a late-night trip to the hospital with my feverish fifteen-month-old; the second is the support we received from the nursery school that my four-year-old attended in the village. Like Callaci, as well as Kelly Dombrowski and others, I want to make more visible the care work (including reproductive labour and Indigenous hospitality) necessary for knowledge production in the human sciences.[6] To think about the work and labour in the southwestern Pacific, it is instructive to highlight West and Aini's description of work and labour in Papua New Guinea, where they see work as, among other things, connected to the effort of participating in social networks of obligation and exchange, while "labour" is the measure and analysis of work in a capitalist system.[7]

CARE IN THE FIELD

Before leaving for Vanuatu with my daughters, we visited a travel doctor in Auckland, where I had spent four months conducting archival research at the Western

Pacific Archives. This doctor told us that, owing to the presence of cerebral malaria in Vanuatu, any high fever in a child that lasts more than four hours needed to be evaluated by a health professional. In Port Vila (but not most of Vanuatu at that time) this would only, and under ideal conditions, be possible during the day. I was two months into my postdoctoral fieldwork in 2010 (which built on three prior trips) when my younger daughter developed a high fever. Knowing that fevers are common occurrences in children the world over, I was conflicted about how seriously to consider it. When the four-hour mark passed and her fever was still high, it was dark and raining and hailing a bus from Pango into town would not have been possible. My neighbour (also my landlady, and part of the family who had welcomed us on previous trips) talked with people in her networks to try to find someone with a vehicle who could take us. She eventually found an off-duty taxi driver with his own vehicle, who quickly agreed to drive us the fifteen minutes into town.

When we ultimately arrived at the hospital, the man who had driven us kindly offered to wait for me, and I took my daughter to the emergency ward. There were rows of ni-Vanuatu women sitting with disconsolate and very sick-looking children. I looked down at my fifteen-month-old, who was more flushed than usual but clearly healthy in comparison. I felt stupid and unequivocally privileged. Without making me wait, the triage nurse waved me to her station. Before looking at my baby or taking a history she said to me: "Thank you for coming," repeating the formalities of hospitality and welcome that I had experienced in a wide variety of social gatherings in Vanuatu. She looked at my daughter and gently felt her forehead with her hand. She took her temperature with a thermometer in her ear. After assessing the fever to be rather low-grade, the nurse asked, "*Fasbon blo yu?*" (Is this your first child?) looking at me rather kindly. I said, "no, *sekenbon.*" But I took her meaning. I was overreacting to a normal common fever as an inexperienced mother would. She patted my baby's head. "Thank you for taking such good care of your baby," she said, "she will come good, she doesn't have malaria. Thank you for coming." She again used the language of hospitality. Ashamed for taking her time, I thanked her profusely and left the hospital.

The taxi driver was waiting for me in the damp darkness of the potholed hospital parking lot. He was surprised how quickly we had emerged. Driving back to the village, he told me I should call him if I ever needed anything, and he would come and help. All of these kindnesses were understood as necessary by my hosts and by me because I was a guest—if I were not, I would not have needed their help. They were offered with the explicit recognition that I was a white mother unfamiliar with common childhood illnesses and without my own support network. Never before, nor since, have I been welcomed and thanked for coming into a medical clinic in Europe or North America.

Very early on after arriving in Pango on this particular trip, as part of the welcoming encounters we had with people in the village, friends came to the house we were renting and invited us to have our four-year-old attend the preschool in the village. This was a far cry from the years-long wait lists and gatekeeping I encountered to access pre-schools at home in Toronto or the litany of forms and interviews we had to complete in Berlin. My four-year-old began attending this school in the mornings, run by a formally trained ni-Vanuatu teacher in her house. There were some nominal fees. In addition to daily routines, the teacher and her assistants prepared celebrations for birthdays, Mother's Day, and International Children's Day. Words cannot express how much this nursery school contributed to our lives and my research. I met other mothers of small children who generously agreed to be interviewed. I learned later that the contact with this welcoming person who led the school would also support my presence and research in the village when others questioned whether it was appropriate for me to be there. The teacher was welcoming to our family and was a generous research interlocutor—she taught me a great deal about the history of the village, about changes to women's access to land, and the changes to women's lives through telling me her own life history. The nursery school swept us into the community in ways that were nurturing to us as people, as well as conducive to my research.

Beyond the hospitality of adults, the children we came to know were also welcoming to us, and to my daughters in particular. The values of hospitality were inculcated in the expectations ni-Vanuatu adults had for children. If I was looking for someone's house, for example, adults would give me general directions and then say, "When you are closer, ask the children around to take you." Hospitality was an expectation. Children were expected to include my four-year-old daughter in their games, to share their toys with this guest, and to speak English (their third or fourth language) with her. Children would accompany us on walks, they greeted us warmly (sometimes following prompting by their parents) when we came to their houses, and when we ran into them around the village. We tried hard to be welcoming and respond in kind too. I was, and remain, incredibly grateful for the hospitality, and I have tried, over several years of precarious employment, to remain in reciprocal relations as a respectful guest.

These two descriptions of encounters in the field involve many types of care. Beyond the reproductive labour carried out by me and my partner on behalf of our children and the project, there was the work of ni-Vanuatu. Though different in kind, this resembles the work that ni-Vanuatu carried out to support the Cambridge anthropologist W.H.R. Rivers and many others during the last century of research in biomedicine and anthropology in Vanuatu.[8] While the knowledge and expertise of research interlocutors is part of research results, their care work is generally not written about in peer-reviewed publications, except in reflexive essays like this one.

REPRODUCTIVE AND HOSPITALITY LABOUR

At its simplest, reproductive labour constitutes "the everyday tasks involved in staying alive and helping others stay alive."[9] Since the 1970s, Marxist feminists (largely in Euro-American contexts) have called attention to the amount of unpaid care work that is required for people to do their paid work. Their analyses have often focused on housework, the labours of raising children, and other care-giving responsibilities, and they have examined what these forms of work have contributed to labour power and the workings of capitalism. They have argued that despite being a form of labour, such care has been made invisible in important statistics like the gross domestic product (GDP) and were routinely taken for granted in social policy.[10] Critiques, often from the Global South, have since been levelled at this scholarship for the lack of attention to the fact that reproductive labour is itself stratified and can take paid and unpaid forms.[11] In response, more recent analyses of reproductive labour have shown how it was structured by inequalities based on race and colonialism—decisively the case in this chapter.

Reproductive labour is often both invisible and yet necessary for scientific work. There are long-standing feminist analyses of how the practice of taking children to the field potentially impacts their health and safety and affects rapport with interlocutors amongst other issues.[12] Moreover, research grants and university fellowships generally only fund the cost of unaccompanied fieldwork, so the fieldwork of someone who brings their family must be subsidized by private funds and unpaid labour.

Reproductive labour is an important part of my story but does not capture all of the care work and skill of hospitality needed for successful field research with children. In this piece, I aim to render visible the ni-Vanuatu hospitality work for me and my family that was essential for my research in a way that extends beyond reproductive labour. Being a good host involves care work of many kinds. This work marks people as guests and hosts in culturally inflected ways and is informed by particular histories.

RESEARCH ENCOUNTERS AND INDIGENOUS HOSPITALITY

Encounters between indigenous people and researchers is central to natural and social scientific knowledge production in the Pacific Islands (as it is in many places).[13] The invisibilities of such encounters and intersubjective relationships are the subject of long-standing critical analyses of the object of knowledge in field sciences, particularly anthropology, and are imbricated in the expansion of colonialism.[14] Indeed, in present-day social anthropology, scholars are highlighting and unpacking such encounters, putting emphasis on the co-production of knowledge by anthropologists and hosts.[15]

Some such encounters can be characterized as part of the relations of hospitality, as Adrian Young has insightfully shown.[16] Acts of hospitality need not be considered "labour," but when hospitality is a culturally important idiom for understanding foreigner–Indigenous relations, it helps to recognize the agency and, indeed, work of those who take on the care and socialization of researchers as guests. And hospitality does take significant labour in the context of such encounters: Paige West and John Aini have shown that the work of hosting visiting NGO workers, filmmakers, and journalists in the New Ireland region of Papua New Guinea amounts to a great deal of undercompensated effort, often invisible to guests.[17] While it could be argued that the nurse, the taxi driver, and the preschool teacher that I was fortunate to meet were "just doing their jobs," hospitality and welcome figured prominently in the way that they went about providing care.

Dictionaries typically define "hospitality" as: "the friendly and generous reception and entertainment of guests, visitors, or strangers." French philosopher Jacques Derrida writes that hospitality captures the aspects of the host's identity: having a home that is open at times to others is at the core of a host's character.[18] In this way, Derrida portrays "the foreigner" as a key figure in metropolitan society, reinforcing the host's identity by partaking in their practices of hospitality (or their refusal). But hospitality must be understood with regard to local idioms and practices, and in the southwestern Pacific, hospitality is a social framework that indicates moral social obligations on both sides in ways that are not universal. In the southwestern Pacific, welcoming strangers is a powerful assertion of sovereignty. Anthropologist Debra MacDougall writes that the universal belonging of strangers in Oceanic places is anathema to Oceanic forms of Indigeneity, but Indigenous sovereignty is necessary for being a good host, to grow food for guests, among other things.[19]

MacDougall's research in the rural Solomon Islands shows us that engagements between host and guest allow "the transformation from stranger to kinsperson"[20] while keeping host's sovereignty over land at the forefront. This is quite different than what Derrida argues for: a radical unconditional hospitality to pair with the conditional hospitality more familiar to modern nation states. MacDougall writes, contra Derrida: "Hospitality is not [simply] neutral tolerance of the stranger in one's midst on ground that already belongs to everyone."[21] Rather, MacDougall argues, in the rural Solomon Islands, the land belongs to the hosts, who require it to fulfil their obligations of hospitality. More specifically relating to Vanuatu, anthropologist Rachel Smith has demonstrated that ni-Vanuatu hospitality to foreigners and even international companies is "moralized and marked by obligations of mutuality or reciprocity."[22] Furthermore, she writes that while the idioms of hospitality might appear to mask exploitative relations, between the ni-Vanuatu seasonal workers interviewees and their New Zealand employers, they can also "be used to make moral claims, asserting mutual obligations and a right to dignity because of the general character of the 'law of hospitality,' and its resonance across political, social, and cultural divides."[23] Smith demonstrates this claim from her observation that ni-Vanuatu have also received their New Zealand employers as guests back in their home villages, showing that these are relationships that involve reciprocity, dignity, and sovereignty. In other words, through hospitality discourses and practices, which figure in relations of paid work, a host and guest's obligations to one another are articulated and reciprocated. While it could be argued that the nurse, the taxi driver, and the preschool teacher that I was fortunate to meet were "just doing their jobs," in the way that they went about their labour and work, hospitality and welcome figured prominently. The lesson for me, then, is that by undertaking hospitality labour in relation to me and my family, my hosts were asserting that I was part of social worlds in guest/host frameworks.

Providing hospitality to a visiting researcher involves socially specific skills, expertise, and values; and it is connected to particular histories, identities, and economies. It requires time and effort. It also allows for particular kinds of control and surveillance of the researcher. As Rosanna Dent (chapter 24) insightfully shows, A'uwẽ-Xavante people in Brazil invest a great deal of time and labour to host foreign researchers, in part to enrol those researchers in A'uwẽ-Xavante political aspirations. Dent's story reminds us that another, reciprocal, aspect of the invisible work of hospitality is the efforts that researchers must make to be a proper guest in relation with hosts. "Being a guest is as much work as being a host," Grandmother and Traditional Knowledge keeper Joanne Dallaire (Cree, Wolf Clan) told Anishnabe scholar Ruth Koleszar-Green. Koleszar-Green is writing about host–guest relations on Turtle Island, primarily on the lands taken by Canada (where this chapter was written), and she insists that protocols of hospitality that include the responsibilities of hosts and guests are specific to each Indigenous nation. Her general point is relevant beyond these places. Indeed it is a small step towards meeting the point that Hawaiian scholar Huanani-Kay Trask wrote in 1993, that anthropologists "exploit the hospitality and generosity of native people."[24] Paying attention to the invisibility of hospitality as care work in the production of knowledge is also to call attention to the work of relationship building that should come from the researcher as well.

Of course there is care work involved in negotiating social relationships in many fieldwork contexts. In this volume, Lara Keuck (chapter 21) demonstrates the impact that the care work of archival gatekeepers has on the circulation of sources and the production of historical knowledge. But particularly in postcolonial contexts, paying attention to the invisible care work of hospitality can concomitantly render colonial legacies and gendered racial politics visible through local practices of how outsiders are treated. I was made aware of my outsider status in the way I was welcomed as a mother as well as by the unearned privilege of being received as a guest in the hospital and the village nursery school.

VISIBLE CARE

The COVID-19 pandemic, during which schools and childcare arrangements collapsed from one day to the next, has made the reproductive labour needed for research highly visible. The invisible labour needed for research to take place—and the gendering of that labour—was made visible by, for example, journal article submission rates: submissions by women were significantly reduced, while men's increased.[25] Care work, and specifically the work that underpins scholarly endeavours, is arguably more visible than ever, and so it is perhaps time for researchers across the social sciences and humanities to consider how that work can be reflected on and analysed in order to make better, and more responsive, social scientific knowledge. Considering hospitality as part of the invisible care work in fieldwork would be part of this endeavor.

Acts of hospitality may not always be thought of as work, but when hospitality is a culturally important idiom for shaping foreign researcher–Indigenous relations, there is a need to recognize the work and culturally valuable skills of those who graciously receive those researchers as guests. Doing so also situates the researcher in a network of reciprocal relations, moral expectations, and recognitions of sovereignty, rather than figuring them as someone who arrives merely to extract material in the pursuit of knowledge. This is not to claim that Indigenous hospitality is the correct register for thinking about *all* encounters in fieldwork contexts. To be sure, there is ample record of the connection between field sciences and colonialism's extractive social forms that combine land alienation, pernicious racisms, and glaringly visible forms of labour exploitation. Hospitality may well have been initially extended to the researchers while in the field, but the historical record of colonialism is replete with guests who have blatantly missed their reciprocal obligations.

I began this chapter with the words that ni-Vanuatu used to narrate their reproductive choices, and I point out that they did so in relation to their reproductive labour and the labour they did not want their children to do. Rivers interpreted this as their "loss of interest in life" rather than as an invitation to him to contribute to making their work visible or reciprocal, though he was opposed to the conditions of "the labour trade" at the time of writing. To my early twenty-first-century sensibility, these ni-Vanuatu words remain a powerful refusal of the colonial extraction of care, labour, work, and perhaps hospitality rendered more poignant here because of its visibility within Rivers' project. Highlighting the centrality and invisibility of reproductive labour and hospitality in research contexts is a small piece of the unfinished work of decolonizing anthropology.

NOTES

1. W. H. R. Rivers, "The Psychological Factor," in *Essays on the Depopulation of Melanesia*, ed. W. H. R. Rivers (Cambridge: Cambridge University Press, 1922), 84–113, 104.
2. Vanuatu is a nation of over eighty islands in the southwestern Pacific. Having achieved independence from Britain and France in 1980, the 292,680 inhabitants speak over 110 languages.
3. Rivers made field trips to these islands in 1908 and 1914.
4. Emily Callaci, "On Acknowledgments," *The American Historical Review* 125 (2020): 126–131.
5. I take my cue in part from historian Lyn Schumaker, who analyses the significance of fieldwork in the history of science, and in the history of social anthropology in particular. Schumaker takes what she calls a "field approach" to examine the scientific and non-scientific practices of fieldwork, including the organization of work and equipment: Lyn Schumaker, "Constructing Racial Landscapes: Africans, Administrators and Anthropologists in Northern Rhodesia," in *Colonial Subjects: Essays on the Practical History of Anthropology*, eds. Peter Pels and Oscar Salemink (Ann Arbour: University of Michigan Press, 2000), 327.
6. Kelly Dombroski, "Awkward Engagements in Mothering: Embodying and Experimenting in Northwest China," in *An Anthropology of Mothering*, eds. Michelle Walks and Naomi McPherson (Toronto: Demeter Press, 2011), 49–66.
7. Paige West and John Aini, "'I Will Be Travelling to Kavieng!': Work, Labour and Inequality in New Ireland, Papua New Guinea," in *Unequal Lives: Gender, Race and Class in the Western Pacific*, eds. Nicholas Bainton, Debra McDougall, Kalissa Alexeyeff, and John Cox, 47–75, 64 (Canberra: ANU Press, 2021).
8. W. H. R. Rivers (1864–1922) was a field-defining social anthropologist for his work with the genealogical method. He was also a psychiatrist and is also well known for coining the term "shell shock" from his research with World War I veterans.

9. Helen Hester and Nick Srnicek, "The Crisis of Social Reproduction," https://www.bbvaopenmind.com/en/articles/the-crisis-of-social-reproduction-and-the-end-of-work/, accessed September 30, 2020.

10. For example, Silvia Federici, *Wages against Housework* (Bristol: Falling Wall, 1973); Maria Mies, *Patriarchy and Accumulation on a World Scale: Women in the International Division of Labour* (New York: Palgrave Macmillan, 1998); Marilyn Waring, *Counting for Nothing: What Men Value and What Women Are Worth*, Second Edition (Toronto: University of Toronto Press, 1999).

11. For example, Michelle Murphy, "Reproduction," in *Marxism and Feminism*, ed. Shahrzad Mojab (New York: Zed Books, 2015), 291–92.

12. For example, Julianna Flinn, Leslie Marshall, and Jocelyn Armstrong, eds. *Fieldwork and Families: Constructing New Models for Ethnographic Research* (Honolulu: University of Hawaii Press, 1998); Fabienne Braukmann, Michaela Haug, Katja Metzmacher, and Rosalie Stolz, eds., *Being a Parent in the Field: Implications and Challenges of Accompanied Fieldwork* (Bielefeld, Germany: transcript Verlag, 2020); Danielle Drozdzewski and Daniel Robinson, "Care-work on Fieldwork: Taking Your Own Children into the Field," *Children's Geographies* 13 (2015): 372–78.

13. For example, Bronwen Douglas and Chris Ballard, eds., *Foreign Bodies: Oceania and the Science of Race 1750–1940* (Canberra: Australian National University Press, 2008); Alexandra Widmer, "From Research Encounters to Metropolitan Debates: The Making and Meaning of the Melanesian 'Race' during Demographic Decline," *Paideuma: Mitteilungen zur Kulturkunde* 58 (2012): 69–93; Adrian Young, "Mutiny's Bounty: Pitcairn Islanders and the Making of a Natural Laboratory on the Edge of Britain's Pacific Empire" (PhD diss., Princeton University, 2016).

14. For example, Talal Asad, ed., *Anthropology and the Colonial Encounter* (London: Ithaca Press, 1973).

15. Lieba Faier and Lisa Rofel, "Ethnographies of Encounter," *Annual Review of Anthropology* 43 (2014): 363–77.

16. Adrian Young, "Putting Anthropology in Its (Hospitable) Place: Harry Shapiro's Fieldwork on Pitcairn Island, 1934–1935," *History and Anthropology*, DOI: 10.1080/02757206.2020.1762591.

17. West and Aini, "'I Will Be Travelling to Kavieng!,'" 47–75, 64.

18. Jacques Derrida, *Of Hospitality: Anne Dufourmantelle Invites Jacques Derrida to Respond* (Stanford: Stanford University Press, 2000).

19. Debra MacDougall, *Engaging with Strangers: Love and Violence in the Rural Solomon Islands* (New York: Berghahn Books, 2016), 239.

20. MacDougall, *Engaging with Strangers*, 22.

21. Ibid., 239.

22. Rachel Smith, "Be Our Guest/Worker: Reciprocal Dependency and Expressions of Hospitality in Ni-Vanuatu Overseas Labour Migration," *Journal of the Royal Anthropological Institute* 25 (2019): 356.

23. Smith, "Be Our Guest/Worker," 364.

24. Cited in Linda Tuhiwai Smith, *Decolonizing Methodologies: Research and Indigenous People*, second edition (London: Bloomsbury, 2012), 70.

25. Colleen Flaherty, "No Room of One's Own," *Inside Higher Ed*, April 21, 2020; Anna Fazackerley, "Women's Research Plummets during Lockdown—But Articles from Men Increase," *The Guardian*, May 12, 2020.

24

Invisible Infrastructures

A'uwẽ-Xavante Strategies to Enrol and Manage *Warazú* Researchers

Rosanna Dent

AN A'UWẼ-XAVANTE FIELD SEMINAR

According to Fabrício Rodrigues dos Santos, when the members of his Genographic field team arrived in the A'uwẽ-Xavante territory of Pimentel Barbosa in Brazil in 2010, it was almost as if they were entering a ready-made training system for scientists.[1] The researchers were working on a large-scale initiative to track pre-historic human migration through the genetic sampling of Indigenous groups, and they had extensive experience with other Indigenous communities. However, in Santos' account, their short stay in Mato Grosso was exceptional. It was more of a crash course in A'uwẽ-Xavante culture than a routine collection of cheek swabs.

Residents of the community of Etênhiritipá stipulated that even to study Y-chromosomes and mitochondrial DNA, the researchers had to understand the A'uwẽ-Xavante way of life. In Santos' account, leader Jurandir Siridiwê Xavante told them, "For you to understand us, you have to live with us."[2] The *warazú* (non-A'uwẽ) visitors were installed in an old schoolhouse and were assigned two village residents as guides and guards to help with daily tasks and to protect them from overly curious children. Rather than days filled with genealogical data and genetic samples, Santos' narrative centred on hunting and fishing trips, a movie night, and incredible star-filled skies. "And not only that," Santos told me, "We participated in rituals with them. Not the rituals they put on for tourists, ones that they were really doing." Pedro Paulo Vieira, another researcher in the field team, echoed this sense of engagement, saying, "The Genographic [Project] was adopted by the Xavante in Brazil. . . . Fabrício, myself, and some other members of our team were even assigned to clans within the village. . . . We became part of the Xavante community."[3] As they tell it, the researchers' time in Pimentel Barbosa was a consuming, immersive experience.

This account may seem a far cry from the Genographic Project historians of science expect. We are likely more familiar with the critiques of the project by scholars such as Kim TallBear and Joanna Radin, who examine the practical and conceptual continuities between the Genographic Project and earlier programs.[4] These critiques resonate with scholarship on the unethical creation, sale, and use of immortalized cell lines of Indigenous groups in South America.[5] Collectively, these works are part of a larger literature that shows that extractive research practices in the human and biomedical sciences have failed to offer participants concrete benefits. They have also created risks by disregarding or contradicting Native cosmologies, perpetuating damaging notions that Indigenous peoples are disappearing, and using Native groups to stand in for human history. However, the A'uwẽ-Xavante of Etênhiritipá are aware of histories of misuse of Indigenous biosamples in South America—they were not naive, and yet they welcomed the Genographic researchers.

How can we make sense of the geneticists' claims that they were accepted, embraced, and adopted, all over the course of a few short visits? Their account of hospitality could be interpreted as a claim to legitimacy—perhaps hoping to preempt my critiques, or to use my work to provide a counter narrative as I turned them "into subjects of historical research" (Radin, chapter 20). But the geneticists' seemingly exceptional account is consistent with a broader pattern. It reflects my experience in the neighbouring community of Pimentel Barbosa and overlaps with the accounts of many recent researchers. This A'uwẽ-Xavante generosity and extension of kinship make sense only when seen in deeper historical context.

Scholars in the history and anthropology of science have shown that human experience of research is rarely clear-cut. Subjects do not necessarily interpret their experiences through dominant discourses—not all consider themselves "vulnerable," for example.[6] In Warwick Anderson's words, research participants have "complex moral sensibilities" and "sensuous apprehension of relationships" that inform their actions.[7] As Susan Lindee has shown, human subjects have been active collaborators in genetics research, their ideas sometimes "incorporated whole into scientific texts."[8] Even when power dynamics are highly unequal, scientific encounters may not be wholly violent.

Over time, A'uwẽ-Xavante communities in Terra Indígena (TI, Indigenous Land) Pimentel Barbosa (figure 24.1) have systematized their approach to enrolling, managing, and training researchers. Although invisible to academic audiences—and at times even to researchers themselves—a great deal of labour is involved in hosting us.[9] A'uwẽ-Xavante leaders have created social infrastructures to mediate and distribute the work and compensation associated with research. I refer to this as an "infrastructure" because it is enduring and undergirds the creation and movement of the materials of science.[10] Without this resilient social system, scientists might not be able to create the DNA samples, ethnographic fieldnotes, photographs, surveys, or other materials of their scientific processes.

Recognizing the A'uwẽ-Xavante systems that make these forms of knowledge production possible, however, is not enough to tell us why they might invest. As A'uwẽ-Xavante perform the extensive labour of caretaking for researchers, they

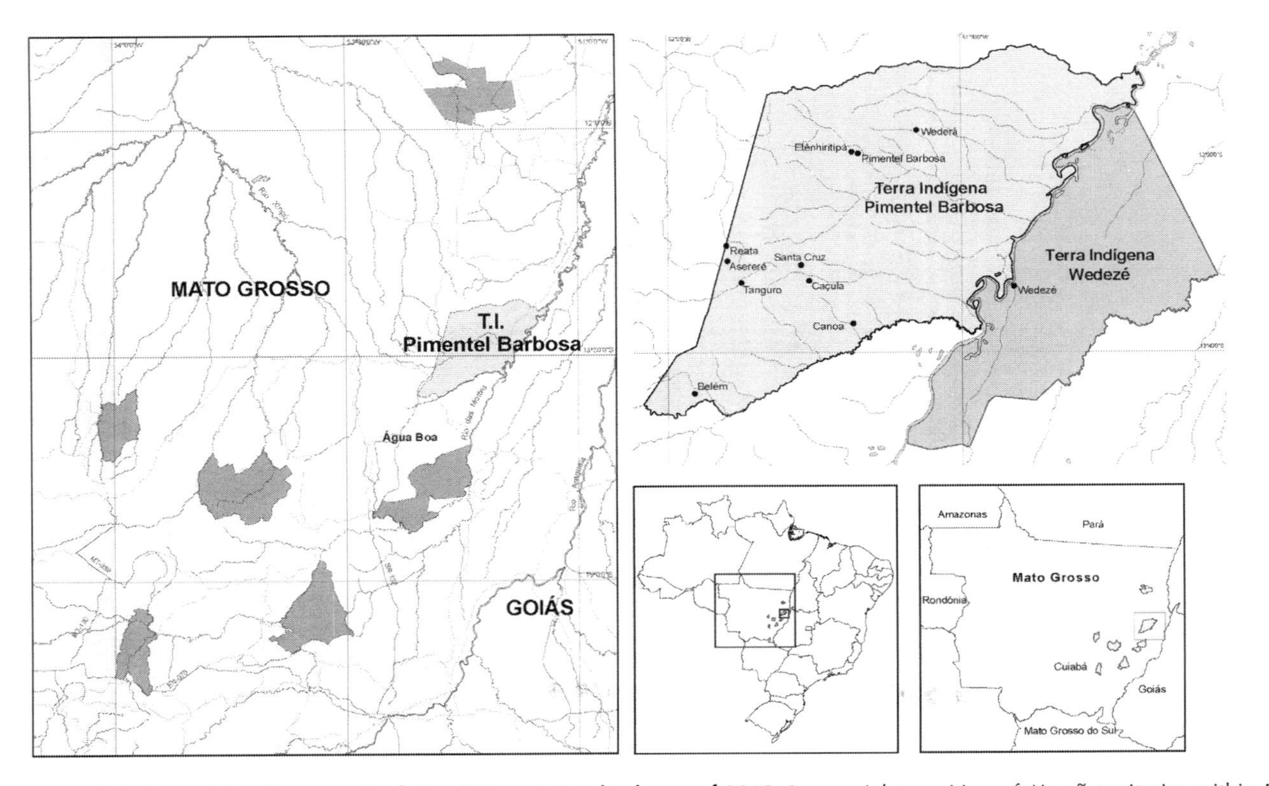

Figure 24.1. Left: Map of legally recognized A'uwẽ-Xavante territories as of 2010. Lower right: position of A'uwẽ territories within Brazil, and in Mato Grosso state. Upper right: close up of TI Pimentel Barbosa and the delimited territory, Wedezé. Adapted from Welch et al. 2013. Copyright James Welch, used with permission.

cultivate relationality that demands reciprocity. They work to make research a "two-way" process of collaboration and mutual benefit. This is one component of larger projects of political visibility in communities' ongoing struggles for land, health care, and education. Their investment makes sense when the double bind of bureaucracy is brought into view: to access the limited rights afforded to Indigenous peoples, A'uwẽ-Xavante need to be represented to the state by those with bureaucratically recognizable expertise. Their invisible labour is in the pursuit of another form of visibility, one that is highly elusive.

AN ENDURING SITE OF RESEARCH

My own enrolment in the neighbouring community of Pimentel Barbosa was a more eyes-wide-open kind of affair than that of the Genographic team. By 2014, when I first met with village leaders, I had spent the better part of a year studying the history of research in A'uwẽ-Xavante territories. I had heard that participating communities valued researchers' presence, camaraderie, and potential political help often enough to mitigate my scepticism. Ricardo Ventura Santos, an anthropologist, public health researcher, and mentor, suggested that my historical work might be of interest as Pimentel Barbosa was building a cultural documentation centre. With his introduction, I set out on the fourteen-hour bus ride from Brasília.

On a hot June day in the small town of Água Boa, Mato Grosso, I sat down with four leaders from Pimentel Barbosa. Above the persistent growl of the air conditioning, I described my research. Within an hour of discussion, the Elders surprised me by professing their interest in the history of science. "We want to know," they told me, "what was said? What was written in those books, in that research?"[11] Their enthusiasm—along with the support of Ventura Santos and his colleagues—led me to arrive in the community a year later with a hard drive of images, scanned from researchers' papers. I had been enrolled to help create a digital repository, the beginning of a project that has taken me back to TI Pimentel Barbosa almost every year since. In doing so, I joined the long line of scholars that the villages have hosted.

The first researchers arrived shortly after the A'uwẽ-Xavante established diplomatic relations with the Brazilian government in the mid-1940s. Anthropologist David Maybury-Lewis completed a major ethnographic study in 1958.[12] Building on this early work and Maybury-Lewis' 1962 collaboration with human geneticists Francisco M. Salzano and James V. Neel,[13] academic visitors followed from all walks of the human sciences.

Initially Maybury-Lewis struggled to carry out his research. He commented that residents "were little inclined for the tedium of instructing a foreigner in their tongue" and that when asked for help "they were not usually of much assistance, for they had no experience at that time either of translation or of paraphrase."[14]

Maybury-Lewis' six-month stay in 1958 established a precedent for inscription activities. Cultural anthropologist Nancy Flowers, then still a graduate student, perceived as much during her fourteen months of fieldwork in the mid-1970s. It was only her third afternoon in the field when a local woman took it upon herself to teach the new *warazú* a lesson in social organization. Even as Flowers carefully repeated back the names of the age sets, the woman scolded her. Flowers noted in her field journal, "She said I should write them down right away like David always did, but I had my cameras with me and not a notebook. Very bad—one should always carry a notebook."[15] Likewise, Flowers observed how "everyone" now took an interest "correcting my Xavante pronunciation and grammar."[16] The Elders already had ideas about what interested anthropologists and opinions about what anthropologists should do. But A'uwẽ-Xavante have not limited their engagement to tutoring individual researchers; they developed infrastructures that go beyond those that were so visible for Flowers.

AUTHORITY AND EXCHANGE IN RESEARCH RELATIONSHIPS

When Maybury-Lewis arrived, members of the village had been in contact with Brazilian society for over ten years. Before their "pacification" of the *warazú*, the A'uwẽ-Xavante had been at constant war with invading settlers.[17] But by 1958 they were accustomed to government agents and received regular visits from officials, journalists, and filmmakers for photo opportunities. Each visit was mediated through extensive gifts. Effectively, material goods became the basis for most interactions with *warazú*, whether researchers or not.[18]

The expectation of material gifts was challenging for researchers, who had to think carefully about how and when to offer compensation for hospitality, food, shelter, tutoring, and time.[19] Influxes of material goods were also complicated for communities. Distribution of the newfound wealth created competition and jealousy along political lines. Extensive pressure from land-invading settlers and outbreaks of disease exacerbated this political conflict, and in the late 1950s and early 1960s, the community fractioned several times.[20]

Anthropologists have long described A'uwẽ-Xavante society as a moiety system, where each individual belongs to one of two groups: *öwawe* or *poreza'õno*. Moiety belonging passes from father to child, and individuals of one moiety marry a member of the other. This institution is one of a number that structure both familial political power and ritual relations. Leaders able to ally themselves with researchers have benefitted from the political prestige accompanying the resulting availability of goods, in ways that have reflected and fuelled internal political competition.[21] With the arrival of outside researchers since the mid-twentieth century, A'uwẽ-Xavante had to balance these tensions in order to both maintain community cohesion and continue to host researchers.

The Genographic researchers' experience in the 2010s provides insight into how the A'uwẽ-Xavante have adapted existing social institutions. In the case of Santos' field research, two individuals participated in each activity. As he explained, "With the Xavante, it's always in twos. One *öwawe* and one *poreza'õno*."[22] That is to say, one member of each moiety is assigned to each task. For our cultural documentation work, the two individuals chosen by community leadership to be trained in photography and film came from each moiety. Including researchers in the moiety system also has helped share labour. Half of Santos' team became *öwawe*, while the other half was identified as *poreza'õno*. Many researchers who spend more than a few days in the village are adopted into a moiety.

Cultural anthropologist James Welch described his initial days of fieldwork in 2004 as a whirlwind of activity: "At least at the beginning, it was just busy, busy, busy. You think, "It's going to be peaceful and calm . . . [but] one person would say 'I'll take you to my garden,' and then the next, 'I'll take you here, I'll take you there.'"[23] In addition to training the anthropologist in A'uwẽ-Xavante ways, these efforts served to establish Welch's place in the kinship system.

Once he was located in a family, age-set, and moiety, this social position informed who would help with his research. His adoptive family answered myriad questions and helped with language. Members of his age-set explained the ritual practices that they performed. But even once situated in a specific social role, Welch was still expected—and explicitly reminded—to share his attention and gifts broadly.[24] This distribution of the work has also resulted in a distribution of the benefits of the researchers' presence.

POLITICAL PRESSURES, POLITICAL DEMANDS

In TI Pimentel Barbosa, A'uwẽ-Xavante think creatively about the future when they find common ground with researchers. Over time, the presence of *warazú* scholars has become a source of political potential. Demands for political engagement date back at least to Nancy Flowers' research in the mid-1970s and have become increasingly nuanced and strategic over time.

Flowers first arrived at a precarious moment. Under unofficial siege during the military dictatorship, page after page of Flowers' field notes document the encroachment of ranchers, and waves of epidemics. Flowers described how Elders asked for help: "The Indians thought that I had a lot more power than I had, of course. Because Warodi [one of the leaders] would say, 'well when you get back to the United States you tell your president . . .'"[25] Flowers left frustrated by what she saw as her inability to intervene in the systems that were stripping A'uwẽ-Xavante of their land, health, and livelihood. But although she initially despaired, her return to the village twelve years later would help fulfil demands she first thought impossible.

In the early 1990s, Flowers returned in collaboration with Ricardo Ventura Santos and Carlos E. A. Coimbra Jr., two young public health researchers from Fundação Oswaldo Cruz (Fiocruz). Building on genetic research from 1962 and Flowers' doctoral research, their study on health and nutrition in Pimentel Barbosa became one of the first comprehensive diachronic Indigenous health studies in Brazil.[26] The initial project also inadvertently grew into a long-term research program that, over the course of twenty-five years, has addressed some A'uwẽ-Xavante political concerns.

Perhaps most significantly, after James Welch joined the Fiocruz team, they conducted a land study for Wedezé, a large area adjacent to Pimentel Barbosa. The months of work included extensive field documentation, mapping, and interviews with Elders and leaders. In addition to extensive community input, they drew essential evidence from Flowers' field notes and dozens of other scholarly sources. Their 2011 report led to the delimitation of 150,000 hectares of A'uwẽ-Xavante land at a time when few new Indigenous territories were being recognized (figure 24.1).[27]

Over the past sixty years, various government institutions in Brazil have been responsible for land demarcation, health care, and education in Indigenous territories, and yet, they have consistently left needs and rights unattended. A'uwẽ-Xavante have responded to this abandonment with a wide variety of political tactics, including engaging researchers. They are not alone in this; Brazilian anthropologists are routinely called upon for demarcation studies, and those with deep relationships prepare stronger reports.

Thinking back to the beginning of his fieldwork, Welch described his hosts' vision as well beyond what he thought he could offer: "I found out later on, that one particular person, when he supported my coming, was hoping that one day I could help with the land fight for Wedezé. . . . [They] had learned that when you invest in a relationship with someone, there is a possibility that it will turn out to be a long-term relationship."[28] A'uwẽ-Xavante political demands, which began in the 1970s with somewhat naive requests for Flowers to talk to her president developed into strategic engagement.

TWO-WAY RESEARCH: AN INCONCLUSION

While the Fiocruz researchers' participation in land demarcation is an inspiration for politically useful academic work, it is not a redemptive story. Rather, it highlights the gross inequities that Indigenous groups face as they navigate a colonial legal system. Within this context, it is not surprising that even a significant political success such as the Wedezé delimitation study is viewed differently across communities and amongst individuals.

For one of my hosts in Pimentel Barbosa community, the study is a testament to trust-building over decades of collaborative work. Leader Tsuptó first suggested that Coimbra, Ventura Santos, and Welch undertake the study. He interpreted their months of unremunerated work as a testament to their character: "Beyond research, they were doing it because they are honest. Because they had gained this trust, friendship. A [different] anthropologist wouldn't do this for free" he told me.[29] When I asked more generally about where researchers have gone wrong, Tsuptó and other residents of his community avoided my questions, offering a few superficial or humorous examples. In my view, these evasions indicated not that everything had been ideal, but rather that community members found value in presenting a positive vision of researchers to and through me.

In neighbouring Etênhiritipá, Jurandir's focus was on the limitations of the Wedezé study. "It's taken years, and Wedezé isn't out [of the courts],"[30] he said, referring to the ongoing legal battle to finalize the demarcation. Instead, his discussion focused on ongoing fights: efforts for Sõrepré, another unrecognized territory; struggles for educational resources; lack of basic health infrastructure.

Jurandir also offered a much more pointed critique. He described much scholarship as benefitting the scholar without serving the community. Reflecting on the work of Maybury-Lewis and Flowers, he said, "It's the fruit of one-way [mão única] work. It does not have two-way results. . . . The guys come here and then they disappear." Ten years after the Genographic Project, Jurandir expressed frustration that Etênhiritipá's hospitality had not instilled reciprocity: "The guys from [the Federal University of] Minas Gerais said National Geographic did the project to understand where man comes from. . . . We talked: 'Oh, cool, we can contribute. But how?' 'Saliva, from your cheeks.' 'Oh, cool. Okay. And where's the *contrapartida* [counterpart]?' That's what we need to discuss."[31] *Contrapartida* is a Portuguese word I could translate as "compensation." But it more precisely connotes a "balancing entry" in accounting. It's the part that corresponds, in reciprocity. For Jurandir, the Genographic team has yet to fulfil their responsibilities in a two-way relationship.

Though his discourse is starkly different from Tsuptó's, I also interpret Jurandir's words as relationship-building. I was the newest addition to his collection of researchers in 2019. His criticisms were accompanied by an invitation to be different. Focusing on the need to develop research collaboratively, he said, "So projects always come top down. . . . What is missing? Dialogue. What is missing? Respect." And yet, emphasizing my disciplinary training, he drew me in: "Historians: they talk to us." he said. This affirmation was also a warning and an exhortation. He was labouring in good faith, with the hope that I would as well.

These leaders' adopted tactics reflected diverging perspectives, but also our different relationships. Tsuptó described his connections with the Fiocruz team after I had been collaborating in Pimentel Barbosa for five years. I had been adopted by his uncle. I had discussed our digital project in numerous community meetings. Jurandir and I had met only once or twice. It was my first formal visit to Etênhiritipá, and we were beginning conversations about the digital archive. Over the following days I would present the project in community meetings for the first time.

In my understanding, what unites their differing discourses is a shared emphasis on rationality.[32] Tsuptó and Jurandir leveraged connection differently but both in service of shifting power dynamics of scholarship to serve community interests. This labour is fundamental. It is part of an enduring social infrastructure of knowledge production. It also reshapes researchers and our practices, even if haltingly, unevenly, or incompletely.

When A'uwẽ-Xavante community members draw a new researcher into their circle, they are imagining futures for engagement that go beyond our narrow research plans. In a certain sense, A'uwẽ-Xavante in TI Pimentel Barbosa engage in our research for the advocacy we might one day offer. Faced with the predicament of relying on state recognition, their scientific interlocutors contribute to making them visible and "protectable." Ultimately, it is a precarious technique that depends on the consistency, dedication, and resources of researchers who may or may not meet A'uwẽ-Xavante expectations. With their time, care, and the labour of research participation, they compel us to try.

NOTES

1. This work was supported by a visiting fellowship at the Max Plank Institute for the History of Science, the Social Science Research Council with funds provided by the Mellon Foundation, and by the Conselho Nacional de Desenolvimento Científico e Tecnológico (CNPq). Thanks to many for thoughtful feedback and especially to Tess Lanzarotta, Orkan Telhan, and Jenny Bangham. In 2010, Terra Indígena (Indigenous Land, TI) Pimentel Barbosa was home to approximately 1,200 A'uwẽ-Xavante distributed in nine villages (James R. Welch, Ricardo Ventura Santos, Nancy M. Flowers, and Carlos E. A. Coimbra, Jr., *Na primeira margem do rio: Território e ecologia do povo Xavante de Wedezé* [Rio de Janeiro, Brazil: Museu do Índio/FUNAI, 2013]; figure 24.1).

2. Fabrício Rodrigues dos Santos, interview with Rosanna Dent, March 6, 2014, Belo Horizonte, MG.

3. Pedro Paulo Vieira, interview with Rosanna Dent, May 7, 2014, Rio de Janeiro.

4. Joanna Radin, *Life on Ice: A History of New Uses for Cold Blood* (Chicago, IL: University of Chicago Press, 2017); Kim TallBear, *Native American DNA: Tribal Belonging and the False Promise of Genetic Science* (Minneapolis: University of Minnesota Press, 2013), especially 143–76. See also: Jenny Reardon, *Race to the Finish: Identity and Governance in an Age of Genomics* (Princeton, NJ: Princeton University Press, 2004).

5. Carlos E. A. Coimbra, Jr. and Ricardo Ventura Santos. "Ética e pesquisa biomédica em sociedades indígenas no Brasil," *Cadernos de Saúde Pública* 12, no. 3 (1996): 417–22.

6. Nancy D. Campbell and Laura Stark, "Making Up 'Vulnerable' People: Human Subjects and the Subjective Experience of Medical Experiment," *Social History of Medicine* 28, no. 4 (2015): 825–48.

7. Warwick Anderson, *The Collectors of Lost Souls: Turning Kuru Scientists into Whitemen* (Baltimore, MD: Johns Hopkins University Press, 2008), 7.

8. Susan Lindee, *Moments of Truth in Genetic Medicine* (Baltimore, MD: Johns Hopkins University Press, 2005), 3.

9. For another take on the labour (or not) of hosting, see Sandra Widmer, this volume.

10. Brian Larkin, "The Politics and Poetics of Infrastructure," *Annual Review of Anthropology* 42 (2013): 327–43.

11. Tsuptó Buprewên Wa'iri Xavante, Sidówi Wai'azase Xavante, Luiz Hipru Xavante, and Agostinho Seseru Xavante, interview with Rosanna Dent, June 4, 2014, Água Boa MT.

12. David Maybury-Lewis, *Akwẽ-Shavante Society*, Second Edition (Oxford: Oxford University Press, 1974 [1967]).

13. James V. Neel, Francisco M. Salzano, Friedrich Keiter, David Maybury-Lewis, and Pedro Clóvis Junqueira, "Studies on the Xavante Indians of the Brazilian Mato Grosso," *American Journal of Human Genetics* 16, no. 1 (1964): 52–140.

14. David Maybury-Lewis, *The Savage and the Innocent*, Second Edition (Boston, MA: Beacon Press, 1988 [1965]), xxi.

15. Nancy M. Flowers, "Field Diaries, 1976–1977," Papers of Nancy M. Flowers, Museu do Índio Archive, Rio de Janeiro.

16. Flowers, "Field Diaries."

17. Seth Garfield, *Indigenous Struggle at the Heart of Brazil: State Policy, Frontier Expansion, and the Xavante Indians, 1937–1988* (Durham, NC: Duke University Press, 2001); Laura R. Graham, *Performing Dreams: Discourses of Immortality among the Xavante of Central Brazil* (Austin: University of Texas Press, 1995).

18. Garfield, *Indigenous Struggle*, 77.

19. Maybury-Lewis, *Savage and the Innocent*; James V. Neel, *Physician to the Gene Pool: Genetic Lessons and Other Stories* (New York: J. Wiley, 1994).

20. Maybury-Lewis, *Akwẽ-Shavante Society*.

21. Garfield, *Indigenous Struggle*, 76–78.

22. Santos, interview.

23. James R. Welch, interview with Rosanna Dent, March 27, 2014, Rio de Janeiro.

24. Welch, interview.

25. Nancy Flowers, interview with Rosanna Dent, August 22, 2013, New York City.

26. Carlos E. A. Coimbra, Jr., Nancy M. Flowers, Francisco M. Salzano, and Ricardo V. Santos, *The Xavánte in Transition: Health, Ecology, and Bioanthropology in Central Brazil* (Ann Arbor: University of Michigan Press, 2002).

27. Welch et al., *Na primeira margem do rio*.

28. Welch, interview.

29. Tsuptó Buprewên Wai'ri Xavante, interview with Rosanna Dent, Pimentel Barbosa, July 10, 2019.

30. Jurandir Siridiwê Xavante, interview with Rosanna Dent, Etênhiritipá, July 4, 2019.

31. Jurandir Siridiwê Xavante, interview.

32. Jurandir and Tsuptó's approaches resonate with Indigenous Studies scholarship from North America, which has shaped my thinking. This work recasts research as something that must be relationally based. See, Kim TallBear, "Standing with and Speaking as Faith: A Feminist-Indigenous Approach to Inquiry," *Journal of Research Practice* 10, no. 2 (2014): Article N17; Nicholas J. Reo, "Inawendiwin and Relational Accountability in Anishnaabeg Studies: The Crux of the Biscuit," *Journal of Ethnobiology* 39, no. 1 (2019): 65–75; and Margaret Kovach, *Indigenous Methodologies: Characteristics, Conversations and Contexts* (Toronto: University of Toronto Press, 2009).

25

Cultivating a Northern Australian Public for Yolŋu Cosmologies

"Keeping Visible" Yolŋu Research Practices and Their Effects

Michaela Spencer, with contributions from *Gawura Wanambi, Joy Bulkanhawuy, Yasunori Hayashi, Rosemary Gundjarranbuy, and Stephen Dhamarrandji*

> Balanda [nonIndigenous people] are panicking because they don't realise we have own ways of doing things properly. They are trying to put the fear into the Yolŋu, but they themselves are panicking and going to the supermarket to buy so many toilet papers. Balanda are panicking and calling Aboriginal people "high risk" [from COVID] and very vulnerable, but we have our own internal power to tackle this pandemic. By using that old law which was always there before colonisation, that is a strong law for all Indigenous people. Following this law shows our original leadership.
>
> —Gawura Wanambi

> The news about COVID comes from the government to the shire council and from the shire council to the clinic. Then the clinic gives everyone a bar of soap—with no story. This is what makes the senior people afraid. No-one has come to them to tell them the story first. They are the ones that should share it with their clan groups and coordinate everyone.
>
> —Rosemary Gundjarranbuy

> Let's find the way where we can find the bush medicine, food, trees, flowers, waters and the food. We as a Yolŋu people, are already strong. We know where to find everything and can show this to the non-Indigenous people, so we can work

together. But this work will come out from cultural awareness and from Yolŋu people's knowledge of our djalkarri-ŋur (foundations).

—Joy Bulkanhawuy

In the weeks following the arrival of COVID-19 in northern Australia, there was a scramble of activity surrounding the delivery of remote health services in Aboriginal communities. Messages about coronavirus were produced in many different Aboriginal languages, and there sprang to life a range of management initiatives oriented around keeping a generation of Elders safe. Those of us working in the "Ground Up" research team of predominantly Yolŋu Aboriginal Australian researchers were swept up in the action and in speaking from the team's implicated experiences within these events, sought to offer a scholarly intervention and response.

Now, sometime later, having moved past this initial phase of COVID activity, I am grappling with the difficulty of keeping this response "visible." I am thinking through how some COVID management responses easily gathered organizational and institutional lives, generated the "publics" necessary to support, and maintained their enactment. While others, by contrast, have seemingly struggled against the grain to make visible alternative commitments within health care practice.

Central to the intervention made by the Yolŋu research team was an insistence on the presence of a "Yolŋu-body" within the COVID crisis and pandemic response. The team were clear that for Yolŋu, healing was and always had been present in the land, and in health care practices that help keep place–people bonds strong and Yolŋu people–places as one. Within Yolŋu cosmologies, Yolŋu bodies and Yolŋu places are not separable. People and places as such do not exist outside of the eternal relationalities of Waŋarr "the Dreaming." But this metaphysical commitment became largely invisible in the context of mainstream COVID management responses, with their predominant focus on discrete biomedical bodies, disease-causing microbes and personal responsibility focused solely on the disease and the microbe. Mainstream responses to COVID brought to life more strongly than ever the social and spatial configurations of epidemiology focused on managing individual bodies as parts of healthy (or unhealthy) populations working to arrange secular bodies in space rather than doing bodies-as-place.[1] The Ground Up researchers, however, sought ways in which Yolŋu cosmological commitments may be made visible and public at this very specific historical moment.[2]

Telling stories of the experiences of our small research team during these COVID times, I am offering a second round of analysis; an "inquiry into our original inquiry," if you like.[3] I am revisiting and retelling our experiences of conducting Yolŋu-led research in the midst of the pandemic, so as to also tell another story about "research impact": its presence within institutional arrangements of university, organizational and public life, and the challenges of producing different possibilities from within these often aligned and mutually reinforcing orderings.[4] In what follows, I offer three overlapping layers of observation and analysis: direct quotations (in pull-out quotes), a description of our original findings and analysis (in italics) and my subsequent and ongoing analyses and reflections (in roman font).

STORY 1: COLLABORATIVE RESEARCH REVEALING ONTOLOGICAL DIFFERENCE

COVID times necessitated new ways of working for the Ground Up team at Charles Darwin University. For this piece of work, researchers, Gawura Wa̲nambi and Joy Bulkanhawuy sat on a Zoom call in their offices and connected to other members of the team—Rosemary Gundjarranbuy and Stephen Dhamarrandji—sitting over 500 km away on Yolŋu country in Galiwin'ku in East Arnhem Land. This group of four made up the core team of Yolŋu researchers—two female and two male, two in the urban centre of Darwin, two in an Aboriginal community where so much activity around COVID was taking place. In the team, there were also two non-Yolŋu researchers—Yasunori Hayashi and myself, Michaela Spencer. Yasunori assisted with translation from Yolŋu languages into English, while I helped with the epistemic translation as the stories told by the Yolŋu researchers would also be interpreted and presented in the webinar format to be heard by other audiences.

The impetus for our work together was research around Yolŋu perspectives on disaster management. This had gained particular prescience in COVID times, and we were expanding the research to reflect on COVID management in Aboriginal communities. We were to present the research via a webinar titled "Don't Panic," which would be broadcast to a diverse audience of government staff, healthcare providers and academics. The work proceeded through a Yolŋu practice of knowledge making, as the researchers speaking only in Yolŋu matha (language) initially worked to negotiate their roles and relations to each other, before moving into telling stories of COVID emerging as a new presence and active participant in the social life, governance and healthcare practice in Aboriginal communities. Gradually, in sharing stories with each other, the research took shape and a series of Zoom video and audio recordings were produced.

Once the "raw materials" of the research had been developed through this collective work, the next phase involved attentive translation and editing by Yasunori and myself, as we pored over the audio files, and spoke back and forth with the Yolŋu researchers, carefully curating a presentation which "brought forward" key points from the research and made them available to another audience. It was in this collaborative work that we settled on a few key concepts grounding the research and webinar presentation. Central to the story were the set of quotes that appear at the beginning of this chapter. The first of these was articulated by Gawura as he pushed back against an assumption of deficit and vulnerability, which he saw as accompanying COVID management practices arriving in remote Aboriginal settlements.

This became the first quote to appear in the webinar presentation, with Gawura speaking the words in Yolŋu matha, while an English translation appeared on the screen. His comments proposed the presence of an extant and orderly process of care, which was present within Yolŋu collective life, and which was the means through which Yolŋu together confront new events and crises.

> Balanda are panicking and calling Aboriginal people "high risk" and very vulnerable, but we have our own internal power to tackle this pandemic. . . . Following this law shows our original leadership.

Another key concept was elucidated through a story told by Rosemary Gundjarranbuy. This pointed to COVID management in her community as a form of governance practice that configured each person as individual and the same, rather than as clan groups with their own hierarchies of authority in which each person had a role and were part of a specific collective.

> The clinic gives everyone a bar of soap—with no story. This is what makes the senior people afraid. . . . They are the ones that should share it with their clan groups and coordinate everyone.

Her story narrated the fears that arose as each community member was given a bar of soap, to help with washing their hands. This bar of soap accompanied the government visit to the community and was provided by the local health care organization. The rationale for the soap, when read in relation to the epidemiological concerns for hygiene and cleanliness of the individual so as to protect the population, made sense. However, in deferring to this politico-epistemic arrangement, and by-passing Yolŋu authority, this action generated confusion and concern amongst Yolŋu Elders.

The italic text describes some of our "research findings," and the way we presented these in the "Don't Panic" webinar. This meeting was well attended by government staff, health care providers, academics, and others, and the Yolŋu researchers were pleased with the collaborative effort and their adherence to correct modes of knowledge-making, which sought to "make visible" or bring to presence, the dhudi dhawu (the underneath story), of COVID management in Aboriginal communities. Emphasizing the specificity of Yolŋu practices for caring for people–place relations, they hoped this form of resilience already present on the ground and within Yolŋu life may be recognized by government and other organizations.

However, in the context of the online webinar, it was very difficult to surmise what the impact of this research may have been. Had the story been heard? Through their own Yolŋu research practices, the team had worked to properly constitute a distributed collective of scholars, each able to contribute and negotiate the process of knowledge-making so as to reach a point where the story was clear and flowing.[5] But the audience of the webinar included frontline government staff and service providers who were caught up in their own norms of organizational practice associated with COVID management. At the same time, the "Don't Panic" webinar initiative was part of an effort by the university to offer public-facing research engaging audiences beyond academia and generating research impact.[6] Might this opportunity help to query default arrangements of authority, and make way for multiplicities in the doing of health care and COVID management?

STORY 2: SHARING OUR RESEARCH ON THE GROUND AND IN THE MODERN ACADEMY

Beyond the initial research process, the team was keen to keep going with the work, and to grow the project beyond these humble beginnings. However, these efforts encountered two notable interruptions. The first was when we were contacted by a staff member from one of the health care providers managing the COVID response in a number of Aboriginal communities. The second was when we attempted to engage with a public-facing academic publisher.

On the day following the webinar, I received a call from an employee of an Indigenous-run healthcare organisation. He wanted to talk about the bar of soap. He seemed a little agitated, and also a little uncertain, and it took a little while for me to understand the issue he wanted to discuss. He was appreciative of the webinar, but also concerned. "Yes, we did hand out bars of soap" he said. "But this wasn't something that we had to do; it was a goodwill gesture." Perturbed that this could be expressed as a cause for concern, he went on to talk about how they had approached an Elder within their organisation and had been given the go-ahead to engage people in the community around COVID.

The point that we had been making wasn't that the soap was inappropriate, but that in prescribing one way of managing COVID, it also articulated particular forms of individual and collective responsibility in the practice of healthcare. This was the effect we wanted to draw attention to, however, in the course of the conversation this point kept getting lost, and I hung up the phone feeling rather concerned. The man on the phone had insisted that in offering the soap all the right processes had been followed in the particular community we referred to (but hadn't identified) in the webinar. He was sure the soap could not be a cause for concern, but rather was something that helped to bring people together connecting them with the clinic, and supporting residents and healthcare organisations to do the work of managing COVID together. From his perspective, our phone call offered a means to correct this misunderstanding and omission, a way to be able to recover the practices, which had worked well in preventing the spread of disease amongst Yolŋu in Arnhem land.

So perhaps the difficult online format of the webinar had meant that this health care worker had not been able to clearly have his say on the day of the presentation. Or perhaps this phone conversation was a polite gesture, a strategy to raise concerns outside the webinar to avoid ruffling the discussion on the day. The emotion in the exchange suggested that something quite significant and visceral was being engaged in the telling of stories (and counter stories) about appropriate community health care practice. At the same time, the relations enacting Yolŋu bodies, kinship, and authority seemed to partly recede as these other arrangements prioritizing biomedical bodies and epidemiological authority were re-asserted.

Following the webinar, there was strong interest on the part of the Yolŋu team to publish their work so it could easily be shown and shared with government staff across local, state and federal tiers of government. An obvious and speedy way to help broadcast their academic work was through a public media outlet.

We drafted a short news-style article and approached an online public academic publisher, to discuss whether they would be interested in supporting the article. We felt that in forums to support these kinds of output, this story should have a good chance of publication. COVID was a topical issue, as was its potential danger within Australian Aboriginal settlements. We hoped that research conducted by scholars able to speak directly of these experiences, and to craft research in a style appropriate to engaging these contexts, would be recognised as significant and valuable.

> *One of the editors saw that this story* was *different, and made time to speak with us on the phone. She was supportive and enthusiastic about the work. But while she helpfully suggested other outlets, she also made clear that it would not be possible for her own team to publish such a piece. She considered that the article was closer to an "opinion" piece than to a defined research project with discrete results to report. As it was, this kind of account did not have a place in the fast paced academic news cycle that promises to deliver "results."*

Here the refusal of the publisher seemed, at least at first glance, to be of a different kind to that expressed by the health care provider. The editor's objection was to the mode and character of the research. The configuration of senior Yolŋu authorities negotiating stories and knowledge claims, working collaboratively with STS scholars interpreting and translating those stories in a manner that might make an impact, was not seen by the publisher as robust scholarly work. Although we were highlighting a story previously invisible within the modern academy and virus management practices, the publisher did not recognize it as a "research finding."

REFLECTING ON THESE STORIES

The question for me as an epistemic translator involved in the Ground Up team is: In what ways did the presentation of the Yolŋu story fail to gain traction in these other settings? How should we read the refusals as political practices of knowledge work?

In their chapter "Gino's lesson on humanity," sociologists Michel Callon and Vololona Rabeharisoa looked back over some of their research around muscular dystrophy on Reunion Island.[7] While hearing many stories from people living with the condition, they were brought up short by one interviewee, "Gino," who refused to "comprehend the lessons of genetics and to become part of the medical and associative networks that implement and diffuse that knowledge." Being met with silence from their potential interview subject, Callon and Rabeharisoa read this refusal as the expression of a particular moral position that had been excluded and made invisible by the sphere of action assumed by their inquiry. Their reading enabled an alternative moral configuration to come into view. They produced an analysis which refused to position the subject Gino as recalcitrant or ignorant, but instead presented Gino as making an active moral response in a situation where the only means to enact that response was through silence.

This is the understanding of refusal that I follow here. In presenting stories of the health care provider challenging our responses to the bar of soap and the publisher refusing the Yolŋu-led research, I consider these stories as presenting moments where the differing commitments articulated within our Ground Up work was denied traction, refused and sealed off within the mainstream discourses of Coronavirus management and research within which they tried to speak. In the events of these refusals, there was a configuration of health care and academia that rendered the work we had produced on the COVID response as "having no place/being out of place."

The health care provider reconfigured the bar of soap as a gift, a "goodwill gesture," an act that transcended the ordinary, and thus resisted acknowledging its role in enacting certain relations of governance and health care that singularized community members and collapsed other hierarchies of authority. The public academic publisher saw the work we had carried out as falling outside the bounds of what could be deemed research—and instead perceived it as "opinion" or perhaps as "mere politics." And what felt so troubling was that these re-concealments seemed to be difficult to resist.

Within a logic of epidemiological health care, the soap arrived seamlessly as part of a regime of management in which hand washing was a means to slow or remove the transmission of germs. The health care worker explained that consultation about this approach had occurred, and through this permission to go ahead had been secured. In re-telling this story after the webinar, it felt that an effort was being made to paper over the quite particular difference that the story had pointed to, effacing our collaborative research process. I was also struck by the similar individualizing propensity of neoliberal academic discourses. The coercive and largely invisible collaboration between biomedical and neoliberal academic orderings made it very difficult for any ontological difference proposed in the research of our team to become visible (and to generate effects) through the publishing process. By contrast, here in the present chapter, as a subsequent analytic endeavour, the character of these authorizing relations can be made more visible, more tangible, and perhaps more open to intervention.

CULTIVATING PUBLICS FOR A YOLŊU COSMOLOGY

In her work on in/visibilities in STS research, Lisa Garforth looks back over the trajectory of STS inquiry into scientific knowledge practices.[8] She queries the emphasis on "epistemologies of vision" that were present in early laboratory studies, as social scientists sought to intervene in abstract accounts of scientific knowledge-making and instead show that science was a "matter of ordinary practice, rather than a special, inaccessible form of cognition."[9] Garforth also noticed an alignment between the emphasis on vision in these STS analyses and the audit culture of contemporary universities, which assess value predominantly in terms of what can be shown, rather than that which cannot. However, she also suggests that to maintain this focus on the easily visible—either as a social scientist researching knowledge-making, or a university auditing its academic practices—is to lose much of the nuance and significance of invisible practices in knowledge work.[10]

In reflecting on a rather different instance of knowledge-making, the stories I have told in this chapter have also highlighted a particular consistency. That is between the conditions of visibility within biomedical practice and COVID management on the ground in northern Australian communities, on the one hand, and the culture of public academia through which impactful research becomes visible, on the other. As foreshadowed by Garforth, I see the academy as itself mirroring and enacting similar interpolations and silences as the practices of coronavirus health care.

Looking back over the stories and resistances presented in this chapter, they can be re-read as expressing the relational arrangements and limits of particular viewing publics. Understood in this way, it becomes easy to see how those involved in grounded health care practices in Aboriginal communities may struggle to see their work as enacting a quite specific epistemics (knowing bodies) and doing quite specific politics (forms of governance authority). Just as for the mainstream public academic media outlet, it may be difficult to see the Ground Up work as configuring specific forms of legitimate research practice, and performing a politics of inclusion and exclusion on this basis. However, such awareness becomes invaluable when considering the work of how to cultivate viewing publics able to "see" Yolŋu research, and cosmology.

Opening up, and looking at, the relations that constitute both the visibilities and invisibilities in Yolŋu research may offer a helpful guide to the careful conceptual and practical work of cultivating publics able to witness ontological difference and dynamically maintain its presence across settings. The activity and labour of attending to the relations by which existing publics are sustained, and the re-configuring of these relations in the performance of new knowing publics, may be transformative work.

In the initial stages of generating the webinar, we attended carefully to the epistemic and political practices by which we could make visible to the audience online differences in the knowing of healthy bodies, the doing of authoritative health care. We worked to generate a visual and scholarly display that showed an image of the research team members as both working together online and variously present in each of our particular places. We continued to present the work in Yolŋu matha, while also adding English translations. We made sure to have a full presentation and to also engage seriously in a long question and answer session at the end to discuss all queries and concerns. However, in cultivating a public that recognizes Yolŋu cosmologies, there are perhaps still further lessons to learn and politico-epistemic practices to be enacted. Such work is unlikely to be possible within the bounds of a single presentation alone, but as part of a continuing performative effort explicitly concerned with the making of in/visibilities in northern Australian publics and collective life.

NOTES

1. Miriam Solomon, *Making Medical Knowledge* (New York: Oxford University Press, 2015), 3.
2. Helen Verran, "Screens and Ontological Work: Four Projects of Yolŋu Aboriginal Australian Meaning-Making Mobilising ICT Machines," in *Gendered Configurations of Humans and Machines*. eds. Jan Büssers, Anja Faulhaber, Myriam Raboldt, and Rebecca Wiesner (Opladen; Berlin; Toronto: Barbara Budrich, 2021), 205–29.
3. John Dewey and Melvin L. Rogers. *The Public and Its Problems: An Essay in Political Inquiry* (University Park, PA: Penn State Press, 2012).

4. Ingunn Moser and John Law. "Good Passages, Bad Passages." *The Sociological Review* 47, no. S1 (1999): 196–219.

5. Timothy Buthimaŋ, "Garmak Gularriwuy" (2010), http://www.cdu.edu.au/centres/yaci/pdf/Buthimang_Gularri.pdf, accessed August 2, 2020.

6. Rebecca Lave, Philip Mirowski, and Samuel Randalls. "Introduction: STS and Neoliberal Science," *Social Studies of Science* (2010): 659–75.

7. Michel Callon and Vololona Rabeharisoa, "Gino's Lesson on Humanity: Genetics, Mutual Entanglements and the Sociologist's Role," *Economy and Society* 33, no. 1 (2004): 1.

8. Lisa Garforth. "In/visibilities of Research: Seeing and Knowing in STS." *Science, Technology, & Human Values* 37, no. 2 (2012): 264–85.

9. Garforth, "In/visibilities of Research," 266.

10. Garforth, "In/visibilities of Research," 282.

Index

Page references for figures are italicized.

About the Editors and Contributors

Elena Aronova is an associate professor at the Department of History, the University of California, Santa Barbara. She has published on the history of environmental data collection, history of the International Geophysical Year, seismology, and the historiography of science. She is the author of *Scientific History: Experiments in History and Politics from the Bolshevik Revolution to the End of the Cold War* (2021), which documents the history of the continuous efforts to integrate scientific knowledge and new technologies—from plant genetics to computers—into historical research. She has co-edited two collections of essays, *Science Studies during the Cold War and beyond: Paradigms Defected* (2016) and *Data Histories* (2017).

Jenny Bangham is a Wellcome University Award lecturer in the School of History, Queen Mary University of London, where she researchers the politics, meanings, and practices of genetics. She is author of *Blood Relations: Transfusion and the Making of Human Genetics* and co-editor (with Emma Kowal and Boris Jardine) of the open access volume, "How Collections End: Objects and Loss in Laboratories and Museums" (*BJHS Themes*, 2019).

Sarah Blacker is an SSHRC postdoctoral fellow in the Department of Anthropology at York University in Toronto. Her current research on environmental health and justice examines the role of data practices in making—and delegitimizing—knowledge claims about the causes of race-based health inequities. She is working on a book manuscript titled *Warding off Disease: Racialization and Health in Settler Colonial Canada.*

Margaret M. Bruchac is an associate professor of anthropology at the University of Pennsylvania, where she is also associate faculty in the Penn Cultural Heritage Center and coordinator of Native American and Indigenous studies. She directs a restorative research project, called "The Wampum Trail," that focuses on the materiality, meaning, and recovery of historical wampum objects. She has received research awards from the American Philosophical Society, Five College Fellowship, Ford Foundation, Mellon Foundation, and the School for Advanced Research, among others. Bruchac is co-editor, with Siobhan M. Hart and H. Martin Wobst, of *Indigenous Archaeologies: A Reader in Decolonization.* Her 2018 book, *Savage Kin: Indigenous Informants and American Anthropologists* received the inaugural Council for Museum Anthropology Book Award.

Elise K. Burton is an assistant professor at the Institute for the History and Philosophy of Science and Technology at the University of Toronto. She is a historian of the life sciences in the modern Middle East, with particular interests in genetics and evolutionary biology, as well as their role in shaping local ideas about race and nationalism. She is the author of *Genetic Crossroads: The Middle East and the Science of Human Heredity.*

Xan Chacko is a lecturer and director of undergraduate studies in the science, technology, and society (STS) program at Brown University. A feminist science studies scholar, her research complicates the taken-for-grantedness of scientific knowledge production to argue for a feminist re-envisioning of science that is committed to justice. She is currently writing a book that situates the emergence of cryogenic seed banking as a response to catastrophic species loss of plant life in the twentieth century. The book demonstrates how concepts like "biodiversity" and "food security" are evoked in a neoliberal era to enable the continuation of extractive colonial practices like plant collecting.

Susannah Chapman is a research fellow in the School of Social Science at The University of Queensland. Trained as an environmental and legal anthropologist, her research explores the intersection of law, science, and society, with a particular focus on transformations in human–plant relations and seed interventions during the twentieth century. She is currently writing a book that traces changes in farmer varietal selection and seed exchange alongside the emergence and elaboration of seed law and seed science in The Gambia. In following historical and contemporary encounters concerning the seed, the book shows how attempts to regulate plants have entailed discipline, assessment, and translation of diverse human and more-than-human worlds.

Sabine Clarke is a senior lecturer in modern history at the University of York. She is a historian of science, technology, and medicine and British imperialism in the twentieth century and a historian of science policy in Britain. Her monograph *Science at the End of Empire: Experts and the Development of the British Caribbean, 1940–1965* was published in 2018. Her current project is called "The Chemical Empire: A New History of Synthetic Insecticides in Britain and Its Colonies, c. 1920–1970" and is funded by a Wellcome Trust Investigator Award.

Rosanna Dent is an assistant professor in the Federated Department of History of New Jersey Institute of Technology and Rutgers–Newark. She is broadly interested in how human interactions unfold in the context of knowledge production, and the implications of these relationships for political and social justice. Her current book manuscript examines the history of human sciences research in A'uwe-Xavante (Indigenous) communities in Central Brazil.

Omnia El Shakry is a professor of history at the University of California, Davis. She specializes in the intellectual and cultural history of the modern Middle East, with a particular emphasis on the history of the human and religious sciences in modern Egypt. She is the author of *The Arabic Freud: Psychoanalysis and Islam in Modern Egypt* and *The Great Social Laboratory: Subjects of Knowledge in Colonial and Postcolonial Egypt*.

Boris Jardine is a researcher at the University of Cambridge and is currently co-investigator on the AHRC-funded project "Tools of Knowledge: Modelling the Creative Communities of the Scientific Instrument Trade, 1550–1914." He works on the history of scientific instrumentation, with a particular focus on the instrument trade in Britain, 1550–1700. Other areas of interest include the history and theory of collections, the nature of scientific architecture and infrastructure, and the relationship between aesthetic modernism and the sciences. Jardine was formerly curator for history of science at the Science Museum in London, and Munby fellow in bibliography at Cambridge University Library.

Judith Kaplan is a historian of the human sciences and a fellow with the Consortium for the History of Science, Technology, and Medicine in Philadelphia. Her work focuses on the rise of the modern linguistics in nineteenth-century Germany and on the subsequent development of comparative and historical approaches. She has published widely on topics from orientalism to sound studies and is currently completing a manuscript titled *Living Language and the Transformation of the Language Sciences, 1871–1918*.

Lara Keuck is a historian and philosopher of medicine at Bielefeld University. She holds a PhD in history, philosophy and ethics of medicine, and worked as a postdoc in philosophy and history departments. In 2015, she was awarded a Branco Weiss Fellowship of ETH Zurich for conducting the project "Learning from Alzheimer's Disease." Since 2021, she leads the independent Max Planck Research Group on "Practices of Validation in the Biomedical Sciences" at the Max Planck Institute for the History of Science in Berlin, Germany.

Whitney Laemmli is an assistant professor in the Department of History at Carnegie Mellon University, where she specializes in the history of modern science and technology with a special interest in the history of the human body. She is currently completing a book that explores how and why human movement became a central object of scientific, political, and popular concern over the course of the twentieth century.

Lan A. Li is an assistant professor in the Department of the History of Medicine at the Johns Hopkins University School of Medicine. Li's work focuses on the history of medicine in global East Asia with particular interest in the ontological and epistemic implications of practiced multiplicity within and across knowledge systems. Li's first book explores the long history of representing meridians and neurophysiological structures, attending to the role that graphic genre played in articulating objects of anatomy. Li's work as a filmmaker, researcher, and director of the Medicine Race Democracy Lab (mrdlab.org) further explores community clinics as sites of health care innovation.

M. Susan Lindee is a historian at the University of Pennsylvania in Philadelphia, where she is the Janice and Julian Bers Professor of the History and Sociology of Science. She studies historical and contemporary questions raised by human and medical genetics, science in the Cold War, nuclear weapons, and radiation genetics. Her books include *Rational Fog: Science, Technology and Modern War*; *Moments of Truth in Genetic Medicine*, *The DNA Mystique*: *The Gene as a Cultural Icon* with the late sociologist Dorothy Nelkin, and *Suffering Made Real: American Science and the Survivors at Hiroshima*. She is a Guggenheim fellow, a Weiler fellow, and the winner of a Burroughs Wellcome Fund 40th Anniversary Award and the Schuman Prize of the History of Science Society.

Stuart McCook is a professor of history at the University of Guelph, Canada. He is the author of *States of Nature: Science, Agriculture, and Environment in the Spanish Caribbean*, and *Coffee Is Not Forever: A Global History of the Coffee Leaf Rust*, which received the Agricultural History Society's Wallace Prize. He is interested in the relations between scientists and farmers in the global tropics, especially the complex and fragmented modernization of coffee farming.

Alexandra Noi is a PhD student at the history department of University of California, Santa Barbara. Her dissertation is a comparative intellectual and social history of forced labour and reeducation in the Soviet Union and in twentieth-century China. She received a doctoral degree in modern Chinese literature from St. Petersburg State University in 2017.

María Fernanda Olarte-Sierra (Mafe) is a medical anthropologist and an anthropologist of science with a focus on ethnographic research. She addresses interactions of health, technology, and the body in highly biomedicalized and technological contexts, including forensic victim identification in armed conflicts, prenatal testing, congenital cardiac diseases, and childhood cancer in Latin America. She recently finished a Marie Skłodowska-Curie fellowship at the University of Amsterdam with a project that addresses the role of judicial and humanitarian forensic knowledge in co-producing collective accounts of violence in Colombia. Currently, she is a postdoctoral researcher at the University of Vienna, in the Institute for Cultural and Social Anthropology in the group of Medical Anthropology and Global Health, where she will work on the search of forcibly disappeared persons as practices of collective care and sites of reconciliation in war-ridden contexts in Latin America.

Joanna Radin is a historian of biomedical futures and associate professor of history of medicine at Yale University in the United States. Her research examines the social and technical conditions of possibility for the systems of biomedicine and biotechnology that we live with today. She is the author of *Life on Ice: A History of New Uses for Cold Blood* and co-editor (with Emma Kowal) of *Cryopolitics: Frozen Life in a Melting World.*

Julia E. Rodriguez teaches history at the University of New Hampshire. She is the author of *Civilizing Argentina: Science, Medicine, and the Modern State* and has published articles in the *American Historical Review, Isis, Science in Context*, and the *Hispanic American Historical Review*. She is also editor of the open-source teaching website HOSLAC: History of Science in Latin America and the Caribbean (www.hoslac.org) that she developed with support from the National Science Foundation (CAREER Award # 0547125). Rodriguez's research focuses on the history of the human sciences in the late nineteenth and early twentieth centuries, including transatlantic scientific networks, the history of racial science and conceptions of humanity and personhood, the roots of cultural relativism, and the ethical dimensions of research on human subjects. She is currently completing a book about the nineteenth-century origins of Americanist anthropology.

Gabriela Soto Laveaga is a professor of the history of science and Antonio Madero Professor for the Study of Mexico at Harvard University. Her current research interests interrogate knowledge production and circulation between Mexico and India, medical professionals and social movements, and science and development projects in the twentieth century. Her first book, *Jungle Laboratories: Mexican Peasants, National Projects and the Making of the Pill*, won the Robert K. Merton Best Book prize in Science, Knowledge, and Technology Studies from the American Sociological Association. She is completing two book manuscripts, both of which approach the theme of invisible labour in hospitals and experiment stations.

Michaela Spencer is a research fellow with the Northern Institute at Charles Darwin University (CDU) in northern Australia. Her current research involves working from the "ground up" with Indigenous knowledge authorities and differing traditions of knowledge and governance. Most often, this is centred around public policy research, which engages and works collaboratively with government agencies, service providers, university staff and Indigenous Elders and researchers in remote communities. Michaela also facilitates CDU's Diploma of Indigenous Research and CDU's Indigenous Community-based Researcher micro-credential programs. Michaela's collaborators Gawura Wanambi, Joy Bulkanhawuy, and Yasunori Hayashi are based at Charles Darwin University; while Rosemary Gundjarranbuy and Stephen Dhamarrandji have a long-term affiliation with "Yalu" Marŋgithinyaraw Research Centre, Elcho Island.

Laura Stark is an associate professor at Vanderbilt University. She is the author of *Behind Closed Doors: IRBs and the Making of Ethical Research* and author of many articles on science, justice, and social-political theory. Her second book is under contract. *The Normals: A People's History* explores how a global market for healthy civilian "human subjects" emerged in law, science, and everyday imagination. It is based on a vernacular archive Stark created with more than 100 "normal control" research subjects and the scientists who experimented with them from 1950 through 1980. The Vernacular Archive of Normal Volunteers is now housed at the Countway Library for the History of Medicine. Her website is www.laura-stark.com.

Mihai Surdu is a sociologist working as a postdoctoral fellow at University Heidelberg, Research Centre on Antigypsyism, Germany. His academic research addresses critically the politics of knowledge production about Roma with a focus on data collection procedures, processes of stigmatization and minoritization, and social consequences resulting from ethnic categorization in various fields. His research has been supported by the Deutsche Forschungsgemeinschaft (DFG), Freiburg Institute of Advanced Studies (FRIAS), the Institute of Advanced Study at Central European University, the Max Planck Institute for the History of Science, and Open Society Foundations. He is author of *Those Who Count: Expert Practices of Roma Classification*. His recent work focuses on genetics and society.

Ana Carolina Vimieiro Gomes is a professor of the history of science in the Department of History at the Federal University of Minas Gerais (UFMG), Brazil. Her main research themes include human biological diversity, racial taxonomies, and genetics. She was a research fellow at the Max Planck Institute for the History of Science, the Escola Nacional de Saúde Pública, Fiocruz, and the Consortium for the History of Science, Technology and Medicine. She has published on the history of body classification, eugenics, and national identity; her current project focuses on the history of human genetics in Brazil.

Alexandra Widmer is an assistant professor of anthropology at York University in Toronto. She has conducted ethnographic and archival research on the impact of British and Australian colonial histories on women's health, nutrition, and reproduction in the southwestern Pacific. She has also published on the global and colonial history of demographic thinking and practice, particularly on its connection with racism(s) and economization. She is co-editor of *Health and Difference: Rendering Human Variation in Colonial Engagements*. She was previously a research scholar at the Max Planck Institute for the History of Science in Berlin.

Caitlin Donahue Wylie is an associate professor of science, technology, and society in the University of Virginia's School of Engineering and Applied Science. She studies how non-scientists, such as technicians, students, and volunteers, contribute to research in science and engineering. Caitlin uses qualitative methods, such as participant observation and interviews, to access the unwritten, everyday work of science and engineering. She teaches undergraduate engineers how to assess the social and ethical dimensions of technology and professional work, so that they can design safe, equitable, and successful sociotechnical systems. Caitlin's book *Preparing Dinosaurs: The Work Behind the Scenes* investigates how research communities construct their social order alongside trustworthy, beautiful fossils that serve as scientific evidence of past life.